SOME PHYSICAL CONSTANTS

Speed of Light	c	3.00×10^8 m/s
Gravitational Constant	G	6.67×10^{-11} Nm²/kg²
Coulomb Constant	k	8.99×10^9 Nm²/C²
Planck's Constant	h	6.63×10^{-34} Js
Boltzmann's Constant	k_B	1.38×10^{-23} J/K
Elementary charge	e	1.60×10^{-19} C
Electron Mass	m_e	9.11×10^{-31} kg
Proton Mass	m_p	1.67×10^{-27} kg
Neutron Mass	m_n	1.68×10^{-27} kg

STANDARD PREFIXES

For Powers of Ten

Power	Prefix	Symbol
10^{18}	exa	E
10^{15}	peta	P
10^{12}	tera	T
10^9	giga	G
10^6	mega	M
10^3	kilo	k
10^{-2}	centi	c
10^{-3}	milli	m
10^{-6}	micro	μ
10^{-9}	nano	n
10^{-12}	pico	p
10^{-15}	femto	f
10^{-18}	atto	a

COMMONLY-USED PHYSICAL DATA:

Gravitational Field Strength g	9.80 J·kg^{-1}·m^{-1} = 9.80 m/s²
Density of Water	1000.0 kg/m³ = 1 g/cm³ *
Density of Air	1.2 kg/m³ *

* at normal pressure, 20°C

USEFUL CONVERSION FACTORS

1 meter = 1 m = 100 cm = 39.4 inches = 3.28 ft
1 mile = 1 mi = 1609 m = 1.609 km = 5280 ft
1 inch = 2.54 cm
1 light-year = 1 ly = 9.46 Pm = 0.946×10^{16} m

1 hour = 1 h = 60 min = 3600 s
1 day = 1 d = 24 h = 86.4 ks = 86,400 s
1 year = 1 y = 365.25 d = 31.6 Ms = 3.16×10^7 s

1 J = 1 kg·m²/s² = 0.239 cal
1 kWh = 3.6 MJ

1.0 rad = 57.3° = 0.1592 rev

1 m/s = 2.24 mi/h
1 mi/h = 1.61 km/h
1 ft³ = 0.02832 m³
1 gallon = 1 gal = 3.79×10^{-3} m³ ≈ 3.8 kg H_2O

1 N = 1 kg·m/s² = 1 J/m = 0.225 lb
1 lb = 4.45 N
weight of 1-kg object near earth = 9.8 N = 2.2 lbs

1 W = 1 J/s
1 horsepower = 1 hp = 746 W

1 rev = 360° = 2π rad = 6.28 radians

USEFUL ASTRONOMICAL DATA

1 AU = mean distance from earth to sun = 1.50×10^{11} m

	Mass	Radius	Mean Orbital Radius	Orbital Period	Eccentricity
Sun	1.99×10^{30} kg	696,000 km	- - -	- - -	- - -
Moon	7.36×10^{22} kg	1740 km	384,000 km	27.3 days	0.055
Mercury	0.0558 M_E	2439 km	0.387 AU	0.241 y	0.206
Venus	0.815 M_E	6060 km	0.723 AU	0.615 y	0.007
Earth	5.98×10^{24} kg ≡ M_E	6380 km	1.000 AU	1.000 y	0.017
Mars	0.107 M_E	3370 km	1.524 AU	1.88 y	0.093
Jupiter	318 M_E	69,900 km	5.203 AU	11.9 y	0.048
Saturn	95.1 M_E	58,500 km	9.539 AU	29.5 y	0.056
Uranus	14.5 M_E	23,300 km	19.182 AU	84.0 y	0.047
Neptune	17.2 M_E	22,100 km	30.058 AU	165 y	0.009
Pluto/Charon	0.0025 M_E	3500/1800 km	39.785 AU	248 y	0.254
Ceres (asteroid)	1.2×10^{21} kg	500 km	2.768 AU	4.61 y	0.077
Halley's comet	1.2×10^{14} kg	≈ 7 km	17.94 AU	76.0 y	0.967

(Based mostly on data in D. Halliday, R. Resnick, *Fundamentals of Physics,* 3/e, New York:Wiley, p. A6.)

SYMBOLS AND THEIR MEANINGS

Symbol	Meaning
$=$	is equal to
$\neq$	is not equal to
$\approx$	is approximately equal to
$>$	is greater than
$<$	is less than
$>>$	is much greater than
$<<$	is much less than
$\equiv$	is defined to be
$\propto$	is proportional to
$\Rightarrow$	implies or therefore
$\int$	integral
∞	infinity
$\cdot$	indicates dot product OR product of units
$\times$	indicates cross product or multiplication by a power of 10
i.e.	*id est* "that is"
e.g.	*exempli gratia* "for example"
etc.	*etcetera* "and so on"
Q.E.D.	*quod erat demonstrandum* "which was to be demonstrated"
$\lvert x \rvert$	absolute value of x
mag()	magnitude of a vector
$\int$	indicates an integral
$'$	indicates a quantity measured in the S′ frame
$\vec{\beta}$	velocity of one frame relative to another
δ	phase constant
Δ	(as a prefix) a largish change in the variable whose symbol follows
Δt	(as a subscript) signifies *average* velocity or acceleration over the interval Δt
ε	orbital eccentricity
θ, ϕ, ψ	angles OR phase constants
$\hat{\theta}$	unit vector in the direction perpendicular to $\hat{r}$
μ_s, μ_k	static and kinetic coefficients of friction
ρ	mass density
$\vec{\tau}$	torque
$\vec{\omega}$	angular velocity
ω	magnitude of angular velocity OR phase rate
$\vec{a}$	acceleration
a	magnitude of $\vec{a}$ OR an arbitrary constant OR the semimajor axis of an elliptical orbit
A	area OR amplitude of an oscillation
$\vec{A}$	acceleration of one frame relative to another
AU	astronomical unit, a unit of distance
b	scalar constant OR unitless variable
c	speed of light OR an arbitrary constant
C	drag coefficient OR an arbitrary constant
C	(not italic) coulomb, the SI unit of charge
CM	(not italic, often as a subscript) center of mass
d	(as a prefix) tiny change in the variable whose symbol follows
đ	(as a prefix) one contribution to a tiny change in the variable following
đ$\vec{p}$	tiny momentum transfer
đK	tiny energy transfer

Symbol	Meaning
d, D	distance
e	charge on a proton OR exponential function
$f()$	function of whatever is in ()
$F()$	antiderivative of f()
$\vec{F}$	force
$\vec{F}_g$	gravitational force (weight)
$\vec{F}_N$	normal force
$\vec{F}_{SF}$	static friction (sticking) force
$\vec{F}_{KF}$	kinetic (sliding) friction force
$\vec{F}_D, \vec{F}_L$	drag force, lift force
$\vec{F}_{Th}, \vec{F}_B$	thrust force, buoyant force
$\vec{F}_{Sp}$	spring force
$\vec{F}_T$	tension force
$\vec{g}$	gravitational field vector
g	gravitational field strength = mag($\vec{g}$)
G	the universal gravitational constant
h	height
i	(as a subscript) means *initial* OR represents an index in a sum
I	moment of inertia
J	(not italic) joule, the SI unit of energy
k	the Coulomb constant
k_s	a spring (stiffness) constant
K	kinetic energy (also KE)
$\vec{L}$	angular momentum
L	length OR magnitude of angular momentum
m	mass
M	mass (usually of a system or large object)
m	(not italic) meter, a unit of distance
n	an arbitrary or unknown integer
N	number of particles in a system
N	(not italic) newton, the SI unit of force
0	(as a subscript) value of the attached variable at time $t = 0$ OR some other original value
O	the origin of a reference frame
$\vec{p}$	momentum
q	electrical charge OR unitless variable
$\vec{q}$	arbitrary vector
$\vec{r}$	a position vector
r	a radius or separation or mag($\vec{r}$)
$\hat{r}$	unit vector in the radial direction
R	a radius (often the fixed outer radius of some object, or a radius distinct from r)
$\vec{R}$	position of one frame relative to another
s	an arclength
t	time
T	period OR constant with units of time
u	$\equiv 1/r$ OR arbitrary and/or unitless variable
$\vec{v}$	velocity [speed $\equiv$ mag($\vec{v}$) $= v$]
$v_\perp$	component of $\vec{v}$ in the $\hat{\theta}$ direction
V	potential energy (also PE)
$\vec{w}, \hat{w}$	arbitrary vector, arbitrary unit vector
x, y, z	position coordinates
x, y, z	(as subscripts) indicates a component of a vector quantity
x-, y-, z-	(as prefixes) indicates a component of a vector quantity

SIX IDEAS THAT SHAPED PHYSICS

Unit N: The Laws of Physics Are Universal

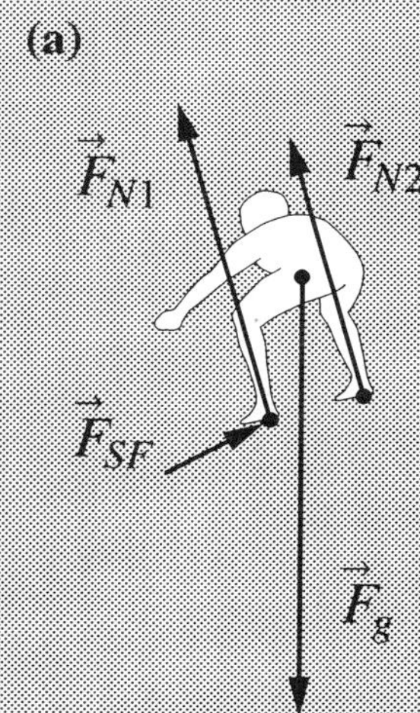

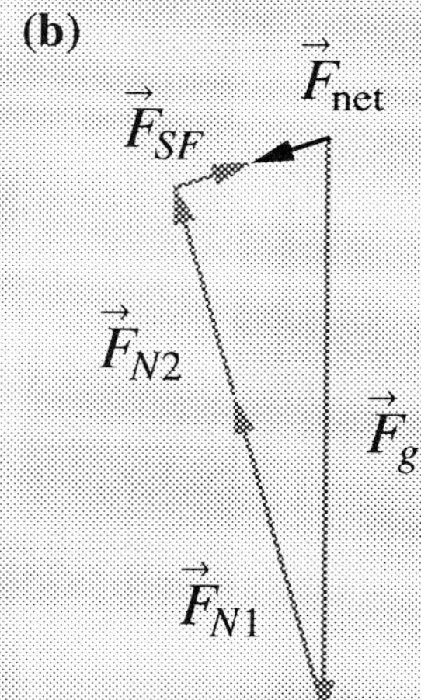

Thomas A. Moore
Pomona College

Boston Burr Ridge, IL Dubuque, IA Madison, WI New York San Francisco St. Louis
Bangkok Bogotá Caracas Lisbon London Madrid
Mexico City Milan New Delhi Seoul Singapore Sydney Taipei Toronto

WCB/McGraw-Hill
A Division of the McGraw-Hill Companies

SIX IDEAS THAT SHAPED PHYSICS/
UNIT N: THE LAWS OF PHYSICS ARE UNIVERSAL

 This book is printed on recycled, acid-free paper containing 10% postconsumer waste.

3 4 5 6 7 8 9 0 QPD/QPD 9 0 9 8

ISBN 0-07-043055-1

Vice president and editorial director: *Kevin T. Kane*
Publisher: *James M. Smith*
Sponsoring editor: *John Paul Lenney*
Developmental editor: *Donata Dettbarn*
Marketing manager: *Lisa L. Gottschalk*
Project managers: *Larry Goldberg, Sheila Frank*
Production supervisor: *Mary E. Haas*
Cover designer: *Jonathan Alpert/SCRATCHworks Creative*
Compositor: *Thomas A. Moore*
Typeface: *9/10 Times Roman*
Printer: *Quebecor Printing Book Group/Dubuque*

Cover photo: © Corel

Library of Congress Catalog Card Number: 97-62302

www.mhhe.com

For Brittany,
whose intuitive understanding of newtonian mechanics is part of what makes her awesome.

CONTENTS

PREFACE

1. INTRODUCTION

Opening comments about this preliminary edition

This volume is one of six that together comprise the PRELIMINARY EDITION of the text materials for *Six Ideas That Shaped Physics,* a fundamentally new approach to the two- or three-semester calculus-based introductory physics course. This course is still very much a work in progress. We are publishing these volumes in preliminary form so that we can broaden the base of institutions using the course and gather the feedback that we need to better polish both the course and its supporting texts for a formal first edition in a few years. Though we have worked very hard to remove as many of the errors and rough edges as possible for this edition, we would greatly appreciate your help in reporting any errors that remain and offering your suggestions for improvement. I will tell you how you can contact us in a section near the end of this preface.

Much of this preface discusses features and issues that are common to all six volumes of the *Six Ideas* course. For comments about this specific unit and how it relates to the others, see section 7.

The course's roots in the Introductory University Physics Project

Six Ideas That Shaped Physics was created in response to a call for innovative curricula offered by the Introductory University Physics Project (IUPP), which subsequently supported its early development. IUPP officially tested very early versions of the course at University of Minnesota during 1991/92 and at Amherst and Smith Colleges during 1992/93. In its present form, the course represents the culmination of over eight years of development, testing, and evaluation at Pomona College, Smith College, Amherst College, St. Lawrence University, Beloit College, Hope College, UC-Davis, and other institutions.

The three basic principles of the IUPP project

We designed this course to be consistent with the three basic principles articulated by the IUPP steering committee in its call for model curricula:

1. **The pace of the course should be reduced** so that a broader range of students can achieve an acceptable level of competence and satisfaction.
2. **There should be more 20th-century physics** to better show students what physics is like at the present.
3. **The course should use one or more "story lines"** to help organize the ideas and motivate student interest.

My additional working principles

The design of *Six Ideas* was also strongly driven by two other principles:

4. **The course should seek to embrace the best of what educational research has taught us** about conceptual and structural problems with the standard course.
5. **The course should stake out a middle ground** between the standard introductory course and exciting but radical courses that require substantial investments in infrastructure and/or training. This course should be useful in fairly standard environments and should be relatively easy for teachers to understand and adopt.

A summary of the course's distinctive features

In its present form, *Six Ideas* course consists of a set of six textbooks (one for each "idea"), a detailed instructor's guide, and a few computer programs that support the course in crucial places. The texts have a variety of innovative features that are designed to (1) make them more clear and readable, (2) teach you *explicitly* about the processes of constructing models and solving complex problems, (3) confront well-known conceptual problems head-on, and (4) support the instructor in innovative uses of class time. The instructor's manual is much more detailed than is normal, offering detailed suggestions (based on many teacher-years of experience with the course at a variety of institutions) about how to structure the course and adapt it to various calendars and constituencies. The instructor's manual also offers a complete description of effective approaches to class time that emphasize active and collaborative learning over lecture (and yet can still be used in fairly large classes), supporting this with day-by-day lesson plans that make this approach much easier to understand and adopt.

In the remainder of this preface, I will look in more detail at the structure and content of the course and briefly explore *why* we have designed the various features of the course the way that we have.

2. GENERAL PHILOSOPHY OF THE COURSE

Problems with the traditional intro course

The current standard introductory physics course has a number of problems that have been documented in recent years. (1) There is so much material to "cover" in the standard course that students do not have time to develop a deep understanding of any part, and instructors do not have time to use classroom techniques that would help students really learn. (2) Even with all this material, the standard course, focused as it is on *classical* physics, does not show what physics is like *today*, and thus presents a skewed picture of the discipline to the 32 out of 33 students who will never take another physics course. (3) Most importantly, the standard introductory course generally fails to *teach physics*. Studies have shown that even students who earn high grades in a standard introductory physics course often cannot

1. apply basic physical principles to realistic situations,
2. solve realistic problems,
3. perceive or resolve contradictions involving their preconceptions, or
4. organize the ideas of physics hierarchically.

What students in such courses *do* effectively learn is how to solve highly contrived and patterned homework problems (either by searching for analogous examples in the text and then copying them without much understanding, or by doing a random search through the text for a formula that has the right variables.) The high pace of the standard course usually drives students to adopt these kinds of non-thinking behaviors even if they don't want to.

The goal: to help students become competent in using the skills listed above

The goal of *Six Ideas* is to help students achieve a meaningful level of competence in each of the four thinking skills listed above. We have rethought and restructured the course from the ground up so that students are goaded toward (and then rewarded for) behaviors that help them develop these skills. We have designed texts, exams, homework assignments, and activity-based class sessions to reinforce each other in keeping students focused on these goals.

The focus is more on skills than on specific content

While (mostly for practical reasons) the course does span the most important fields of physics, the emphasis is *not* particularly on "covering" material or providing background vocabulary for future study, but more on developing problem-solving, thinking, and modeling skills. Facts and formulas evaporate quickly (particularly for those 32 out of 33 that will take no more physics) but if we can develop students' abilities to think like a physicist in a variety of contexts, we have given them something they can use throughout their lives.

3. TOPICS EXPLORED IN THE COURSE

The six-unit structure

Six Ideas That Shaped Physics is divided into six units (normally offered three per semester). The purpose of each unit is to explore in depth a single idea that has changed the course of physics during the past three centuries. The list below describes each unit's letter name, its length (1 d = one day ≡ one 50-minute class session), the idea, and the corresponding area of physics.

First Semester (37 class days excluding test days):
Unit *C* (14 d) *Conservation Laws Constrain Interactions* (conservation laws)
Unit *N* (14 d) *The Laws of Physics are Universal* (forces and motion)
Unit *R* (9 d) *Physics is Frame-Independent* (special relativity)

Second Semester (42 class days excluding test days):
Unit *E* (17 d) *Electromagnetic Fields are Dynamic* (electrodynamics)
Unit *Q* (16 d) *Particles Behave Like Waves* (basic quantum physics)
Unit *T* (9 d) *Some Processes are Irreversible* (statistical physics)

(Note that the spring semester is assumed to be longer than fall semester. This is typically the case at Pomona and many other institutions, but one can adjust the length of the second semester to as few as 35 days by omitting parts of unit *Q*.)

Dividing the course into such units has a number of advantages. The core idea in each unit provides students with motivation and a sense of direction, and helps keep everyone focused. But the most important reason for this structure is that it makes clear to students that some ideas and principles in physics are more important than others, a theme emphasized throughout the course.

Comments about the non-standard order

The non-standard order of presentation has evolved in response to our observations in early trials. **[1]** Conservation laws are presented first not only because they really are more fundamental than the particular theories of mechanics considered later but also because we have consistently observed that students understand them better and can use them more flexibly than they can Newton's laws. It makes sense to have students *start* by studying very powerful and broadly applicable laws that they can also understand: this builds their confidence while developing thinking skills needed for understanding newtonian mechanics. This also delays the need for calculus. **[2]** Special relativity, which fits naturally into the first semester's focus on mechanics and conservation laws, also ends that semester with something both contemporary and compelling (student evaluations consistently rate this section very highly). **[3]** We found in previous trials that ending the second semester with the intellectually demanding material in unit *Q* was not wise: ending the course with Unit *T* (which is less demanding) and thus more practical during the end-of-year rush.

Options for adapting to a different calendar

The suggested order also offers a variety of options for adapting the course to other calendars and paces. One can teach these units in three 10-week quarters of two units each: note that the shortest units (*R* and *T*) are naturally paired with longest units (*E* and *Q* respectively) when the units are divided this way. While the first four units essentially provide a core curriculum that is difficult to change substantially, omitting either Unit *Q* or Unit *T* (or both) can create a gentler pace without loss of continuity (since Unit *C* includes some basic thermal physics, a version of the course omitting unit *T* still spans much of what is in a standard introductory course). We have also designed unit *Q* so that several of its major sections can be omitted if necessary.

Using the volumes alone or in different orders

Many of these volumes can also stand alone in an appropriate context. Units *C* and *N* are tightly interwoven, but with some care and in the appropriate context, these could be used separately. Unit *R* only requires a basic knowledge of mechanics. In addition to a typical background in mechanics, units *E* and *Q* require only a few very basic results from relativity, and Unit *T* requires only a very basic understanding of energy quantization. Other orders are also possible: while the first four units form a core curriculum that works best in the designed order, units *Q* and *T* might be exchanged, placed between volumes of the core sequence, or one or the other can be omitted.

Superficially, the course might seem to involve quite a bit *more* material than a standard introductory physics course, since substantial amounts of time are devoted to relativity and quantum physics. However, we have made substantial cuts in the material presented in the all sections of the course compared to a standard course. We made these cuts in two different ways.

The pace was reduced by cutting whole topics...

First, we have omitted entire topics, such as fluid mechanics, most of rotational mechanics, almost everything about sound, many electrical engineering topics, geometric optics, polarization, and so on. These cuts will no doubt be intolerable to some, but *something* has to go, and these topics did not fit as well as others into this particular course framework.

... and by streamlining the presentation of the rest

Our second approach was to simplify and streamline the presentation of topics we *do* discuss. A typical chapter in a standard textbook is crammed with a variety of interesting but tangential issues, applications, and other miscellaneous factons. The core idea of each *Six Ideas* unit provides an excellent filter for reducing the number density of factons: virtually everything that is not *essential* for developing that core idea has been eliminated. This greatly reduces the "conceptual noise" that students encounter, which helps them focus on learning the really important ideas.

Because of the conversational writing style adopted for the text, the total page count of the *Six Ideas* texts is actually similar to a standard text (about 1100 pages), but if you compare typical chapters discussing the same general material, you will find that the *density* of concepts in the *Six Ideas* text is much lower, leading to what I hope will be a more gentle perceived pace.

Choosing an appropriate pace

Even so, this text is *not* a "dumbed-down" version of a standard text. Instead of making the text dumber, I have tried very hard to challenge (and hopefully enable) students to become *smarter*. The design pace of this course (one chapter per day) is pretty challenging considering the sophistication of the material, and really represents a maximum pace for fairly well-prepared students at reasonably selective colleges and universities. However, I believe that the materials *can* be used at a much broader range of institutions and contexts at a lower pace (two chapters per three sessions, say, or one chapter per 75-minute class session). This means either cutting material or taking three semesters instead of two, but it can be done. The instructor's manual discusses how cuts might be made.

Part of the point of arranging the text in a "chapter-per-day" format is to bee clear about how the pace should be *limited*. Course designs that require covering *more* than one chapter per day should be strictly avoided: if there are too few days to cover the chapters at the design pace, than chapters will *have* to be cut.

4. FEATURES OF THE TEXT

The texts are designed to serve as students' primary source of new information

Studies have suggested that lectures are neither the most efficient nor most effective way to present expository material. One of my most important goals was to develop a text that could essentially replace the lecture as the primary *source* of information, freeing up class time for activities that help students *practice* using those ideas. I also wanted to create a text that not only presents the topics but goads students to develop model-building and problem-solving skills, helps them organize ideas hierarchically, encourages them to think qualitatively as well as quantitatively, and supports active learning both inside and outside of class.

A list of some of the texts' important features

In its current form, the text has a variety of features designed to address these needs, (many of which have evolved in response to early trials):

1. **The writing style is expansive and conversational**, making the text more suitable to be the primary way students learn new information.
2. **Each chapter corresponds to one (50-minute) class session**, which helps guide instructors in maintaining an appropriate pace.
3. **Each chapter begins with a unit map and an overview** that helps students see how the chapter fits into the general flow of the unit.
4. **Each chapter ends with a summary** that presents the most important ideas and arguments in a hierarchical outline format.
5. **Each chapter has a glossary** that summarizes technical terms, helping students realize that certain words have special meanings in physics.

6. **The book uses "user-friendly" notation and terminology** to help students keep ideas clear and avoid misleading connotations.
7. **Exercises embedded in the text** (with provided answers) help students actively engage the material as they prepare for class (providing an active alternative to examples).
8. **Wide outside margins** provide students with space for taking notes.
9. **Frequent *Physics Skills* and *Math Skills* sections** explicitly explore and summarize generally-applicable thinking skills.
10. **Problem-solving frameworks** (influenced by work by Alan van Heuvelan) help students learn good problem-solving habits.
11. **Two-minute problems** provide a tested and successful way to actively involve students during class and get feedback on how they are doing.
12. **Homework problems** are generally more qualitative than standard problems, and are organized according to the general thinking skills required.

5. ACTIVE LEARNING IN AND OUT OF CLASS

The *Six Ideas* texts are designed to support active learning both inside and outside the classroom setting. A properly designed course using these texts can provide to students a rich set of active-learning experiences.

Active learning using two-minute exercises

The *two-minute exercises* at the end of each chapter make it easy to devote at least part of each class session to active learning. These mostly conceptual questions do not generally require much (if any) calculation, but locating the correct answer does require careful thinking, a solid understanding of the material, and (often) an ability to apply concepts to realistic situations to answer correctly. Many explicitly test for typical student misconceptions, providing an opportunity to expose and correct these well-known stumbling blocks.

I often begin a class session by asking students to work in groups of two or three to find answers for a list of roughly three two-minute problems from the chapter that was assigned reading for that class session. After students have worked on these problems for some time, I ask them to show me their answers for each question in turn. The students hold up the back of the book facing me and point to the letter that they think is the correct answer. This gives me instant feedback on how well the students are doing, and provides me with both grist for further discussion and a sense where the students need the most help. On the other hand, students cannot see each others' answers easily, making them less likely to fear embarrassment (and I work very hard to be supportive).

Active demonstrations

Once everyone gets the hang of the process, it is easy to adapt other activities to this format. When I do a demonstration, I often make it more active by posing questions about what will happen, and asking students to respond using the letters. This helps everyone think more deeply about what the demonstration really shows and gets the students more invested in the outcome (and more impressed when the demonstration shows something unexpected).

The exercises support active reading

The in-text exercises and homework problems provide opportunities for active learning *outside* of class. The exercises challenge students to test their understanding of the material as they read it, helping them actively process the material and giving them instant feedback. They also provide a way to get students through derivations in a way that actively involves them in the process and yet "hides" the details so that the structure of the derivation is clearer. Finally, such exercises provide an active alternative to traditional examples: instead of simply displaying the example, the exercises encourage students to work through it.

The types of homework problems

The homework problems at the end of each chapter are organized into four types. ***Basic*** problems are closest to the type of problems found in standard texts: they are primarily for practicing the application of a single formula or concept in a straightforward manner and/or are closely analogous to examples in the text. ***Synthetic*** problems generally involve more realistic situations, require students to apply *several* concepts and/or formulas at once, involve creating or applying models, and/ or require more sophisticated reasoning. ***Rich-Context*** problems are synthetic problems generally cast in a narrative framework where either too much or too little information is given and/or a non-numerical question is posed (that nonetheless requires numerical work to answer). ***Advanced*** problems usually explore subtle theoretical issues or mathematical derivations beyond the level of the class: they are designed to challenge the very best students and/or remind instructors about how to handle subtle issues.

Collaborative recitation sessions

The rich-context problems are especially designed for collaborative work. Work by Heller and Hollenbaugh has shown that students solving standard problems rarely collaborate even when "working together", but that a well-written rich-context problem by its very open-ended nature calls forth a discussion of physical concepts, requiring students to work together to create useful models. I typically assign one such problem per week that students can work in a "recitation" section where can they work the problem in collaborative groups (instead of being lectured to by a TA).

The goal of the course is that the majority of students should ultimately be able to solve problems at the level of the *synthetic* problems in the book. Many of the rich-context problems are too difficult for individual students to solve easily, and the advanced problems are meant to be beyond the level of the class.

The way that a course is structured can determine its success

In early trials of *Six Ideas*, we learned that whether a course succeeds or fails depends very much the details of how the course is *structured*. This text is designed to more easily support a productive course structure, but careful work on the course design is still essential. For example, a "traditional" approach to assigning and grading homework can lead students to be frustrated (rather than challenged) by the richer-than-average homework problems in this text. Course structures can also either encourage or discourage students from getting the most out of class by preparing ahead of time. Exams can support or undermine the goals of the course. The instructor's manual explores these issues in much more depth and offers detailed guidance (based on our experience) about how design a course that gets the most out of what these books have to offer.

6. USE OF COMPUTERS

Using computers

The course, unlike some recent reform efforts, is *not* founded to a significant degree on the use of computers. Even so, a *few* computer programs are deployed in a few crucial places to support a particular line of argument in the text, and unit *T* in particular comes across significantly better when supported by a relatively small amount of computer work.

The most current versions of the computer programs supporting this course can be downloaded from my web-site or we will send them to you on request (see the contact information in section 8 below).

7. NOTES ABOUT UNIT *N*

The purpose and place of this unit in the course

This particular unit is primarily focused on Newton's second law and its application to both terrestrial and celestial physics. Its goal is to help students appreciate the power and breadth of the newtonian perspective as well as the historical importance of Newton's work.

This unit is structured on the premise that students have already studied unit *C*, and indeed it draws on ideas from almost all of the chapters of that unit. It in turn is needed as basic background for all of the other units in the course.

The unit's spiral structure

This unit is designed to teach newtonian mechanics using a "spiral learning" approach. The first five chapters provide a mostly qualitative introduction to the concepts and techniques of newtonian mechanics, while the remaining chapters explore quantitative applications of these ideas in depth. Instructors can help students get the most out of this approach by helping them see the connections between the earlier and later spirals through the given material.

An unusual feature of the first part of this text is the exploration of motion using *motion maps* and *trajectory diagrams*. Both of these tools are designed to deepen students' intuitive understanding of motion, and trajectory diagrams in particular are a powerful tool for qualitatively predicting an object's motion in advance of using mathematics. If students spend enough time practicing the use of both of these tools, their understanding of newtonian mechanics will become much deeper and more flexible.

Motion maps and trajectory diagram

I have also discussed the trajectory diagram in such depth because it provides an excellent conceptual basis for computer programs that calculate trajectories. I am in the process of trying to develop a user-friendly program that automates the trajectory-construction process. When this program is done, it could be used to in the latter part of the course to help students develop a more intuitive understanding of projectile, oscillatory, and planetary motion, and greatly enhance the range of applications that they can explore (for example, projectile or planetary motion with drag). Keep your eyes on the *Six Ideas* web site (see the next section for the URL) for news about this and other supporting computer programs.

The most difficult part of the unit for many students is the material on computing the radial and tangential components of acceleration in chapters N9 and N14. If chapter N14 is to make any sense to students (and this chapter *is* the capstone of the unit) special care and time should be taken to make sure that students understand the material in chapter N9. This material has been deliberately placed long before chapter N14 so that students will have sufficient time to absorb it before using it in chapter N14.

Make sure students understand the material in chapter N9

Unit *N*, like unit *C*, is a mostly indivisible whole. Chapter N6 (which looks at torque and statics problems) could probably be omitted if cuts are absolutely necessary: it is not essential for anything else in the course. Chapters N14 and N8 also cover material that not needed in the rest of the course, but I would recommend against omitting these units: chapter N8 is very important for developing students' understanding of Newton's third law and how linked objects interact, and dropping chapter N14 would mean that students would not see the fulfillment of the unit's "great idea." In short, if cuts need to be made, start with chapter N6, but all of the other chapters have important roles to play.

One can omit chapter N6 but not much else

Please see the instructor's manual for more detailed comments about this unit and suggestions about how to teach it effectively.

8. HOW TO COMMUNICATE SUGGESTIONS

As I said at the beginning of this preface, this is a preliminary edition that represents a snapshot of work in progress. I would greatly appreciate your helping me make this a better text by telling me about errors and offering suggestions for improvement (words of support will be gratefully accepted too!). I will also try to answer your questions about the text, particularly if you are an instructor trying to use the text in a course.

Please help me make this a better text!

McGraw-Hill has set up an electronic bulletin board devoted to this text. This is the primary place where you can converse with me and/or other users of the text. Please post your comments, suggestions, criticisms, encouragement, error reports, and questions on this bulletin board. I will check it often and respond to whatever is posted there. The URL for this bulletin board is:

The *Six Ideas* bulletin board

```
http://mhhe.com/physsci/physical/moore
```

Before you send in an error or ask a question, please check the error postings and/or FAQ list on my *Six Ideas* web site. The URL for this site is:

The *Six Ideas* web site

```
http://pages.pomona.edu/~tmoore/sixideas.html
```

Visiting this site will also allow you to read the latest information about the *Six Ideas* course and texts on this site, download the latest versions of the supporting computer software, and visit related sites. You can also reach me via e-mail at `tmoore@pomona.edu`.

How to get other volumes or ancillary materials

Please refer questions about obtaining copies of the texts and/or ancillary materials to your WCB/McGraw-Hill representative or as directed on the *Six Ideas* web-site.

9. APPRECIATION

Thanks!

A project of this magnitude cannot be accomplished alone. I would first like to thank the others who served on the IUPP development team for this project: Edwin Taylor, Dan Schroeder, Randy Knight, John Mallinckrodt, Alma Zook, Bob Hilborn and Don Holcomb. I'd like to thank John Rigden and other members of the IUPP steering committee for their support of the project in its early stages, which came ultimately from an NSF grant and the special efforts of Duncan McBride. Early users of the text, including Bill Titus, Richard Noer, Woods Halley, Paul Ellis, Doreen Weinberger, Nalini Easwar, Brian Watson, Jon Eggert, Catherine Mader, Paul De Young, Alma Zook, and Dave Dobson have offered invaluable feedback and encouragement. I'd also like to thank Alan Macdonald, Roseanne Di Stefano, Ruth Chabay, Bruce Sherwood, and Tony French for ideas, support, and useful suggestions. Thanks also to Robs Muir for helping with several of the indexes. My editors Jim Smith, Denise Schanck, Jack Shira, Karen Allanson, Lloyd Black, and JP Lenney, as well as Donata Dettbarn, David Dietz, Larry Goldberg, Sheila Frank, Jonathan Alpert, Zanae Roderigo, Mary Haas, Janice Hancock, Lisa Gottschalk, and Debra Drish, have all worked very hard to make this text happen, and I deeply appreciate their efforts. I'd like to thank reviewers Edwin Carlson, David Dobson, Irene Nunes, Miles Dressler, O. Romulo Ochoa, Qichang Su, Brian Watson, and Laurent Hodges for taking the time to do a careful reading of various units and offering valuable suggestions. Thanks to Connie Wilson, Hilda Dinolfo, and special student assistants Michael Wanke, Paul Feng, and Mara Harrell, Jennifer Lauer, Tony Galuhn, Eric Pan, and all the Physics 51 mentors for supporting (in various ways) the development and teaching of this course at Pomona College. Thanks also to my Physics 51 students, and especially Win Yin, Peter Leth, Eddie Abarca, Boyer Naito, Arvin Tseng, Rebecca Washenfelder, Mary Donovan, Austin Ferris, Laura Siegfried, and Miriam Krause, who have offered many suggestions and have together found many hundreds of typos and other errors. Finally, very special thanks to my wife Joyce and to my daughters Brittany and Allison, who contributed with their support and patience during this long and demanding project. Heartfelt thanks to all!

Thomas A. Moore
Claremont, California
November 25, 1997

N1

INTRODUCTION

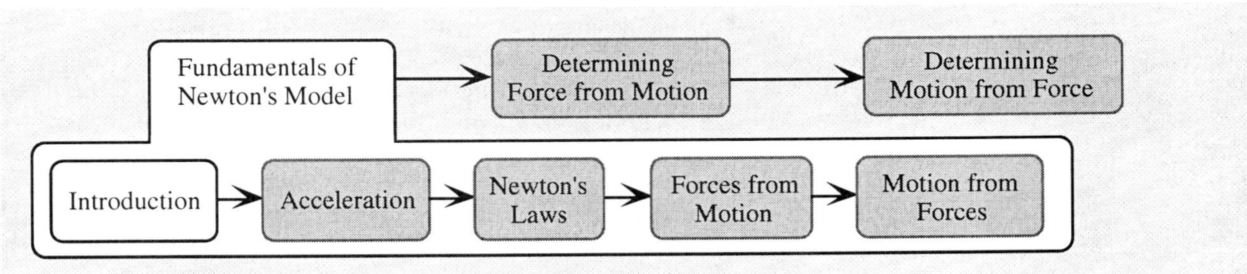

N1.1 OVERVIEW

In the last unit, we explored the constraints that the three great conservation laws (conservation of *linear momentum*, conservation of *energy*, and conservation of *angular momentum*) put on the behavior of systems of interacting objects. Part of the power of these laws is that we can apply them without knowing much about the detailed nature of the interactions involved or exactly how they act to modify the objects' motions.

Our task in this unit is to delve into the details. This is because our ultimate goal in this unit is to understand the **newtonian synthesis**: that is, how Newton's model of mechanics is able to explain celestial as well as terrestrial motion. In order to appreciate this, we *have* to understand exactly how interactions modify an object's motion. In this unit, then, we will sequentially develop the analysis skills we need to illuminate (first) terrestrial physics and (finally) celestial physics at the end of the unit.

This chapter starts the process by first setting the historical context for the newtonian synthesis (so that we can better understand its importance) and then introducing the basic principles of mathematical analysis of motion. Here is a summary of the sections in this chapter.

N1.2 *ARISTOTELIAN PHYSICS* describes the view of physics accepted by Western thinkers before Galileo, a theory that assumed a sharp distinction between the causes of celestial and terrestrial motion.

N1.3 *THE ARISTOTELIAN WORLD-VIEW CRUMBLES* describes how this perspective began to fall apart when confronted with new ideas and new observations in the late 1500s and early 1600s.

N1.4 *THE NEWTONIAN SYNTHESIS* explores the significance and impact of Newton's work in this context.

N1.5 *OVERVIEW OF UNIT N* describes the goals and organization of the unit and how we will explore the Newtonian synthesis.

N1.6 *THE TIME-DERIVATIVE OF A VECTOR* begins our exploration of the mathematics of motion by defining the time-derivative of a vector.

N1.7 *VELOCITY* applies this definition to formally and rigorously define what we mean by the *instantaneous velocity* of an object.

**** ***MATH SKILLS:*** *DERIVATIVES* reviews the definition of the derivative and various useful theorems (such as the sum and product rules).

N1.2 ARISTOTELIAN PHYSICS

One of the first people to think about the laws of motion in a systematic way was the Greek philosopher Aristotle (384-322 BC). In his treatises entitled *Physica* (Physics) and *De Caelo* (On the Heavens), Aristotle described laws of motion for terrestrial and celestial objects that represented a thoughtful classification and presentation of common-sense notions of motion and change gathered from experience. Aristotle's work in these treatises provided the foundation for Western thought on the physics of motion until well into the Renaissance.

In considering Aristotle's ideas about mechanics, it is important to recognize that Aristotle had goals and aims in writing these treatises that seem irrelevant to us now. Aristotle was seeking to understand motion in the context of *change* in general, and was interested in how such changes were connected to an object's intrinsic nature. Aristotle's thinking about these issues was subtle, careful, and insightful, and to isolate the small parts of these works that address what we would now think of as being "laws of mechanics" is to take them out of context in a way that does a certain amount of violence to the coherence of Aristotle's thought.

Aristotle's common-sense model of physics

Having said this, let me try to summarize Aristotle's thoughts on motion, particularly as they were interpreted by later thinkers. Objects move in certain ways according to their *natures*. *Heavy* objects have a tendency to move toward the center of the universe (i.e. the earth) and come to rest there because such objects are constructed of earth and thus seek to join the earth. *Light* objects (like fire) tend to move away from the center of the earth toward the heavens. Celestial bodies move endlessly in the heavens in perfect circles by their very nature.

Thus, according to Aristotle, the natural state of a *heavy* object is to be at rest on the surface of the earth. If such an object is in motion, there must be a *cause* for that motion. (In modern language, we might think of this cause as being some kind of *force*, though this is not exactly what Aristotle had in mind.) The object's motion will continue as long as the cause continues, and when the cause is no longer present, motion ceases. The speed at which an object moves depends on the strength of that which causes the motion.

The motion of celestial bodies, however, is quite different. The natural state of a celestial body is endless circular motion. This motion does not require a special cause: it is a reflection of the intrinsic character of these objects.

Let me make two points about Aristotle's thinking on these subjects. The first is that his thoughts on motion are quite consistent with what we know to be true from our daily experience. Heavy objects *do* have a tendency to seek the center of the earth, and the heavier the object is, the stronger the tendency. You have to press on the accelerator to cause a car to move, and when you release the accelerator, the car stops moving. The speed of the car depends on how strongly you push the accelerator. Aristotle's concepts of motion seem natural and intuitive to us. Indeed, research has shown that if people entering a physics class are questioned about the laws of motion, they express themselves in terms of concepts that are reminiscent of Aristotle's.

Cartoons often play on our deeply-held aristotelian ideas about motion. For example, a cartoon character who runs off the edge of a cliff is depicted as traveling forward horizontally for a short time, then coming to rest as the "cause" of its forward motion dissipates, and then falling vertically downward, seeking the ground. This kind of motion is a logical extension of Aristotle's ideas as described above. (The actual motion of an object sliding off of a horizontal surface is very different! Try rolling a ball off of the edge of your desktop and observe its subsequent motion very closely: what does the ball *really* do?)

The second point that I want to make is that in Aristotle's view, celestial physics is very different from terrestrial physics. The endless motion of celestial bodies in perfect circles does not require a cause, as the motion of terrestrial objects does. The laws of celestial motion are thus completely disconnected from the laws of terrestrial motion.

Ptolemy extends Aristotle's celestial model

As I mentioned before, Aristotle's ideas provided the starting point for all of western thought on *natural philosophy* (as physics was then known) until the late 1500s. This didn't mean that his ideas went unquestioned. Aristotle's simple image of celestial objects moving in perfect circles around the earth was quickly seen to be inadequate, and in about 140 AD the Alexandrian astronomer Ptolemy published a more sophisticated (but still essentially aristotelian) geocentric model of the universe that involved a complicated nesting of purely circular motions. Ptolemy's book was so encyclopedic and brilliantly written that it became the accepted text on astronomy until the 1600s (9th century Arab astronomers referred to this book as *Almagest*, i.e. "The Greatest"). Even though this book extended and modified Aristotle's ideas, its popularity did much to spread around the basic aristotelian concepts of motion.

Medieval thinkers extend Aristotle's terrestrial ideas

Medieval thinkers questioned other aspects of Aristotle's work. For example, in Aristotle's scheme, a heavier object seeks the earth more strongly than a lighter object. This would seem to imply that a heavier object should fall more rapidly than a light object (indeed, one would think that the speed of a falling object should be proportional to its weight). This prediction was known to be false by the early Middle Ages. Another problem discussed in the medieval west was the flight of an arrow. The release of the bow clearly caused the motion of an arrow initially, but what kept it going once it left the bow? Aristotle proposed that the motion of the arrow set up a kind of "wake" in the air that continued to propel the arrow forward after it left the bow, but medieval scholars argued that this explanation was inadequate. Consideration of these problems led medieval scholars to develop extensions to Aristotle's ideas to solve these puzzles.

But the fact remains that until the time of Galileo (the late 1500s and early 1600s), no one seriously questioned the most basic assumptions of aristotelian mechanics, specifically (1) that continuing motion of terrestrial bodies requires a continuing *cause* of that motion, (2) that the speed of such a moving body is proportional to the strength of its cause, and (3) that celestial physics was fundamentally distinct from terrestrial physics.

N1.3 THE ARISTOTELIAN WORLD-VIEW CRUMBLES

The Copernican model of the solar system

The aristotelian world-view began to fall apart in the mid-1500s when the Polish scholar Nicolaus Copernicus proposed a model of the solar system that turned Ptolemy's system inside out, putting the sun at the center of the universe (as the solar system was considered then) rather than the earth. Copernicus had hoped that this model would make it possible to explain the observed behavior of the planets in terms of perfectly circular orbits without the complicated nesting of circular motions required by Ptolemy's model. While this hope proved vain, astronomers found that the model was so handy for doing calculations that they found Copernicus' book indispensable, even if they were unable to endorse the crucial heliocentric hypothesis, which by this time was facing mounting opposition from both the church and other scholars.

The problem with the copernican solar system was that it contradicted not only the accepted ptolemaic model but the foundations of aristotelian physics. The aristotelian scheme of "natural" motion toward the earth or toward the heavens rested firmly on the assumption that the earth was the fixed center of the universe. Rejection of the ptolemaic solar system amounted to rejection of aristotelian physics as well, and no reasonable replacement seemed available.

Galileo criticizes the aristotelian model

In the late 1500s and early 1600s, the Italian physicist Galileo Galilei became convinced of the validity of the copernican model, and began to outline a new way of looking at mechanics consistent with this model. Central to Galileo's perspective was the hypothesis that smooth and constant motion was the natural state of all objects and required no explanation in terms of a cause. Galileo needed this hypothesis to answer one of the most basic arguments against the hypothesis of a moving earth.

In *Almagest*, Ptolemy had argued that if the earth were moving, then an object dropped vertically from rest would appear to someone on the Earth to be swept backwards in a direction opposite to the earth's motion, since there was nothing to cause the object to move along with the Earth. Since this is not observed, Ptolemy concluded, the earth must be at rest. Galileo argued that if constant motion needed no "cause", then the earth and all objects upon it would move naturally together, and an object dropped by someone on a horizontally moving earth would continue to move horizontally along with the Earth, and thus appear to someone on the earth to fall straight downward, making the motion of the earth unobservable.

Galileo also was on the leading edge of several hot new trends in the "natural philosophy" of his time: (1) the desire to make science *quantitatively predictive* as well as qualitatively descriptive, and (2) the development of an experimental outlook that sought to actually test scientific hypotheses using careful observations of nature under controlled circumstances. (Previously, philosophers like Aristotle had been content to present careful and painstakingly logical but completely *theoretical* arguments based on everyday observations of nature, without posing questions to nature itself in the form of experiments.) Galileo himself performed experiments that enabled him to make the first quantitative description of the motion of falling objects (among other things).

Other developments undermined Aristotle's model

Several other crucial developments in the early 1600s laid the foundations for the new mechanics proposed by Newton. In 1609, Johannes Kepler was able to show (to his surprise and dismay) that the copernican model was quantitatively consistent with the best available astronomical data only if it was hypothesized that the planets moved in *ellipses* instead of perfect circles, and by 1618 had published three empirical laws that provided a complete quantitative description of planetary motion, making the copernican model as accurate as the ptolemaic system and conceptually much simpler. At about the same time, the newly invented telescope made it possible to see the actual disks of the nearby planets, and the observation of the phases and the apparent size of Venus by Galileo and others meant that the planet Venus (at least) had to orbit the Sun instead of the Earth, putting the final nail in the coffin of the ptolemaic model.

Finally, the French philosopher René Descartes began in the mid 1600s to express the hope that it should be possible to quantitatively explain *all* of the phenomena of nature (terrestrial and celestial) in terms of matter in motion. While Descartes was unable to achieve his aim of quantitatively explaining planetary motion using a vortex model of the solar system (where the planets were swept around in their orbits by a vast whirlpool centered on the sun), his hope for a purely mechanical explanation of natural phenomena became a guiding principle for physicists at the time (in much the same way as the quest for the unified field theory guides physicists in this century).

The stage was set for the birth of a fundamentally new vision of mechanics.

N1.4 THE NEWTONIAN SYNTHESIS

In 1661, the 19-year old Isaac Newton arrived in Cambridge to begin his college education. Like most undergraduates of his time, Newton learned basic aristotelian philosophy, but the radical new ideas of Galileo and Descartes were circulating unofficially. A set of Newton's papers dating from 1664 showed that he had carefully studied the work of Descartes and other proponents of the new philosophy of science. At about the same time, Newton was developing his skills as a mathematician, again starting with the work of Descartes.

In 1665, just after Newton received his bachelor's degree, the university was closed because of the Black Plague. Newton spent the following two years at his home in rural Lincolnshire. During this time, Newton essentially invented calculus, and applied it to the motion of the moon and the planets. He was able to show crudely that these objects appeared to move as if they were attracted to the centers of their orbits by a force whose strength varied with the inverse square of the distance, but he did not publish any of these findings at the time.

Publication of the *Principia*

In 1679, after he had returned to Cambridge as a teacher, correspondence with a rival physicist (Robert Hooke) goaded Newton to renew his explorations of planetary motion. With the additional encouragement of the astronomer Edmund Halley (after whom Halley's Comet is named), Newton published in 1687 a complete description of his work on planetary motion, a work entitled *Philosophiae Naturalis Principia Mathematica* ("Mathematical Principles of Natural Philosophy"). This book turned out to be one of the most important single works in the history of science.

In the *Principia* (which is how people usually refer to this work) Newton provided for the first time a complete and quantitative theory of the motion of all bodies (terrestrial or celestial) based on three simple laws of motion and the principle of universal gravitation, which asserted that every object is attracted to every other object in the universe by a gravitational force that varies quantitatively as the inverse square of the distance between the objects. Using his newly-invented techniques of calculus, Newton was able to show mathematically that these hypotheses implied that planets must orbit the sun in ellipses having exactly the characteristics that Kepler described seventy years previously on the basis of observational data.

Newton's theory of mechanics was more than just a new theory of celestial mechanics. It embraced what was quantitatively known about the motion of terrestrial objects as well. It yielded predictions regarding falling bodies near the earth that were consistent with Galileo's quantitative measurements, and provided a framework for understanding a variety of other terrestrial phenomena. Moreover, it provided for the first time a physical basis for understanding the moon's influence on the tides.

In short, Newton's theory of mechanics was the first that was able to explain both terrestrial *and* celestial phenomena within the same theoretical framework. This unification of terrestrial and celestial physics is called the **newtonian synthesis**, and stands as one of the most amazing intellectual achievements of all time.

Indeed, the *Principia* was so impressive that nearly everyone doing physics at the time dropped what they were doing and began to explore the implications of this new perspective on mechanics. The *Principia* forged from a diversity of competing schools and perspectives a *single* community of scholars dedicated to the refinement and extension of this wonderful new theory, transforming physics from a branch of natural philosophy into a quantitative science.

N1.5 AN OVERVIEW OF UNIT *N*

Purpose of this unit

In the previous unit, you have already been introduced to the basic principles of newtonian physics in the context of conservation laws. The purpose of this unit is to help you learn enough about how to analyze motion in the newtonian model to understand and appreciate the newtonian synthesis.

This unit is comprised of three subunits. The first provides an introduction to the mathematical description of motion using calculus and the basic features of the newtonian model. In the next subunit, we will learn how to solve problems involving (mostly terrestrial) objects that are constrained to move along certain kinds of paths. Our task in such problems is to use our knowledge about the object's motion to determine the magnitude and directions of one or more

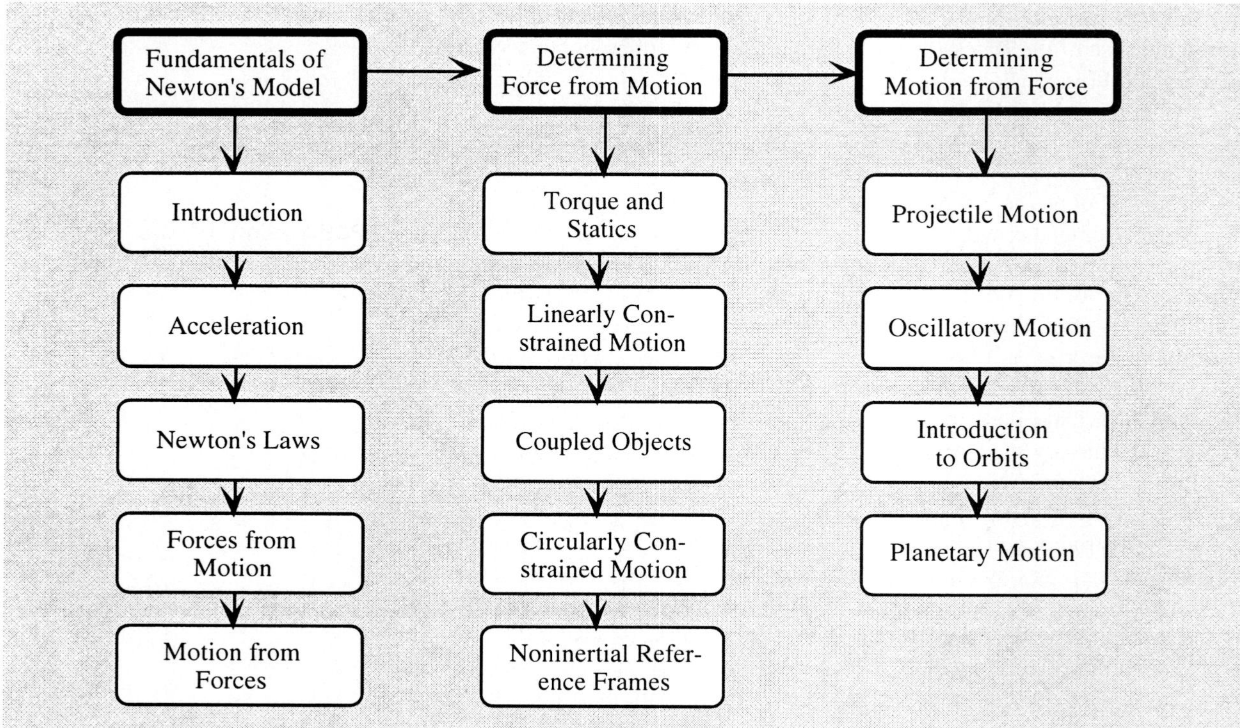

Figure N1.1: The structure of Unit *N*.

unknown forces acting on the object and deduce their effects and/or implications. In the third subunit, we will face the more complicated problem of determining how an object moves knowing only the forces that act on it. This subunit concludes with an argument that planets attracted toward the sun by an inverse-square gravitational force will follow elliptical orbits (as observed by Kepler).

Three major subunits

Figure N1.1 illustrates the structure of the unit. The three boxes with dark outlines at the top represent the three major subunits; the boxes strung vertically below each represent the chapters in that subunit.

In *this* subunit, the remainder of this chapter and the next are devoted to exploring how motion can be described using the language of calculus. The middle chapter describes Newton's famous three laws of motion and how they are related to the ideas discussed in Unit *C*. The fourth and fifth chapters discuss in general (and mostly qualitative) terms how we can apply Newton's laws to determine either (1) things about the forces acting on an object when we know how it moves, or (2) how an object will move if we know about the forces acting on it. These chapters thus provide an overview of the issues that we will explore in chapters N6–N10 and N11–N14, respectively.

Beware of your aristotelian intuition!

A problem that you will face throughout the unit (even more than in Unit *C*) is that *everyone's* common-sense, intuitive ideas about physics start out being more aristotelian than newtonian, and so are at odds with the ideas in this unit. Your task is not to *ignore* but rather *transform* your physical intuition, so that you begin to naturally look at the world through newtonian eyes. This is a tricky process that almost everyone finds challenging. The keys to transforming your intuition are to (1) be *aware* of what your physical intuition is saying in a given situation, but (2) *distrust* it until you can convince yourself that what it is saying is also logically consistent with the newtonian model.

The reward is that if you really learn to look at the world through fully newtonian eyes, you will be awestruck by both the simplicity and breadth of your vision. The power of the newtonian world-view is that it includes so much within a simple conceptual framework. The aristotelian world-view seems altogether parochial in comparison!

N1.6 THE TIME-DERIVATIVE OF A VECTOR

Calculus is the key to understanding the detailed link between interactions and motion: indeed, Newton had to *invent* calculus to work out the implications of his model. There are only a handful of applications of Newton's model that can be studied in depth without calculus. The purpose of the remainder of this chapter and the next is to explore how we can use calculus to describe motion.

Time-derivative of an ordinary mathematical function

Consider a quantity $f(t)$ that varies with time t. The **time derivative** df/dt of such a quantity at a given instant of time t is defined as follows

$$\frac{df}{dt} \equiv \lim_{\Delta t \to 0} \frac{f(t+\Delta t) - f(t)}{\Delta t} \tag{N1.1}$$

In words, this equation says that we compute the change $\Delta f = f(t+\Delta t) - f(t)$ in the quantity's value during a time interval Δt starting at time t, calculate the value of the ratio $\Delta f/\Delta t$, and then take the limiting value of this ratio as Δt approaches zero. The derivatives of ordinary numerical functions of a single variable and some of their applications are typically described in an introductory course in calculus. (For a more detailed review of this definition and the limit-taking process, see the *Math Skills* section at the end of this chapter.)

Time-derivative of a vector

We discovered in the last unit that many important quantities in physics (such as force, velocity, momentum, and the like) are described not by ordinary numbers but by *vectors*. Let $\vec{q}$ be some vector quantity that varies with time t. In analogy to equation N1.1, the time derivative $d\vec{q}/dt$ of $\vec{q}$ is defined to be

$$\frac{d\vec{q}}{dt} \equiv \lim_{\Delta t \to 0} \frac{\vec{q}(t+\Delta t) - \vec{q}(t)}{\Delta t} \tag{N1.2}$$

In words, this equation says that we compute the *change* $\Delta\vec{q} = \vec{q}(t+\Delta t) - \vec{q}(t)$ in the vector during the time interval Δt starting at time t, calculate the ratio $\Delta\vec{q}/\Delta t$ (which is a vector), and determine the limiting value of this ratio as Δt approaches zero.

Components of the time-derivative of a vector

By the component definition of the difference between two vectors, the change $\Delta\vec{q} = \vec{q}(t+\Delta t) - \vec{q}(t)$ in $\vec{q}$ can be written

$$\Delta\vec{q} = \begin{bmatrix} q_x(t+\Delta t) \\ q_y(t+\Delta t) \\ q_z(t+\Delta t) \end{bmatrix} - \begin{bmatrix} q_x(t) \\ q_y(t) \\ q_z(t) \end{bmatrix} = \begin{bmatrix} q_x(t+\Delta t) - q_x(t) \\ q_y(t+\Delta t) - q_y(t) \\ q_z(t+\Delta t) - q_z(t) \end{bmatrix} \tag{N1.3}$$

Thus

$$\frac{d\vec{q}}{dt} \equiv \lim_{\Delta t \to 0} \frac{\Delta\vec{q}}{\Delta t} = \lim_{\Delta t \to 0} \frac{1}{\Delta t} \begin{bmatrix} q_x(t+\Delta t) - q_x(t) \\ q_y(t+\Delta t) - q_y(t) \\ q_z(t+\Delta t) - q_z(t) \end{bmatrix} \tag{N1.4}$$

By the definition of multiplication of a vector by a scalar (in this case $\Delta\vec{q}$ by $1/\Delta t$) and since the limit applies to each component individually, this becomes

$$\frac{d\vec{q}}{dt} \equiv \begin{bmatrix} \lim_{\Delta t \to 0} \frac{1}{\Delta t}[q_x(t+\Delta t) - q_x(t)] \\ \lim_{\Delta t \to 0} \frac{1}{\Delta t}[q_y(t+\Delta t) - q_y(t)] \\ \lim_{\Delta t \to 0} \frac{1}{\Delta t}[q_z(t+\Delta t) - q_z(t)] \end{bmatrix} \equiv \begin{bmatrix} \frac{dq_x}{dt} \\ \frac{dq_y}{dt} \\ \frac{dq_z}{dt} \end{bmatrix} \tag{N1.5}$$

In words, this says that the components of the time derivative vector $d\vec{q}/dt$ are simply the ordinary time derivatives of the components of $\vec{q}$.

Exercise N1X.1: Imagine that the position vector of a certain object is a function of time as follows:

$$\vec{r}(t) = \begin{bmatrix} (3.0 \text{ m/s}^2)t^2 \\ 1.2 \text{ m} \\ (2.5 \text{ m/s})t \end{bmatrix} \tag{N1.6}$$

Find the components of the time derivative $d\vec{r}/dt$ of this vector, and evaluate these components at time $t = 2.0$ s. (See the *Math Skills* section for a review of how to take the time derivative of simple powers of t.)

N1.7 VELOCITY

In Unit *C* we defined an object's velocity vector $\vec{v}$ at a time t to be

The definition of velocity that we used in Unit *C*

$$\vec{v} \equiv \frac{d\vec{r}}{dt} \equiv \frac{\text{small displacement}}{\text{short time interval}} \tag{N1.7}$$

where $d\vec{r}$ is the object's displacement during the time interval dt, which in turn (1) encloses the instant t in question, and (2) is "sufficiently short" that the velocity doesn't change significantly during the interval.

This definition is reasonably intuitive and was sufficiently precise for our purposes in Unit *C*. However, this definition really is a bit fuzzy. What constitutes a "sufficiently short" dt is not very clearly defined, partly because the main criterion refers to the quantity (velocity) that we are trying to define! While we may have an *intuitive* sense of what this criterion means, it is not very rigorous and thus cannot be used as a basis for any mathematically precise analysis of motion. The lack of a mathematically precise definition of velocity at an *instant* was a significant stumbling block for natural philosophers until Newton.

The concept of the *time derivative of position* provides a natural way to put the concept of velocity on a mathematically firm foundation. An object's **instantaneous velocity** vector at a given instant of time t is defined to be

The formal definition of *instantaneous velocity*

$$\vec{v} \equiv \frac{d\vec{r}}{dt} \equiv \lim_{\Delta t \to 0} \frac{\Delta \vec{r}}{\Delta t} \equiv \lim_{\Delta t \to 0} \frac{\vec{r}(t+\Delta t) - \vec{r}(t)}{\Delta t} \tag{N1.8}$$

This definition is essentially the same as given by equation N1.7 except that *no* interval Δt is considered "sufficiently short": we define $\vec{v}$ by the *limiting value* that $\Delta\vec{r}/\Delta t$ approaches as Δt goes to *zero*. This definition thus entirely avoids the problem of defining when an interval is "sufficiently short". It also makes it mathematically clear what we can possibly mean by an object's velocity *at an instant*, an idea that superficially seems to be incompatible with the definition of velocity as the object's *displacement* during a (nonzero!) *interval* of time divided by the duration of that interval.

According to equation N1.5, this definition implies that

$$\vec{v} \equiv \frac{d\vec{r}}{dt} \equiv \begin{bmatrix} \lim_{\Delta t \to 0} \frac{1}{\Delta t}[x(t+\Delta t) - x(t)] \\ \lim_{\Delta t \to 0} \frac{1}{\Delta t}[y(t+\Delta t) - y(t)] \\ \lim_{\Delta t \to 0} \frac{1}{\Delta t}[z(t+\Delta t) - z(t)] \end{bmatrix} \equiv \begin{bmatrix} \frac{dx}{dt} \\ \frac{dy}{dt} \\ \frac{dz}{dt} \end{bmatrix} \tag{N1.9}$$

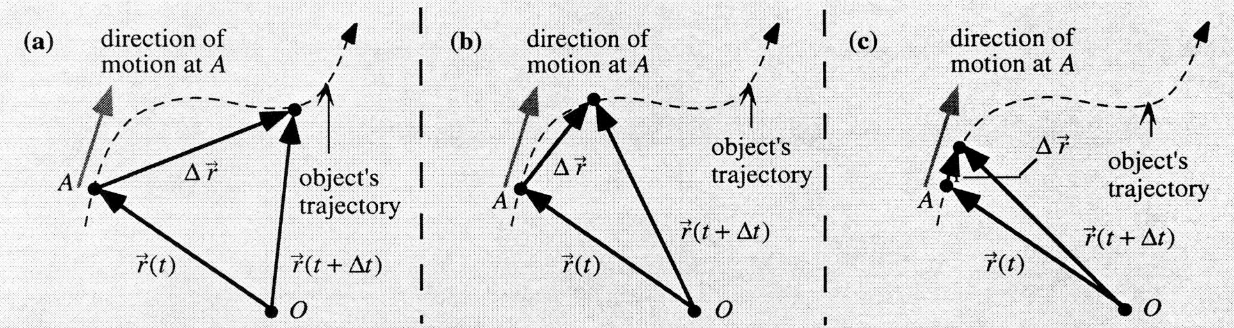

Figure N1.2: The shorter the time interval Δt, the closer that the direction of $\Delta\vec{r}/\Delta t$ (which is the same as that of $\Delta\vec{r}$) becomes to the direction of the object's velocity $\vec{v}$ at time t (which is the direction of the object's motion as it passes point A).

We will still call the components $v_x \equiv dx/dt$, $v_y \equiv dy/dt$, and $v_z \equiv dz/dt$ the object's ***x*-velocity**, ***y*-velocity**, and ***z*-velocity** at time t respectively. The object's speed at time t is

$$v \equiv \text{mag}(\vec{v}) \equiv \sqrt{v_x^2+v_y^2+v_z^2} = \sqrt{\left(\frac{dx}{dt}\right)^2+\left(\frac{dy}{dt}\right)^2+\left(\frac{dz}{dt}\right)^2} \quad \text{(N1.10)}$$

EXAMPLE N1.1

Problem: The components of an object's position as a function of time are given by $x(t) = at^2 + b$, $y(t) = ct$, and $z(t) = 0$, where $a = 2.0$ m/s^2, $b = 5.2$ m, and $c = 1.6$ m/s. What is the object's speed at time $t = 0$?

Solution Taking the time derivative of each of these position components (using the methods discussed in the *Math Skills* section), we find that the components of the object's velocity (as functions of time) are $v_x(t) = dx/dt = 2at$, $v_y(t) = dy/dt = c$, and $v_z(t) = dz/dt = 0$. Evaluating these at time $t = 0$, we find that $v_x(0) = 2a\cdot 0 = 0$, $v_y(0) = c$, and $v_z(0) = 0$. Therefore, the object's speed at time $t = 0$ is $v(0) = [v_x^2+v_y^2+v_z^2]^{1/2} = [0 + c^2 + 0]^{1/2} = c = 1.6$ m/s.

Exercise N1X.2: The components of an object's position vector as functions of time are given by $x(t) = at^2 + b$, $y(t) = -ct$, and $z(t) = -at^2 + ct$, where $a = 1.5$ m/s^2, $b = 3.0$ m, $c = 4.0$ m/s. Find the components of this object's velocity and its speed at $t = 0$ and $t = 2.0$ s.

Average velocity $\Delta\vec{r}/\Delta t \approx$ **instantaneous velocity**

As Figure N1.2 illustrates, the ratio $\Delta\vec{r}/\Delta t$ computed for any nonzero time interval Δt starting at time t is only *approximately* equal the object's instantaneous velocity $\vec{v}$ *at* time t (the approximation gets better as Δt gets smaller). We will call $\vec{v}_{\Delta t} \equiv \Delta\vec{r}/\Delta t$ the object's **average velocity** during the time interval Δt to distinguish it from the object's instantaneous velocity. (The subscript on the symbol $\vec{v}_{\Delta t}$ specifies the interval used to calculate it and distinguishes it from the symbol $\vec{v}$, which from now on we will use exclusively for *instantaneous* velocity.)

As Figure N1.3 shows, $\vec{v}_{\Delta t} \equiv \Delta\vec{r}/\Delta t$ for a nonzero Δt generally most closely approximates the object's instantaneous velocity at an instant t halfway through the interval Δt. (We will use this idea extensively in the next chapter.)

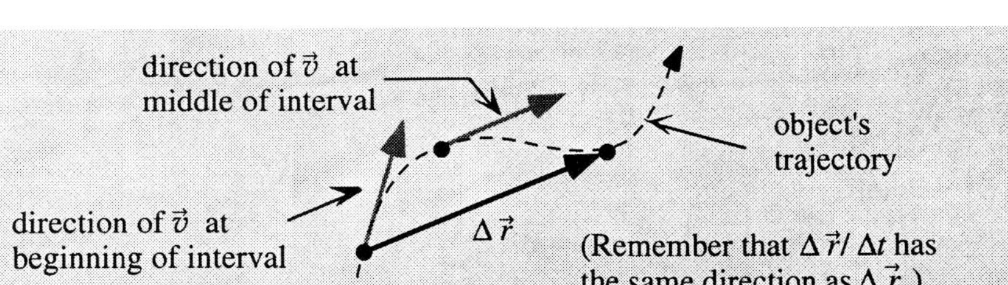

Figure N1.3: The direction of the average velocity $\Delta\vec{r}/\Delta t$ of an object during an interval Δt is generally closer to the direction of its instantaneous velocity $\vec{v}$ at the *middle* of the interval than at the beginning of the interval (the same statement also applies to the *magnitudes* of the vectors).

EXAMPLE N1.2

Problem: The diagram below is a top view of a ball rolling toward the right along an inclined track, showing the ball's position every 0.1 s. Is the ball's speed increasing, decreasing or staying constant? Estimate its speed at $t = 0.3$ s.

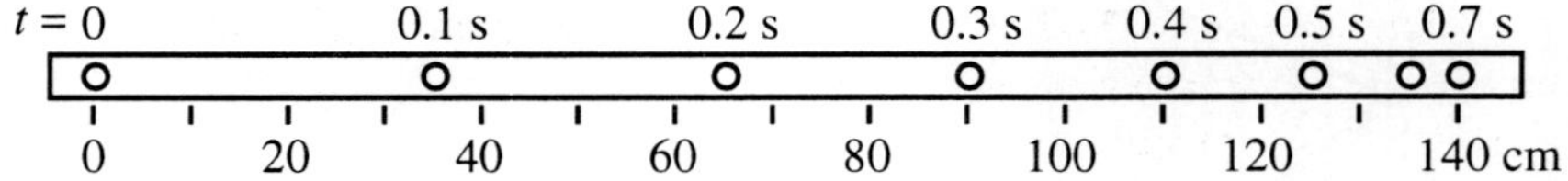

Solution Since the ball's displacement during successive intervals gets smaller, the ball must be slowing down. Its speed at $t = 0.3$ s will be most closely approximated by the magnitude of $\Delta\vec{r}/\Delta t$ for the shortest interval having that instant as its midpoint. Between $t = 0.2$ s and $t = 0.4$ s, the ball travels 45 cm, so $v(0.3\text{ s}) \approx \text{mag}(\Delta\vec{r}/\Delta t)$ during this interval $\approx 45\text{ cm}/0.2\text{ s} = 225\text{ cm/s}$.

MATH SKILLS: Derivatives

The purpose of this section is to review (in the relatively informal language used by physicists) some basic principles of differential calculus, principles that we will use repeatedly in the remainder of this course.

Consider a function of $f(t)$, where t is some *arbitrary* variable (not *necessarily* time, though functions of time will be our main concern in this unit). The t derivative of $f(t)$ evaluated at t is defined to be

Definition of the derivative

$$\frac{df}{dt} \equiv \lim_{\Delta t \to 0} \frac{f(t + \Delta t) - f(t)}{\Delta t} \tag{N1.11}$$

that is, the derivative is the limit as Δt goes to zero of the change in f during the interval Δt starting at t divided by the value of Δt.

Examples of evaluating the limit as Δt goes to zero.

The best way to show what we mean by "the limit as Δt goes to zero" here is to do an example. Imagine that $f(t) = t^2$. The derivative of $f(t)$ in this case is

$$\frac{df}{dt} \equiv \frac{d}{dt}[t^2] \equiv \lim_{\Delta t \to 0} \frac{(t + \Delta t)^2 - t^2}{\Delta t} \tag{N1.12}$$

Multiplying out the binomial in the numerator, we find that

$$\frac{(t + \Delta t)^2 - t^2}{\Delta t} = \frac{t^2 + 2t\Delta t + \Delta t^2 - t^2}{\Delta t} = \frac{2t\Delta t + \Delta t^2}{\Delta t} = 2t + \Delta t \tag{N1.13}$$

As Δt becomes smaller, this clearly gets closer and closer to the value $2t$. Thus

$$\text{if f}(t) = t^2\text{, then } \frac{df}{dt} \equiv \lim_{\Delta t \to 0}[2t + \Delta t] = 2t \tag{N1.14}$$

Here is a second example. Imagine that $f(t) = 1/t$. Its derivative is

$$\frac{df}{dt} \equiv \frac{d}{dt}\left[\frac{1}{t}\right] \equiv \lim_{\Delta t \to 0} \frac{1}{\Delta t}\left(\frac{1}{t + \Delta t} - \frac{1}{t}\right) \tag{N1.15}$$

Putting the expression in parentheses over a common denominator, we find that

$$\frac{1}{\Delta t}\left(\frac{1}{t + \Delta t} - \frac{1}{t}\right) = \frac{1}{\Delta t}\left(\frac{t - (t + \Delta t)}{t(t + \Delta t)}\right) = \frac{1}{\Delta t}\left(\frac{-\Delta t}{t(t + \Delta t)}\right) = \frac{-1}{t(t + \Delta t)} \tag{N1.16}$$

In the limit that Δt becomes very small, the value of $t + \Delta t$ approaches t, so

$$\text{if } f(t) = \frac{1}{t}\text{, then } \frac{df}{dt} = \lim_{\Delta t \to 0}\left[\frac{-1}{t(t + \Delta t)}\right] = -\frac{1}{t^2} \tag{N1.17}$$

In general, one can show using the same general approach that

$$\text{if } f(t) = t^n, \text{ then } \frac{df}{dt} = nt^{n-1} \tag{N1.18}$$

The derivative of $f(t) = t^n$

for any integer $n \neq 0$. (Though the proof is more complicated, this equation also applies to *non-integer* values of n as well.)

If $f(t) = c$ (a constant), taking the limit is easy:

$$\text{If } f(t) = c, \text{ then } \frac{df}{dt} = \lim_{\Delta t \to 0} \frac{c-c}{\Delta t} = \lim_{\Delta t \to 0} \frac{0}{\Delta t} = 0 \tag{N1.19}$$

The derivative of a constant is zero

This case makes it abundantly clear that the *limit* of $0/\Delta t$ as Δt approaches zero (which is zero here) is NOT the same as the *value* of $0/\Delta t$ when $\Delta t = 0$ (which is undefined, as are the ratios in equations N1.12 and N1.15). The limit is the value that the function *approaches* as Δt gets small: since $0/\Delta t$ is zero for all $\Delta t \neq 0$, we have to say that $0/\Delta t$ approaches (indeed *is exactly*) zero as Δt gets small.

The following useful theorems follow from the definition of the derivative: if $f(t)$, $g(t)$ and $h(t)$ are arbitrary functions of t, then

$$\text{if } f(t) = g(t) + h(t), \text{ then } \frac{df}{dt} = \frac{dg}{dt} + \frac{dh}{dt} \tag{N1.20}$$

The sum rule

$$\text{if } f(t) = g(t)h(t), \text{ then } \frac{df}{dt} = \frac{dg}{dt}h + g\frac{dh}{dt} \tag{N1.21}$$

The product rule

$$\text{if } f(t) = \frac{1}{g(t)}, \text{ then } \frac{df}{dt} = -\frac{1}{[g(t)]^2}\frac{dg}{dt} \tag{N1.22}$$

The inverse rule

Equations N1.22 and N1.20 together imply that if c is a constant

$$\text{if } f(t) = c\,h(t), \text{ then } \frac{df}{dt} = c\frac{dh}{dt} \tag{N1.23}$$

The constant rule

Example application

Equations N1.18 through N1.23 make it possible to evaluate the derivative of any polynomial in t. For example, if $f(t) = at^3 + bt + c$, then

$$\frac{df}{dt} = \frac{d}{dt}(at^3) + \frac{d}{dt}(bt) + \frac{d}{dt}(c) \quad \text{(by the sum rule)}$$

$$= a\frac{d}{dt}(t^3) + b\frac{d}{dt}(t) + 0 \quad \text{(by the constant rule and N1.19)}$$

$$= a(3t^2) + b(1) = 3at^2 + b \quad \text{(by equation N1.18)} \tag{N1.24}$$

Exercise N1X.3: Use the rules described above to calculate the t-derivative of the function $f(t) = [at+b]^{-1}$, where a and b are constants.

SUMMARY

I. ARISTOTELIAN PHYSICS
   A. Aristotle (384-322 BC)
      1. Objects have *natural* motions according to their composition:
         a) heavy objects constructed of earth and so seek to rejoin earth
         b) light objects (like fire) tend to move toward heavens
         c) celestial objects move endlessly in perfect circles around earth
      2. Continuing *unnatural* motions must have continuing cause
   B. Ptolemy corrects and extends Aristotle's celestial model (140 AD) (describing planetary motion around earth in terms of *nested* circles)
   C. Basic aristotelian ideas accepted by almost all scholars before late 1500s
      1. continuing motion of terrestrial body required continuing cause
      2. the speed of such a body is proportional to strength of cause
      3. celestial physics based on circular motion, unlike terrestrial physics

II. THE ARISTOTELIAN WORLD-VIEW CRUMBLES
   A. Copernicus proposes a sun-centered model of solar system in late 1500s
   B. Galileo Galilei, at the cutting edge of hot trends in natural philosophy,
      1. sought to make physics *quantitative* as well as qualitative,
      2. used *experiments* to test hypotheses (not just pure reason),
      3. argued that constant motion is natural, requiring *no* cause (this is why things are not swept backwards as earth moves)
   C. Other discoveries and ideas that undermined the aristotelian model
      1. Kepler finds that planets move in *ellipses*, not circles
      2. telescopic observations of Venus show it must *orbit* the sun
      3. Descartes seeks to unify terrestrial and celestial physics

III. THE NEWTONIAN SYNTHESIS
   A. Newton publishes the *Principia* in 1687
      1. It provided a complete theory of terrestrial and celestial mechanics
         a) it predicted planetary orbits consistent with Kepler's laws
         b) it explained a variety of things about terrestrial physics also
      2. It was so impressive it captured the entire physics community
   B. this unification of physics is called the *newtonian synthesis*

IV. AN OVERVIEW OF UNIT *N*
   A. Our focus: how the interactions between objects affect their motion
   B. This unit is divided into three subunits, describing (respectively)
      1. the mathematics of motion and the basics of the newtonian model
      2. how we can use our knowledge of motion to determine forces
      3. how we can use our knowledge about forces to determine motion
   C. Beware of aristotelian ideas buried in your intuition!

V. VELOCITY AND THE CALCULUS OF MOTION
   A. The definition of the time-derivative of a time-dependent *vector* $\vec{q}(t)$ is (in analogy to the definition of an ordinary derivative):

$$\frac{d\vec{q}}{dt} \equiv \lim_{\Delta t \to 0} \frac{\vec{q}(t+\Delta t) - \vec{q}(t)}{\Delta t} = \left[\frac{dq_x}{dt}, \frac{dq_y}{dt}, \frac{dq_z}{dt}\right] \quad \text{(N1.2,5)}$$

   B. An application: the definition of *instantaneous velocity* at some time t

$$\vec{v} \equiv \frac{d\vec{r}}{dt} \equiv \lim_{\Delta t \to 0} \frac{\vec{r}(t+\Delta t) - \vec{r}(t)}{\Delta t} = \left[\frac{dx}{dt}, \frac{dy}{dt}, \frac{dz}{dt}\right] \quad \text{(N1.8,9)}$$

      1. This definition avoids the ambiguity of the definition in Unit *C*
      2. It makes it clear what velocity *at a given instant of time* means
   C. *Average velocity* $\vec{v}_{\Delta t} \equiv \Delta\vec{r}/\Delta t$ approximates instantaneous velocity
      1. this approximation gets better as the interval Δt gets smaller
      2. it most accurately estimates the instantaneous $\vec{v}$ at *middle* of Δt

GLOSSARY

Newtonian synthesis: the unification of celestial and terrestrial physics accomplished by Newton's mechanics.

time derivative $d\vec{q}/dt$ (of a vector $\vec{q}$): the limiting value as Δt goes to zero of the change $\Delta\vec{q}$ in $\vec{q}$ divided by the time interval Δt during which the change takes place:

$$\frac{d\vec{q}}{dt} \equiv \lim_{\Delta t \to 0} \frac{\Delta\vec{q}}{\Delta t} \equiv \lim_{\Delta t \to 0} \frac{\vec{q}(t+\Delta t)-\vec{q}(t)}{\Delta t} \qquad \text{(N1.2)}$$

instantaneous velocity $\vec{v}$ (of an object): the time derivative of the object's position vector:

$$\vec{v} \equiv \frac{d\vec{r}}{dt} \equiv \lim_{\Delta t \to 0} \frac{\Delta\vec{r}}{\Delta t} \equiv \lim_{\Delta t \to 0} \frac{\vec{r}(t+\Delta t)-\vec{r}(t)}{\Delta t} \qquad \text{(N1.8)}$$

This formally defines in a mathematically supportable way what we mean by velocity at a single instant of time.

***x*-velocity** v_x, ***y*-velocity** v_y, ***z*-velocity** v_z: the components of an object's instantaneous velocity. Note that $v_x = dx/dt$, $v_y = dy/dt$, $v_z = dz/dt$.

average velocity (of an object during an interval Δt): the object's displacement $\Delta\vec{r}$ during that interval divided by the duration Δt of the interval. An object's average velocity during an interval is approximately equal to its instantaneous velocity at any instant during the interval: the approximation improves as Δt gets smaller. For any finite Δt, the average velocity during an interval generally best approximates the instantaneous velocity at the *middle* of the interval.

the sum rule: if $f(t) = g(t) + h(t)$, then

$$\frac{df}{dt} = \frac{dg}{dt} + \frac{dh}{dt} \qquad \text{(N1.20)}$$

the product rule: if $f(t) = g(t)h(t)$, then

$$\frac{df}{dt} = \frac{dg}{dt}h + g\frac{dh}{dt} \qquad \text{(N1.21)}$$

the inverse rule: if $f(t) = 1/g(t)$, then

$$\frac{df}{dt} = -\frac{1}{g^2}\frac{dg}{dt} \qquad \text{(N1.22)}$$

the constant rule: if $f(t) = c\,h(t)$, then

$$\frac{df}{dt} = c\frac{dh}{dt} \qquad \text{(N1.23)}$$

TWO-MINUTE PROBLEMS

N1T.1 Classify the following statements about the physical world as expressing an aristotelian viewpoint (A) or newtonian viewpoint (B).

(a) If you push on something it moves. If you push twice as hard, it moves twice as fast.
(b) A heavy object falls faster than a light object.
(c) The moon is held in its orbit around the earth by the force of the earth's gravity.
(d) If you push on an object, it moves. After you release it, it gradually comes to rest because friction drains away the force of the initial push.
(e) The speed of a falling object increases as it falls.

N1T.2 Imagine that you drop a tennis ball while you walk fairly rapidly at a constant speed. Where will the ball hit relative to your position at the instant it hits?

A. somewhat in front of you
B. directly at your feet
C. somewhat behind you

N1T.3 Two marbles are rolling along parallel tracks (which may or may not be inclined). A "stroboscopic photograph showing a top view of the positions of the marbles at equally spaced instants of time looks like this:

t_A t_B t_C t_D t_E t_F t_G

t_A t_B t_C t_D t_E t_F t_G

At what instant(s) of time do the marbles have the same instantaneous velocity?

A. at time t_B	D. some time between t_C and t_D
B. at time t_G	E. some time between t_D and t_E
C. at both t_B and t_G	F. roughly time t_D

N1T.4 Two marbles are rolling along parallel tracks (which may or may not be inclined). A "stroboscopic photograph" showing a top view of the positions of the marbles at equally spaced instants of time looks like this:

t_A t_B t_C t_D t_E t_F t_G

t_A t_B t_C t_D t_E t_F t_G

At what instant(s) of time do the marbles have the same instantaneous velocity?

A. at time t_B	D. some time between t_C and t_D
B. at time t_F	E. some time between t_C and t_E
C. at both t_B and t_F	F. roughly time t_D

N1T.5 An object travels halfway around a circle at a constant instantaneous speed v. What is the magnitude of its average velocity during this time interval?

A. v	D. $v/1.41$
B. $1.41v$	E. $v/1.57$
C. $1.57v$	F. other (specify)
	T. not enough information

N1T.6 An object's position as a function of time is given by $\vec{r}(t) = [-at^2,\ bt+c,\ at^2]$, where a, b, and c are constants with values 2.0 m/s^2, −5.0 m/s, and 2.5 m respectively. What is the object's speed at time $t = 0$?

A. zero	D. −2.5 m/s
B. −5.0 m/s	E. +2.5 m/s
C. +5.0 m/s	F. +5.7 m/s
	T. other (specify)

HOMEWORK PROBLEMS

BASIC SKILLS

N1B.1 What are the derivatives of the following functions of time (where a, b, and c are constants)?
(a) $at + c$ **(b)** $at^2 + (bt)^{-2}$ **(c)** $(at + b)^2$

N1B.2 What are the derivatives of the following functions of time (where a, b, and c are constants?)
(a) $ct(at^2+b)$ **(b)** $(at+b)^{-2}$ **(c)** $(at^3+b)/t$

N1B.3 An object's position as a function of time is given by $\vec{r}(t) = [-a/t,\ bt^2+c,\ qt]$, where a, b, c, and q are constants with values $a = 2.0$ m·s, $b = 1.0$ m/s^2, $c = 5.0$ m, and $q = 2.0$ m/s. Calculate the components of this object's instantaneous velocity at time $t = 2.0$ s.

N1B.4 An object's position as a function of time is given by $\vec{r}(t) = [-at^3+bt,\ c,\ +qt^4]$, where a, b, c, and q are constants with values $a = 0.5$ m/s^3, $b = 2.0$ m/s, $c = 2.0$ m, and $q = 1.0$ m/s^4. Calculate the components of this object's instantaneous velocity at time $t = 1.0$ s.

N1B.5 An object travels exactly once around a circle of radius 5.0 m at a constant instantaneous speed of 3.0 m/s. What is its average velocity during this interval?

N1B.6 Show that equation N1.23 in the *Math Skills* section follows from equations N1.21 and N1.19.

SYNTHETIC

N1S.1 Using the basic approach illustrated by equations N1.12 through N1.17, *prove* that if $f(t) = at$ (where a is a constant) then $df/dt = a$. (Do *not* use equation N1.18.)

N1S.2 Using the basic approach illustrated by equations N1.12 through N1.17, *prove* that if $f(t) = t^{-2}$ then $df/dt = -2t^{-3}$. (Do *not* use equation N1.18.)

N1S.3 An object's position as a function of time is given by $\vec{r}(t) = [c(bt+1)^{-1},\ cbt,\ at^2]$, where a, b, and c are constants. **(a)** What are the units of a, b, and c in the SI units system? **(b)** Find an expression for the object's *speed* as a function of time.

N1S.4 An object's position as a function of time is given by $\vec{r}(t) = [a(t+b)^{-2},\ ct,\ cb]$, where a, b, and c are constants. **(a)** What are the units of a, b, and c in the SI units system? **(b)** Find an expression for the object's *speed* as a function of time.

N1S.5 Between $t = 3.0$ s and $t = 4.0$ s, an object moving along a straight line moves a distance of 2.0 m in the $+x$ direction. Between $t = 4.0$ s and $t = 5.0$ s, the object moves a distance of 4.0 m in this same direction. *Estimate* the object's instantaneous speed at $t = 4.0$ s, and explain how you arrived at this estimate and *why* it is an estimate.

N1S.6 Prove that the *sum rule* (see equation N1.20 in the *Math Skills* section) follows from the definition of the derivative. (You may assume without proof that the limit as $\Delta t \to 0$ of a sum of two quantities is the same as the sum of the limits of each quantity separately.)

N1S.7 Prove that the *inverse rule* (see equation N1.22 in the *Math Skills* section) follows from the definition of the derivative. (You may assume without proof that the limit as $\Delta t \to 0$ of the product of two quantities is the same as the products of the limits of each quantity separately.)

N1S.8 Prove that the *product rule* (see equation N1.21 in the *Math Skills* section) follows from the definition of the derivative. (*Hint:* It is easiest if you start with the right side of equation N1.21 and prove that it is equal to the left side rather than vice versa. You may assume without proof that the limit as $\Delta t \to 0$ of a sum of two quantities is the same as the sum of the limits of each quantity separately.)

RICH-CONTEXT

N1R.1 If you look at the planet Venus through even a small telescope, you can see that at some times it looks like a tiny crescent moon and at other times looks more like a tiny full moon. Galileo noted that whenever Venus looked crescent shaped it also always looked significantly larger than it does when it looks like a full moon. Argue (with the help of some diagrams) that this makes sense if Venus goes around the *sun*, but not if it goes around the earth. Contrast this situation with the moon, which *does* go around the earth.

ADVANCED

N1A.1 Prove that equation N1.18 is true for all n that are nonzero positive integers. (*Hint:* this is easiest to prove by *induction*: show directly that equation N1.18 is true for $n = 1$ and $n = 2$, and then show using the product rule that if it is true for $n = m$, where m is any given integer > 1, then it must be true for $n = m+1$ as well. This effectively proves its validity for all integer values of $n > 1$.)

ANSWERS TO EXERCISES

N1X.1 $d\vec{r}/dt$ = [12 m/s, 0, 2.5 m/s]

N1X.2 At an arbitrary time t, the components of this object's velocity are

$$\vec{v} = \begin{bmatrix} dx/dt \\ dy/dt \\ dz/dt \end{bmatrix} = \begin{bmatrix} 2at \\ -c \\ -2at+c \end{bmatrix} \qquad \text{(N1.25)}$$

At $t = 0$, the velocity is $\vec{v}(0)$ = [0, –4.0 m/s, +4.0 m/s], and the corresponding speed is 5.7 m/s. At $t = 2.0$ s the velocity is $\vec{v}(2.0\text{ s})$ = [6.0 m/s, –4.0 m/s, –2.0 m/s] and the corresponding speed is 7.5 m/s.

N1X.3 Using the inverse rule, we have

$$\frac{df}{dt} = -\frac{1}{[at+b]^2}\frac{d}{dt}[at+b] = -\frac{a}{[at+b]^2} \qquad \text{(N1.26)}$$

N2

ACCELERATION

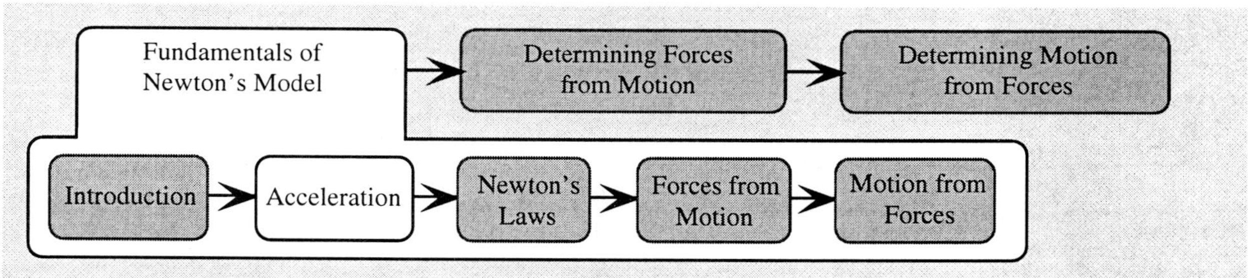

N2.1 OVERVIEW

In the last chapter, we learned how to take the derivative of a vector and used this idea to define rigorously what we mean by an object's velocity at a mathematical instant of time, even though actually measuring the object's velocity involves determining its displacement over a nonzero *interval* of time.

Our task in this chapter is to apply the same mathematical ideas to the concept of *acceleration*. Acceleration is the crucial link in newtonian mechanics between an object's motion and the forces that cause that motion to change.

One of the classic barriers to learning the newtonian model is that *acceleration* is a more abstract concept than velocity. While we can in a certain sense *see* how fast an object is moving and what direction it is going, the magnitude and particularly the direction of an object's acceleration is not something that we intuitively see: we have to *think* about an object's motion to perceive its acceleration. One of the aims of this chapter is to make the idea of acceleration as vivid and intuitive as possible, partly with the help of visual aids.

Here is an overview of the chapter's sections.

N2.2 *THE DEFINITION OF ACCELERATION* carefully discusses the technical meaning of the word *acceleration* in physics.

N2.3 *AVERAGE ACCELERATION* defines the idea of *average acceleration* and discusses how we can use it in a variety of circumstances to estimate an object's instantaneous acceleration without using calculus.

N2.4 *MOTION DIAGRAMS* illustrates a different kind of visual aid that we can use to represent the motion (and qualitatively determine the acceleration) of objects moving in one or two dimensions.

N2.5 *NUMERICAL RESULTS FROM MOTION DIAGRAMS* explains how we can read the actual magnitudes of an object's velocity and acceleration from a carefully constructed motion diagram.

N2.6 *UNIFORM CIRCULAR MOTION* shows how we can use a quantitative motion diagram to help us understand an object's acceleration when it moves at a constant speed in a circular path.

An understanding of acceleration is an *essential* foundation for the rest of this unit: the newtonian approach to analyzing motion uses this concept extensively. Be *sure* that you get your questions answered about this material!

N2.2 THE DEFINITION OF ACCELERATION

We saw in unit *C* that when an object participates in an interaction, the interaction transfers momentum to or from the object, which generally changes the object's total momentum and thus its velocity. Interactions are thus linked to changes in velocity. In order to quantify how interactions affect motion, then, we have to understand how we can mathematically describe the *rate* at which an object's velocity changes, that is, its *acceleration*. The purpose of this chapter is to offer a careful introduction to the concept of acceleration.

The word *acceleration* in everyday English carries a connotation of "speeding up." Its meaning in *physics* is both more general and more precise:

Physics definition of acceleration in words

An object's **acceleration** at an instant is a vector that expresses *how rapidly* and *in what direction* its velocity vector is changing at that instant.

In physics, then, we use the word *acceleration* to describe *any* change in the magnitude or direction of an object's velocity. A car that is speeding up is indeed accelerating (because the magnitude of its velocity is increasing), but according to the physics definition of the word, a car that is slowing down is *also* accelerating, because the magnitude of its velocity vector is changing (in this case decreasing). Even a car moving at a constant speed can be accelerating if it is going around a bend in the road (in this case because the *direction* of its velocity is changing). If an object's velocity vector changes in *any* way, the object is accelerating in the technical meaning of the word.

Mathematically, we define an object's acceleration vector $\vec{a}$ at a given instant of time t as follows

Mathematical definition of the acceleration vector

$$\vec{a} \equiv \lim_{\Delta t \to 0} \frac{\Delta \vec{v}}{\Delta t} \equiv \lim_{\Delta t \to 0} \frac{\vec{v}(t+\Delta t) - \vec{v}(t)}{\Delta t} \equiv \frac{d\vec{v}}{dt} \qquad \text{(N2.1)}$$

In words, this equation says that an object's acceleration at time t is the limiting value as Δt goes to zero of the *change* $\Delta\vec{v}$ in the object's instantaneous velocity between times t and $t+\Delta t$, divided by the duration Δt of that interval. Since velocity is measured in meters per second, the SI units of acceleration are m/s^2.

The components of the acceleration vector

By the definition of the difference between two vectors, the components of an object's acceleration in a given reference frame are:

$$\begin{bmatrix} a_x \\ a_y \\ a_z \end{bmatrix} \equiv \lim_{\Delta t \to 0} \frac{1}{\Delta t}\left(\begin{bmatrix} v_x(t+\Delta t) \\ v_y(t+\Delta t) \\ v_z(t+\Delta t) \end{bmatrix} - \begin{bmatrix} v_x(t) \\ v_y(t) \\ v_z(t) \end{bmatrix} \right)$$

$$= \begin{bmatrix} \lim_{\Delta t \to 0} \dfrac{v_x(t+\Delta t) - v_x(t)}{\Delta t} \\ \lim_{\Delta t \to 0} \dfrac{v_y(t+\Delta t) - v_y(t)}{\Delta t} \\ \lim_{\Delta t \to 0} \dfrac{v_z(t+\Delta t) - v_z(t)}{\Delta t} \end{bmatrix} = \begin{bmatrix} \dfrac{dv_x}{dt} \\ \dfrac{dv_y}{dt} \\ \dfrac{dv_z}{dt} \end{bmatrix} \qquad \text{(N2.2)}$$

Therefore, if we know the components of an object's velocity vector as a function of time, we can compute the components of its acceleration simply by taking the time derivatives of the corresponding velocity components.

Exercise N2X.1: Imagine that the components of an object's velocity vector are $v_x(t) = q$, $v_y(t) = 0$, and $v_z(t) = bt + q$, where $q = 5.0$ m/s and $b = -10$ m/s^2. Find the components of the object's acceleration at time $t = 0$ and $t = 2$ s.

N2.3 AVERAGE ACCELERATION

Just as an object's average velocity $\vec{v}_{\Delta t} \equiv \Delta\vec{r}/\Delta t$ for a given nonzero (but fairly short) interval of time Δt is a good approximation for its instantaneous velocity midway through the interval, so an object's **average acceleration** $\vec{a}_{\Delta t} \equiv \Delta\vec{v}/dt$ during a given nonzero (reasonably short) interval Δt is a good approximation to its instantaneous acceleration midway through the interval:

$$\vec{a}(t+\tfrac{1}{2}\Delta t) \approx \vec{a}_{\Delta t} \equiv \frac{\Delta\vec{v}}{\Delta t} \equiv \frac{\vec{v}(t+\Delta t)-\vec{v}(t)}{\Delta t} \qquad \text{(N2.3)}$$

Definition of the average acceleration

Therefore, if we know an object's velocity at two instants of time, we can estimate its acceleration halfway through the interval by computing the difference $\Delta\vec{v}$ between the velocities and dividing by Δt. This approximation gets better as Δt becomes very small compared to the time required for the object's acceleration to change significantly. If the object's acceleration is *constant* during the interval, then the instantaneous acceleration throughout the interval is the *same* as the average acceleration.

In situations where we need only to *estimate* an object's acceleration, equation N2.3 provides a quick way to connect an object's acceleration to the change in its velocity *without* doing any calculus. The following examples illustrate the application of this idea in several contexts.

EXAMPLE N2.1

Problem: Starting from rest, a car reaches a velocity of 15 m/s eastward after accelerating along a straight stretch of road for 5 s. What was the magnitude and direction of its average acceleration during this period?

Solution Since the car's initial velocity is zero, its *change* in velocity $\Delta\vec{v}$ in this case is simply equal to its final velocity of 15 m/s eastward. The car's average acceleration is simply this change in velocity divided by the duration of the time interval: $\vec{a}_{\Delta t}$ = (15 m/s eastward)/(5.0 s) = 3.0 m/s^2 eastward.

EXAMPLE N2.2

Problem: A driver driving eastward on a straight road at 20 m/s (44 mi/h) sees the brake lights of the person in front go on and so applies the brakes for 1.5 s. This slows the car to 14 m/s. What was the magnitude and direction of the car's average acceleration during this period?

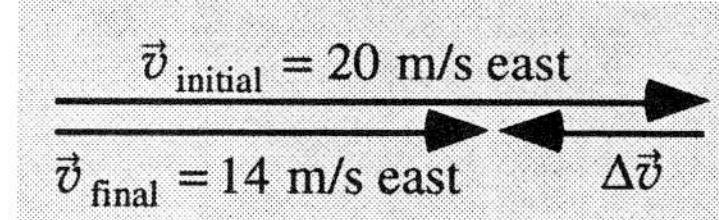

Figure N2.1: The car's *change* in velocity in this example points westward, according to the definition of the vector difference.

Solution The car's initial velocity vector is 20 m/s eastward; its final velocity vector is only 14 m/s eastward. As the vector construction in Figure N2.1 illustrates, this means that the car's change in velocity is $\Delta\vec{v}$ = 6.0 m/s *westward.* (Remember that difference $\Delta\vec{v}$ between the final and initial velocities is the vector that we would have to add to the initial velocity to get the final velocity.) The car's average acceleration in this case is thus $\vec{a}_{\Delta t}$ = (6.0 m/s west)/(1.5 s) = 4.0 m/s^2 westward.

EXAMPLE N2.3

Problem: A bicyclist travels at a constant speed of 12 m/s around a bend in the road during a 34-s interval of time. If the bicyclist was traveling northward at the beginning of the bend and westward at the end, what is the magnitude and direction of the bicyclist's average acceleration during this interval?

Solution Figure N2.2 shows that $\Delta\vec{v}$ in this case is a vector pointing *southwest.* According to the pythagorean theorem, the magnitude of this vector is

$$\text{mag}(\Delta\vec{v}) = \sqrt{(12\text{ m/s})^2+(12\text{ m/s})^2} = 17\text{ m/s} \qquad \text{(N2.4)}$$

The magnitude of the bicyclist's average acceleration is therefore mag($\Delta\vec{v}$)/Δt = (17 m/s)/(34 s) = 0.50 m/s^2. So, the cyclist's average acceleration during this interval is 0.5 m/s^2 southwest (even though the cyclist's speed is constant!).

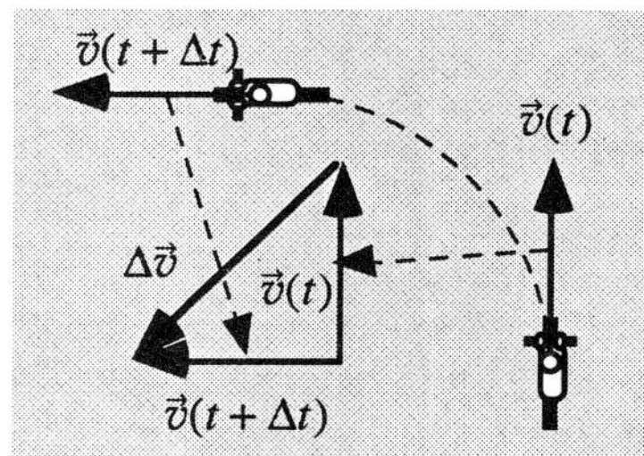

Figure N2.2: The bicyclist's change in velocity.

Exercise N2X.2: A car starting from rest travels forward at in a straight line with an average acceleration of 3.0 m/s^2 for 8.0 s. What is the car's final speed?

Exercise N2X.3: Imagine that you throw a ball vertically into the air. The ball leaves your hand traveling upward at a speed of 15 m/s. Three seconds later, the ball passes you at the same speed on its way downward. What are the magnitude and direction of the ball's average acceleration during this time interval?

N2.4 MOTION DIAGRAMS

Why motion diagrams are useful

When we apply the newtonian model in many practical situations (as we will see), it is often helpful to know the *direction of an object's acceleration vector* before we can even start. A **motion diagram** is a tool that both vividly illustrates an object's motion in one or two dimensions and also provides a way to visually determine the direction of an object's acceleration. The purpose of this section is to discuss how to construct and interpret motion diagrams.

Steps in drawing a motion diagram

Before drawing such a diagram, it helps to visualize what a *strobe photograph* of the object would look like (such a photograph is made using a pulsing flash to record on a single picture multiple images of a moving object at equally spaced instants of time). Figure N2.3a illustrates such a "strobe photograph" of a car that is braking to a stop. Note that the distance between successive images of the car get smaller as time passes, because as the car's speed decreases, its displacement between equally-spaced instants of time gets smaller.

The first step in drawing a motion diagram is to draw a *single* image for the object (to show what object we are considering and how it is oriented). Then we draw a sequence of dots, starting below (or perhaps alongside) that image that represent the positions of object's *center of mass* at equally-spaced instants of time, as shown in Figure N2.3b. Let's label these dots 1, 2, 3, ... in sequence.

We then draw arrows from one dot to the next, as shown in Figure N2.3c. Each such arrow corresponds to the *displacement* $\Delta\vec{r}$ of the object's center of mass as it moves from the dot at the arrows tail to the dot at its head. The object's average velocity during this time interval is $\vec{v}_{\Delta t} = \Delta\vec{r}/\Delta t$, so the arrow is *also* equivalent to the vector $\vec{v}_{\Delta t}\,\Delta t$. We conventionally give the arrow that we draw between dots 1 and 2 the label $\vec{v}_{12}\Delta t$, the arrow that we draw between dots 2 and 3 the label $\vec{v}_{23}\,\Delta t$, and so on (see Figure N2.3c). The subscripts "12", "23", and so on take the place of the Δt subcript that we have been using to indicate the average velocity, but serve the same purpose of specifying the time interval over which we are determining the average velocity.

Note that if the time interval Δt between adjacent dots is always the same, then the arrows $\vec{v}_{12}\Delta t$, $\vec{v}_{23}\,\Delta t$, and so on have the same direction as and are proportional in magnitude to the object's actual average velocity vectors $\vec{v}_{12}$, $\vec{v}_{23}$, and so on, respectively. These "velocity arrows" therefore vividly depict the object's changing velocity as it moves.

Indeed, the object's average velocity $\vec{v}_{12}$ between instants 1 and 2 approximates the object's *instantaneous* velocity most closely at an instant halfway

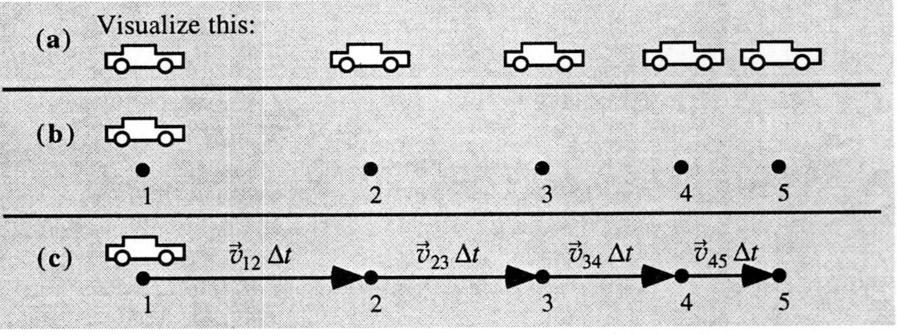

Figure N2.3: How to draw a motion diagram of a car that is braking to a stop. **(a)** Visualize how a "strobe photograph" of the moving object might look. **(b)** Draw a single image of the object, and then draw dots to indicate the position of the object's center of mass at equally-spaced instants of time. **(c)** Draw arrows from each dot to the next. These arrows are proportional to the object's average velocity between the instants depicted.

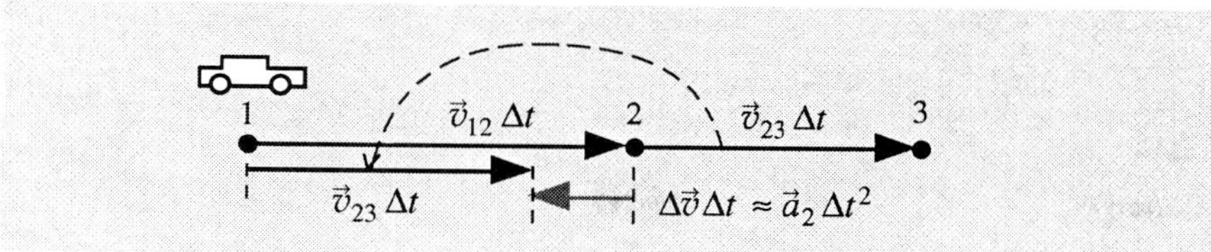

Figure N2.4: How to construct an arrow that represents the object's (approximate) acceleration at the instant when the object passes position 2. The curved dotted arrow shows how we move the $\vec{v}_{23}\,\Delta t$ to perform the subtraction.

between 1 and 2, an instant we might call "1.5". Similarly, the object's average velocity $\vec{v}_{23}$ approximates its instantaneous velocity at the instant halfway between 2 and 3 (instant "2.5"). So the velocity arrows drawn in Figure N2.3c also (approximately) depict the directions and the relative magnitudes the object's *instantaneous* velocities at instants 1.5, 2.5 and so on.

Constructing arrows representing the acceleration

The final step in drawing a motion diagram is to use these velocity arrows to construct arrows representing the object's *acceleration* as it passes the numbered dots. Since the instants 1.5 and 2.5 are also separated by a time interval of Δt, the object's average *acceleration* between instants 1.5 and 2.5 is

$$\vec{a}_{\Delta t} = \frac{\Delta\vec{v}}{\Delta t} = \frac{\vec{v}(2.5) - \vec{v}(1.5)}{\Delta t} \approx \frac{\vec{v}_{23} - \vec{v}_{12}}{\Delta t} \tag{N2.5}$$

The arrows drawn on the diagram *actually* correspond to $\vec{v}_{12}\Delta t$ and $\vec{v}_{23}\,\Delta t$. If we multiply both sides of equation N2.5 by Δt^2, we see that

$$\vec{a}_{\Delta t}\,\Delta t^2 \approx \vec{v}_{23}\,\Delta t - \vec{v}_{12}\,\Delta t \tag{N2.6}$$

The average acceleration $\vec{a}_{\Delta t}$ in turn approximates the object's *instantaneous* acceleration most closely at the instant halfway between the instants 1.5 and 2.5, which is the instant that the object's center of mass passes point 2. Therefore, the arrow representing the vector *difference* $\vec{v}_{23}\Delta t - \vec{v}_{12}\,\Delta t$ between adjacent velocity arrows on the diagram is a good approximation to the vector $\vec{a}_2\,\Delta t^2$, where $\vec{a}_2$ is the object's instantaneous acceleration as it passes point 2:

$$\vec{a}_2\,\Delta t^2 \approx \vec{a}_{\Delta t}\,\Delta t^2 \approx \vec{v}_{23}\,\Delta t - \vec{v}_{12}\,\Delta t \tag{N2.7}$$

Figure N2.4 shows how we can construct this arrow on a motion diagram (for an object moving in one dimension) we:

1. Draw dashed vertical lines down from each dot,
2. Redraw the arrow $\vec{v}_{23}\,\Delta t$ slightly below the arrow $\vec{v}_{12}\Delta t$ so that the tail of $\vec{v}_{23}\,\Delta t$ lines up with the vertical line coming down from dot 1,
3. Then draw an arrow from the tip of the initial velocity arrow $\vec{v}_{12}\Delta t$ to the tip of the final velocity $\vec{v}_{23}\,\Delta t$: this constructed arrow corresponds to the vector difference $\vec{v}_{23}\,\Delta t - \vec{v}_{12}\,\Delta t = \Delta\vec{v}\,/\,\Delta t \approx \vec{a}_2\,\Delta t^2$.

Using the dashed lines as guides for displacing the arrows a bit away from each other (instead of piling them on top of each other) makes the diagram clearer. (Note that this process conveniently puts the tail end of $\vec{a}_2\,\Delta t^2$ at point 2.)

We can construct the acceleration arrows $\vec{a}_3\,\Delta t^2, \vec{a}_4\,\Delta t^2$, and so on analogously. Again, since Δt^2 is a scalar and has the same value for all such arrows on a given diagram, these acceleration arrows have the same direction and are proportional in length to the object's instantaneous acceleration $\vec{a}$ as it passes the numbered dots on the diagram. These arrows on a motion diagram thus usefully *depict* the object's vector acceleration during its motion.

A summary of items in a motion diagram

A complete motion diagram thus shows the following items:

1. A single image of the object in question,
2. a set of dots that represent the positions of the object's center of mass at equally-spaced instants of time,
3. (average) velocity arrows drawn between these dots,
4. acceleration arrows centered on each dot (except the first and last).

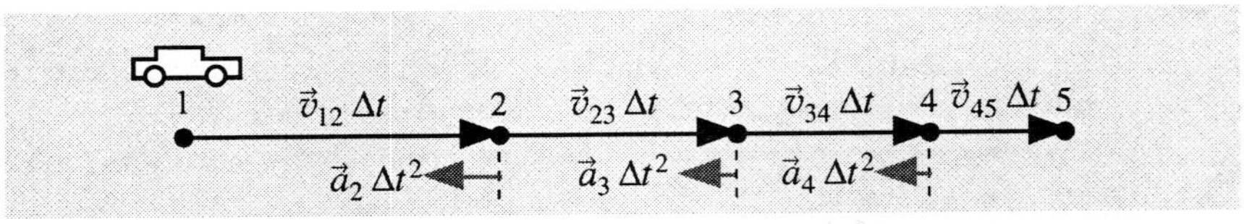

Figure N2.5: A complete motion diagram for a braking car (with optional labels).

Figure N2.5 illustrates a complete motion diagram for the braking car. In a motion diagram that you are using for purely *qualitative* purposes, you do not need to include labels for the points or arrows. (Note that you can tell the difference between the velocity and acceleration arrows even without labels, because the velocity arrows always stretch between the dots while the acceleration arrows generally do not.) You also do not *need* to show how you constructed the acceleration arrows (if you can do the construction in your head), but if you include such diagrams showing how you constructed the acceleration arrows, draw them below or alongside the motion diagram as shown in Figure N2.4.

EXAMPLE N2.4

Problem: A ball rolls up an incline, then back down. Draw a motion diagram of this situation.

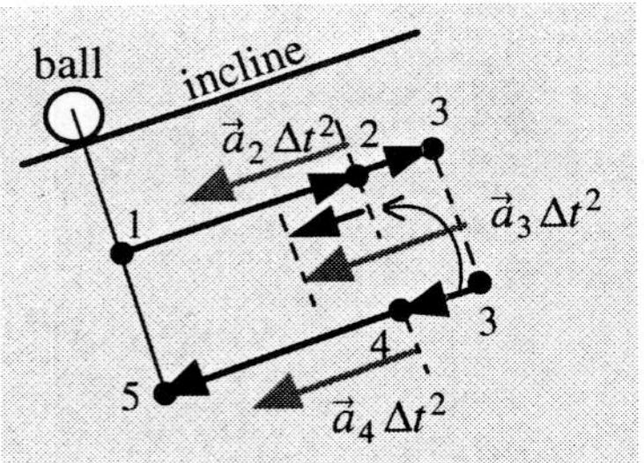

Figure N2.6: Motion diagram of a ball rolling first up and then down an incline.

Solution Figure N2.6 shows an appropriate motion diagram for this situation, showing the ball slowing down as it rolls up the incline and speeding up as it rolls back down the incline. The tricky thing here is that the ball's positions as it rolls up the incline overlap with its positions as it rolls down. This would be very confusing unless we offset the position points for the up and down motions as shown in the diagram. The two points labeled 3 at the right of the diagram both represent the ball's position at the *single* instant when it is at rest at the top of the incline (this instant is common to both the up and down motions).

The diagram explicitly shows the construction of the ball's acceleration arrow $\vec{a}_3\,\Delta t^2$ as the ball passes that position, which is also when its instantaneous velocity passes through zero as it changes direction. (The solid thin curved line shows how I have moved the arrow $\vec{v}_{34}\,\Delta t$ to make its tail coincide with the tail of the arrow $\vec{v}_{23}\,\Delta t$ so that I can do the subtraction yielding $\vec{a}_3\,\Delta t^2$.) This construction clearly shows that the object's acceleration is nonzero even at that instant. This may seem counterintuitive to you, but just because an object's velocity is *passing through* zero at a certain instant doesn't mean that the *rate of change* of its velocity has to be zero at that instant! Indeed, the ball's acceleration seems to be nonzero and directed down the incline throughout its motion.

Exercise N2X.4: Draw a qualitative motion diagram of a car speeding up.

Exercise N2X.5: Draw a qualitative motion diagram of a basketball falling toward the floor and then rebounding from it.

One can often determine the direction (and even estimate the length) of the acceleration arrows by eye on a *one*-dimensional motion diagram. When drawing a motion diagram for an object moving in *two* dimensions, though, you should always show the construction of the acceleration arrow explicitly. This is usually pretty easy, since the arrows to be subtracted usually do *not* lie on top of each other (and so do not have to be displaced in complicated ways as in Figure N2.6). To construct the acceleration arrow at a given point (say, point 2), you

Drawing a motion diagram for two-dimensional motion

1. Take the velocity arrow following the point in question (point 2 here) and move it (without changing its direction) until its tail end coincides with the tail end of the previous velocity arrow; then

2. Draw an arrow from the head of the previous velocity arrow to the head of the arrow you just moved: by definition of vector subtraction the arrow that you have just constructed is $\vec{v}_{23}\,\Delta t - \vec{v}_{12}\,\Delta t \;=\; \Delta\vec{v}\,\Delta t \approx \vec{a}_2\,\Delta t^2$.

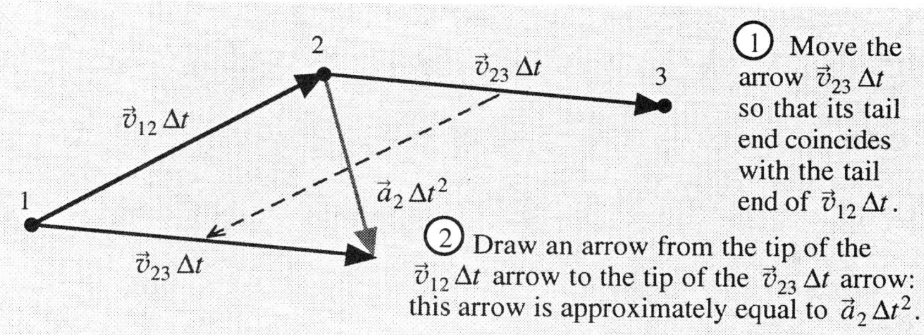

Figure N2.7: How to construct an object's acceleration arrow on a motion diagram for an object moving in two dimensions.

(see Figure N2.7). Note that this method also automatically leaves the acceleration arrow with its tail end at the point to which it applies.

Exercise N2X.6: Draw a qualitative motion diagram of a car going over the crest of a hill at a constant speed. Is its acceleration nonzero?

N2.5 NUMERICAL RESULTS FROM MOTION DIAGRAMS

A hastily-sketched motion diagram can give us important *qualitative* information about the direction and approximate relative magnitudes of an object's acceleration as it follows the motion depicted. However, a carefully-drawn motion diagram can allow us to actually compute the approximate *quantitative magnitudes* of the object's velocity and acceleration at various points along its trajectory by measuring the lengths of the arrows on the diagram.

For example, if we know that the time interval Δt between dots on a motion diagram is 0.10 s and we measure the velocity arrow $\vec{v}_{12}\,\Delta t$ on the diagram to have a length of 2.5 cm, then the actual magnitude of the object's average velocity during the time interval between dots 1 and and 2 is

$$v_{12} \equiv \text{mag}(\vec{v}_{12}) = \frac{\text{mag}(\vec{v}_{12}\,\Delta t)}{\Delta t} = \frac{2.5\text{ cm}}{0.10\text{ s}} = 25\text{ cm/s} \qquad \text{(N2.8a)}$$

Similarly, if we measure the acceleration arrow $\vec{a}_2\,\Delta t^2$ on the diagram to have a length of 1.2 cm, the magnitude of the object's acceleration at point 2 is roughly

$$a_2 \equiv \text{mag}(\vec{a}_2) = \frac{\text{mag}(\vec{a}_2\,\Delta t^2)}{\Delta t^2} = \frac{1.2\text{ cm}}{(0.10\text{ s})^2} = 120\text{ cm/s}^2 \qquad \text{(N2.8b)}$$

EXAMPLE N2.5

Problem: Assume that the time between dots in Figure N2.7 is 0.20 s. What is the magnitude of the object's average velocity between points 1 and 2? What is the magnitude of the object's acceleration at point 2?

Solution I measure the arrow $\vec{v}_{12}\,\Delta t$ on Figure N2.7 to be 4.1 cm long. (You may get a different result if the drawing's size is not exactly preserved during printing and duplication.) According to equation N2.8a, this means that v_{12} = (4.1 cm)/(0.20 s) = 20.5 cm/s. I also measure the arrow $\vec{a}_2\,\Delta t^2$ to be ≈ 2.4 cm long, so $a_2 = (2.4\text{ cm}) / (0.20\text{ s})^2 = 60\text{ cm/s}^2$.

Exercise N2X.7: The drawing below shows an unfinished motion diagram for a button sliding on a tabletop. If $\Delta t = 0.05$ s, what is v_{23}? What is a_3?

1

2

3

4

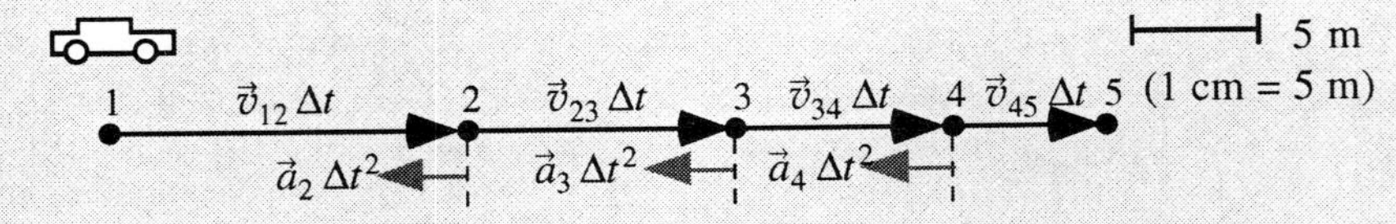

Figure N2.8: A motion diagram for a car braking to a stop that shows the scale used.

Handling scale diagrams

Of course, the examples that we have been considering assume that the motion diagram has been drawn to actual size. Many of the motion diagrams that we will draw will not be actual size, but rather *scale* drawings of the object's motion (if they are quantitatively accurate at all). For example, let us assume that the incomplete motion diagram for the braking car in Figure N2.8 is drawn to a scale such that 1 cm on the diagram corresponds to 5 m of actual distance. In such a case, we have to convert the distances we measure on the diagram to actual physical distances before we can compute velocities or accelerations.

For example, the drawn arrow representing $\vec{v}_{12}\,\Delta t$ in Figure N2.8 is 2.9 cm long, corresponding to

$$2.9\,\cancel{\text{cm}}\left(\frac{5\text{ m}}{1\,\cancel{\text{cm}}}\right) = 14.5\text{ m} \tag{N2.9}$$

of actual displacement for the car. If the time interval between position dots is 0.50 s, then the magnitude of the car's average velocity between points 1 and 2 is (14.5 m)/(0.50 s) = 29 m/s.

Exercise N2X.8: Similarly, find the magnitude of $\vec{v}_{23}$ and $\vec{a}_2$ for the car shown in Figure N2.8 (assuming that the time between dots is still 0.5 s).

N2.6 UNIFORM CIRCULAR MOTION

Circular motion (or approximately circular motion) is a very common kind of motion, applicable to a wide variety of physical situations such as atomic orbitals, bicycle wheels, rotary engines, hurricanes, satellite orbits, and galactic rotation. Historically, one of the most important ideas that helped Newton build his theory was growing recognition in the middle 1600s that even an object that is moving along a circular path (even if it has a constant speed) is accelerating, because the *direction* of its velocity is changing. The purpose of this section is to use motion diagrams to learn what we can about the acceleration of an object moving in a circular path.

Definition of uniform circular motion

Consider an object moving at a constant speed around a circle of radius R: this kind of motion is called **uniform circular motion.** What is the direction and magnitude of the object's acceleration in this case?

A motion diagram for uniform circular motion

Figure N2.9a shows a motion diagram of an object traveling around a circle with a radius of 7.7 cm. For the sake of argument, let's assume that the time interval between position dots is 0.10 s. Note that the position dots are equally spaced, consistent with the idea that the object is moving with a constant speed. Using the techniques discussed in the previous two sections, I have constructed arrows representing the average velocities $\vec{v}_{12}\,\Delta t$ and $\vec{v}_{23}\,\Delta t$ and the arrow representing the object's acceleration as it passes point 2.

Note that when this arrow is attached to point 2, it points directly toward the center of the circle. Since there is nothing special about point 2 (all points on the circle are equivalent) this result applies generally: *the acceleration of an object moving at a constant speed around a circle points directly toward the center of the circle at every instant*. Note that this result applies only if the object is moving at a constant speed: you can see from the diagram that if either velocity arrow is longer than the other, the acceleration arrow will *not* point toward the center.

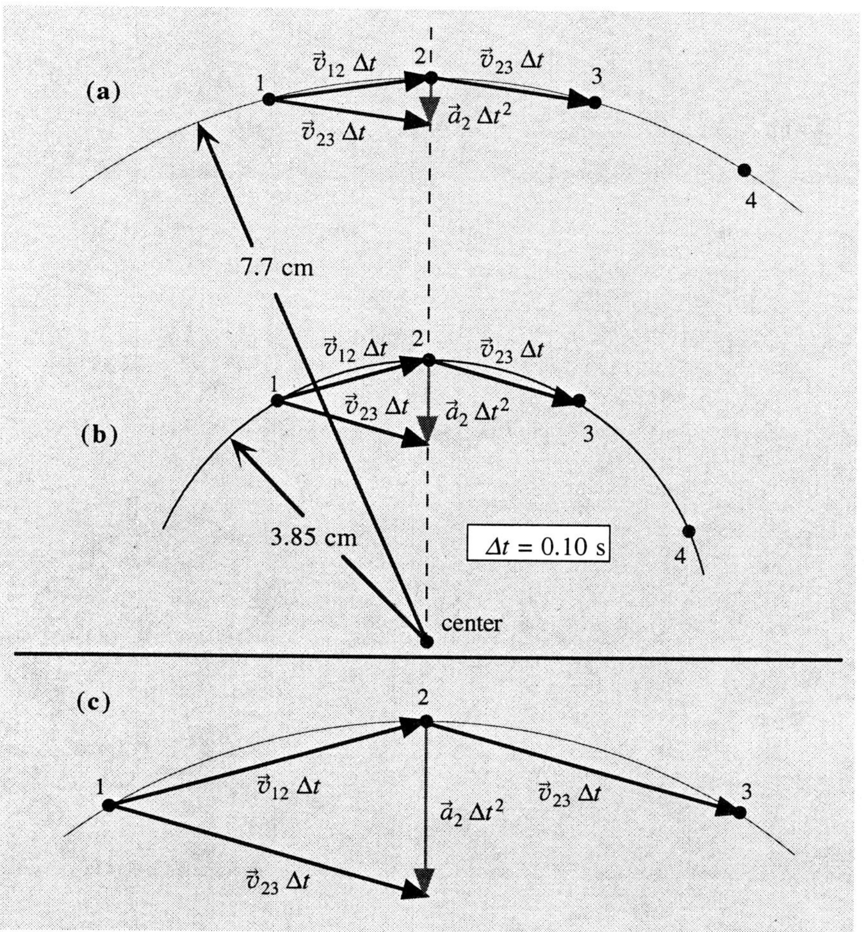

Figure N2.9: (a) An actual-size motion diagram for an object moving at about 22 cm/s in a circle 7.7 cm in radius. Note that the object's acceleration vector as it passes point 2 points directly at the center of the circle.

(b) If we decrease the radius by a factor of two but keep the object's speed the same, the magnitude of its acceleration doubles. This is because the velocity vectors change direction twice as fast as the object goes around the smaller circle.

(c) If we keep the radius the same but double the speed, not only does the direction of the velocity vector change twice as fast but each velocity vector is twice as long. This has the effect of quadrupling the acceleration.

For future reference, let's determine the actual magnitude of the velocities and acceleration in this case. I measure the arrows $\vec{v}_{12}\,\Delta t$ and $\vec{v}_{23}\,\Delta t$ in Figure N2.9a to be 2.2 cm long on the diagram. Since $\Delta t = 0.10$ s here, the actual magnitude of the velocities is (2.2 cm)/(0.10 s) = 22 cm/s. I measure the $\vec{a}_2\,\Delta t^2$ arrow to be about 0.6 cm long on the diagram, corresponding to an actual acceleration of $(0.6 \text{ cm}) / (0.10 \text{ s})^2 = 60 \text{ cm/s}^2$.

What happens if we vary *R* and *v*?

Now, the *magnitude* of the object's acceleration can only depend on the radius of the circle and the object's speed as it travels around its circular path: these two quantities completely determined the object's motion. What happens if we vary the values of these two quantities?

Figure N2.9b shows that if we keep the object's speed unchanged while we decrease R to *half* of its original value, the lengths of the velocity arrows are unchanged, but the angle between them *doubles* (since the object covers twice as great a fraction of the circle during each time interval Δt). This has the effect of *doubling* the length of the acceleration arrow.* We see that the magnitude of an object's acceleration in uniform circular motion must therefore depend inversely on the radius of its circular path: $a = \text{mag}(\vec{a}) \propto 1/R$.

Figure N2.9c shows that if we keep the radius of the circle unchanged while we double the object's speed, not only does the *angle* between the velocity vectors double (since the object goes around the circle twice as fast as before) but their length doubles too. The result is an acceleration vector that is four times as large as it was in Figure N2.9a. We see that the magnitude of an object's acceleration in uniform circular motion depends on the square of its speed: $a \propto v^2$.

* Technically, the acceleration arrow *exactly* doubles in size only in the limit that Δt (and thus the angle between the velocity arrows) goes to zero. See problem N2A.1.

The simplest formula for the magnitude of the object's acceleration in uniform circular motion that embraces both the observation that $a \propto 1/R$ and the observation that $a \propto v^2$ is the following:

A formula for the magnitude of the acceleration of an object moving in a circle at a constant speed

$$a \equiv \text{mag}(\vec{a}) = \frac{\text{mag}(\vec{v})^2}{R} = \frac{v^2}{R} \tag{N2.10}$$

We could multiply the right side of this equation by any constant and get another equation that is *still* consistent with $a \propto 1/R$ and $a \propto v^2$, but it happens that equation N2.10 is correct as it stands. We can check this as follows. We saw earlier that the object in Figure N2.9a moves with a speed of 22 cm/s. According to equation N2.10, the object in that figure should have an acceleration of magnitude $a = v^2/R = (22 \text{ cm/s})^2/(7.7 \text{ cm}) = 63 \text{ cm/s}^2$. This compares reasonably well with the result of 60 cm/s^2 that we found earlier by measuring the length of the acceleration arrow on the diagram (particularly considering the uncertainties associated with the various measurements and in the construction of the arrow).

To summarize, we have learned (by studying a quantitatively accurate motion diagram) that if an object moves at a constant speed v around a circle of radius R (that is in uniform circular motion), its acceleration always points toward the center of the circle and has a magnitude $a = v^2/R$. This is, as we will see, a very important and useful result! Perhaps you can see that motion diagrams can be a powerful tool for analyzing and understanding acceleration, particularly when objects move in two dimensions.

Exercise N2X.9: Construct an arrow on Figure N2.9a that represents the object's acceleration as it passes point 3. When this arrow is attached to that point, does it point toward the circle's center?

Exercise N2X.10: A car traveling at a constant speed of 20 m/s ($\approx$ 44 mi/h) goes round a circular curve to the left in the road whose radius is 600 m. At an instant when the car is going due north, what is the magnitude and direction of the car's acceleration?

SUMMARY

I. THE DEFINITION OF ACCELERATION

A. An object's *acceleration* vector at an instant expresses *how fast* and in *what direction* its velocity is changing at that instant:

$$\vec{a} \equiv \lim_{\Delta t \to 0} \frac{\Delta \vec{v}}{\Delta t} \equiv \lim_{\Delta t \to 0} \frac{\vec{v}(t+\Delta t) - \vec{v}(t)}{\Delta t} \equiv \frac{d\vec{v}}{dt} \qquad \text{(N2.1)}$$

B. The components are $[a_x, a_y, a_z] = [dv_x/dt, dv_y/dt, dv_z/dt]$ (N2.2)

C. The *average acceleration* during an interval is $\vec{a}_{\Delta t} \equiv \Delta\vec{v}/\Delta t$

1. this is ≈ the instantaneous acceleration if Δt is small
2. it most closely approximates that acceleration at the *middle* of Δt
3. it is useful when only an *estimate* of the acceleration is needed

II. MOTION DIAGRAMS

A. Motion diagrams are a useful general tool for

1. visualizing an object's motion in a detailed way
2. estimating the magnitude and direction of its motion

B. Parts of a motion diagram:

1. a single image of object (to show its characteristics and orientation)
2. dots that represent the positions of the object's center of mass at successive equally-spaced instants of time
3. arrows drawn between the dots depicting the object's average velocity (actually $\vec{v}_{\Delta t}\,\Delta t$) during the interval between the given instants
4. an acceleration arrow (which actually represents the vector $\vec{a}\,\Delta t^2$) attached to each dot (except first and last)

C. To draw acceleration arrows in a motion diagram (Figs. N2.4 and N2.7)

1. copy the velocity arrow $\vec{v}_{\Delta t}\,\Delta t$ following a given dot and place it tail to tail with the velocity arrow preceeding the dot
2. draw the vector difference of these two arrows: the difference is a good approximation to $\vec{a}\,\Delta t^2$ as the object passes the given dot

D. To read quantitative magnitudes from motion diagram: measure the lengths of the $\vec{v}_{\Delta t}\,\Delta t$ and $\vec{a}\,\Delta t^2$ arrows on the diagram and use:

1. $v_{\Delta t} = \text{mag}(\vec{v}_{\Delta t}\,\Delta t)/\Delta t$ (N2.8a)
2. $a = \text{mag}(\vec{a}\,\Delta t^2)/\Delta t^2$ (N2.8b)

E. if a diagram is a *scale* drawing, convert measured lengths to real distances *first* and then apply equations N2.8a or N2.8b.

III. UNIFORM CIRCULAR MOTION

A. The importance of circular motion

1. it is a good model of many common situations
2. an understanding of circular motion was crucial for the development of Newton's theory

B. An object in *uniform circular motion* moves in a circle with constant v

1. the object accelerates because the *direction* of its velocity changes
2. mag($\vec{a}$) can only depend on the object's v and the circle's radius R

C. A motion diagram shows that in the case of uniform circular motion:

1. the object's acceleration points always toward the circle's center
2. if R decreases, then mag($\vec{a}$) increases in direct proportion to $1/R$ (because the object covers a greater fraction of circle in given Δt and thus the direction of its velocity changes $1/R$ times faster)
3. if v increases, mag($\vec{a}$) increases in direct proportion to v^2, because
   a) a faster speed leads to a faster change in velocity direction
   b) the size of velocity arrows thus is also proportionally larger
   c) so the *change* in $\vec{v}$ during given (small) Δt increases as v^2

D. The simplest formula consistent with these results is $a = v^2/R$ (N2.10)

1. this formula could *in principle* have unitless constant in front
2. but it is correct as it stands (as a quantitative diagram shows)

GLOSSARY

acceleration $\vec{a}$: a vector quantity that describes how fast and in what direction an object's velocity vector is changing at an instant:

$$\vec{a} \equiv \lim_{\Delta t \to 0} \frac{\Delta \vec{v}}{\Delta t} \equiv \lim_{\Delta t \to 0} \frac{\vec{v}(t+\Delta t) - \vec{v}(t)}{\Delta t} \equiv \frac{d\vec{v}}{dt} \quad \text{(N2.1)}$$

average acceleration (during an interval) $\vec{a}_{\Delta t}$: the change in an object's velocity during an interval of time Δt divided by that interval: $\vec{a}_{\Delta t} \equiv \Delta\vec{v}/\Delta t$. This quantity is most closely equal to the object's instantaneous acceleration at the *middle* of the interval. This approximation becomes better as Δt gets smaller.

motion diagram: a diagram that depicts an object's motion and (at least the direction of) its acceleration at various instants during its motion. A complete motion diagram of a moving object contains the following

1. a single image of the object (to show its orientation)
2. dots representing positions of the object's center of mass at equally-spaced instants of time
3. arrows from dot to dot (depicting average velocities)
4. acceleration arrows attached to each dot.

uniform circular motion: the motion of an object following a circular path at a constant speed.

TWO-MINUTE PROBLEMS

N2T.1 An object can have a constant speed and still be accelerating (T or F).

N2T.2 An object's acceleration vector always points in the direction that it is moving (T or F).

N2T.3 An object's x-velocity is given by $v_x(t) = bt^2$, where b is a constant. The object's x-acceleration is

A. $(1/3)bt^3$
B. $(1/3)bt^3 + C$
C. bt
D. $2bt$
E. $2bt + C$
F. $(1/2)bt$
T. other (specify)

N2T.4 A car starting from rest reaches a speed of 25 m/s (about 55 mi/h) in 5 s. If the car travels in a straight line, what is its average acceleration during this time interval?

A. 5 m/s^2 forward
B. 5 m/s^2 backward
C. 1 m/s^2 forward
D. 1 m/s^2 backward
E. other (specify)

N2T.5 An object falling vertically at a speed of 20 m/s lands in a snowbank and comes to rest 0.5 s later. What is the object's average acceleration during this interval?

A. 10 m/s^2 up
B. 10 m/s^2 down
C. 40 m/s^2 up
D. 40 m/s^2 down
E. 5 m/s^2 up
F. 80 m/s^2 down
T. other (specify)

N2T.6 An object can have zero velocity (at an instant of time) and still be accelerating (T or F).

N2T.7 The x-velocity of an object can be positive at the same time that its x-acceleration is negative (T or F).

N2T.8 A boat hits a sandbar and slides some distance before coming to rest. Which of the arrows shown below best represents the direction of the boat's acceleration as it is sliding? (*Hint*: Draw a motion diagram.)

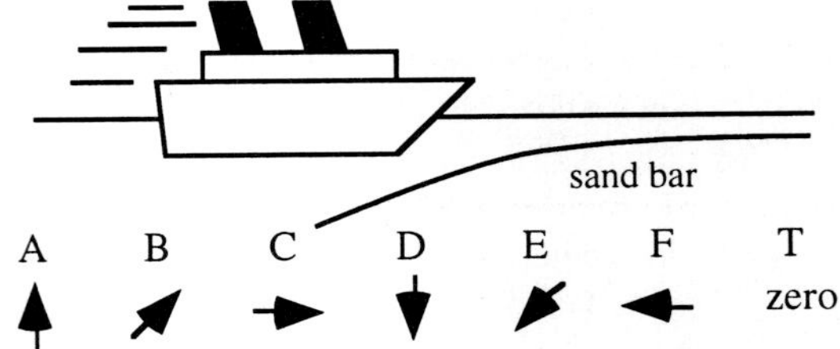

N2T.9 A car moving at a constant speed travels past a valley in the road, as shown below. Which of the arrows shown most closely approximates the direction of the car's acceleration at the instant that it is at the position shown? (*Hint*: draw a motion diagram.)

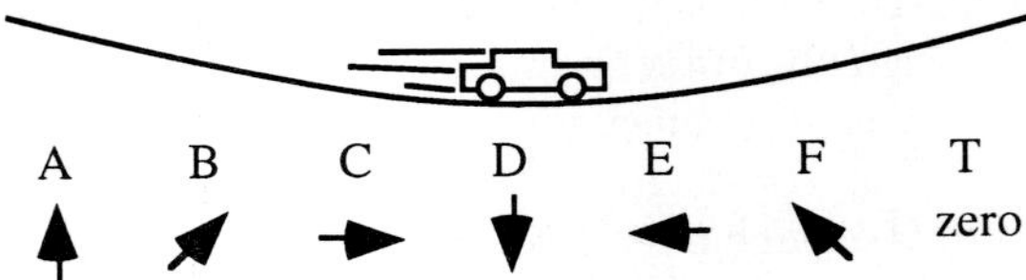

N2T.10 A bike (shown in a top view in the diagram below) travels around a curve with its brakes on, so that it is constantly slowing down. Which of the arrows shown below most closely approximates the direction of its acceleration at the instant that it is at the position shown? (*Hint*: Draw a motion diagram.)

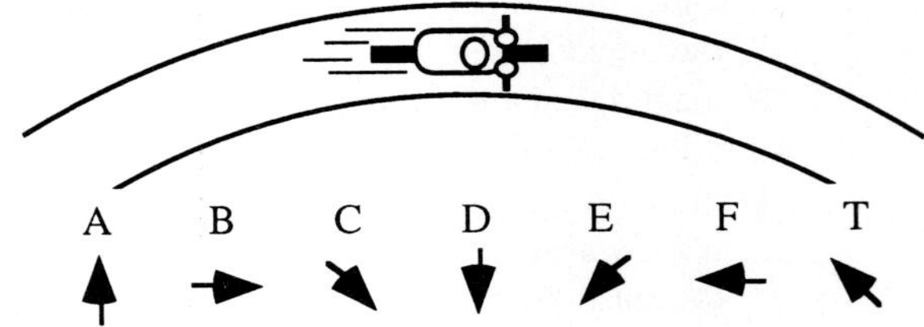

HOMEWORK PROBLEMS

BASIC SKILLS

N2B.1 The driver of a car moving at 24 m/s due east sees something on the road and applies the brakes. The car comes to rest 4.0 s later. What is the magnitude and direction of the car's average acceleration during this interval?

N2B.2 A person hits a trampoline moving downward with a speed of 5.0 m/s and rebounds a short time later with roughly the same speed upward. If the person is in contact with the trampoline for about 1.8 s, what is the magnitude and direction of the person's average acceleration during this time interval?

N2B.3 A certain jet plane has to reach a speed of 50 m/s (110 mi/h) to take off. If it starts at rest and moves with an average forward acceleration of 2.5 m/s, how long must the plane roll along the runway?

N2B.4 A spaceship accelerates from rest with a constant average acceleration of 15 m/s^2 (about the maximum that a human being can tolerate indefinitely). About how long will it take the ship to reach half the speed of light? (Hint: 1 Ms = 10^6 s ≈ 12 days).

N2B.5 Consider the unfinished motion diagram shown below. What is the direction of the object's acceleration at the instant that it passes point 4? Show how you arrived at your answer.

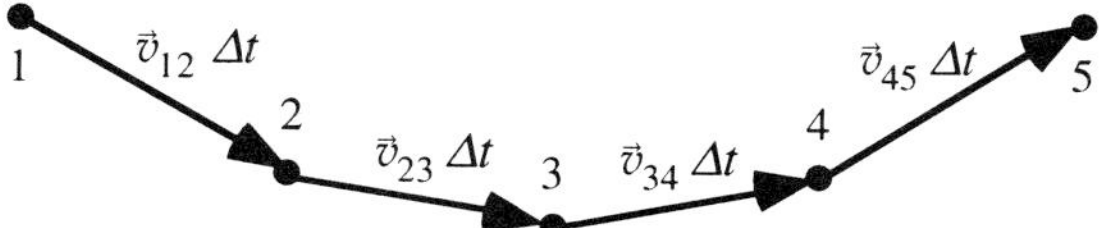

N2B.6 Consider the unfinished motion diagram shown below. What is the general direction of the object's acceleration? Explain your reasoning.

N2B.7 A bicyclist is riding at a constant speed of 10 m/s counterclockwise around a circular track that has a radius of 50 m. At an instant when the cyclist is moving directly south, what is the magnitude and direction of his or her acceleration?

N2B.8 A jet plane flies at a constant speed of 120 m/s (260 mi/h) clockwise in a circular holding pattern of radius 8.0 km. What is the magnitude and direction of the plane's acceleration at an instant when it is traveling due east?

SYNTHETIC

N2S.1 A certain object's velocity vector has components $v_x(t) = bt^2 + c$, $v_y(t) = qt$, $v_z(t) = 0$, where $b = 10$ m/s^3, $c =$ 5 m/s, and $q = -2.0$ m/s^2. What is the magnitude of the object's acceleration at time $t = 0$? At time $t = 3.0$ s?

N2S.2 A certain object's velocity vector has components $v_x(t) = b/(t+c)$, $v_y(t) = q$, $v_z(t) = 0$,, where $b = 10$ m, $c =$ 2.0 s, and $q = -5.0$ m/s. What is the magnitude of the object's acceleration at time $t = 0$? At time $t = 2.0$ s?

N2S.3 An automobile starts from rest and increases its speed for 15 s as it moves along a straight stretch of road. If the magnitude of the car's average acceleration during this interval is 1.2 m/s^2, what is its final speed in mi/h?

N2S.4 A spaceship initially traveling at a speed of $0.1c$ (where $c = 3.0 \times 10^8$ m/s = the speed of light) decreases its speed smoothly to zero in 30 minutes as it approaches a space station. Assuming that the spaceship travels in a straight line, what is the magnitude of its average acceleration during this time interval? Compare to 60 m/s^2, which is roughly the maximum acceleration that a human being can tolerate for a short period of time.

N2S.5 A jogger runs at a constant speed halfway around a circular ring 120 m in diameter in 60 s. If the jogger was initially running east, what is the magnitude and direction of the jogger's average acceleration during this interval?

N2S.6 A car traveling at a constant speed of 20 m/s (about 44 mi/hr) that is initially traveling due northwest rounds a corner so that after 10 s, the car is traveling due northeast. What is the magnitude and direction of the car's average acceleration during this interval of time? (*Hint*: It really helps if you draw a picture.)

N2S.7 Consider the motion of an object swinging back and forth on the end of a string. Draw a qualitatively accurate motion diagram (with at least seven position points) of a single swing (left to right) of such a pendulum.

N2S.8 Draw a qualitatively accurate motion diagram of a ball thrown vertically into the air. (Your diagram should include at least seven position points.)

N2S.9 Figure N2.10 shows a set of dots indicating a certain object's position every 0.10 s. After tracing these dots to your own sheet of paper, draw a complete motion diagram for this object and compute the magnitude of its acceleration at point 3.

N2S.10 Figure N2.11 shows a set of dots indicating a certain object's position every 0.05 s. After tracing these dots to your own sheet of paper, draw a complete motion diagram for this object and compute the magnitude of its acceleration at point 4.

Figure N2.10 (for problem N2S.9).

Figure N2.11 (for problem N2S.10)

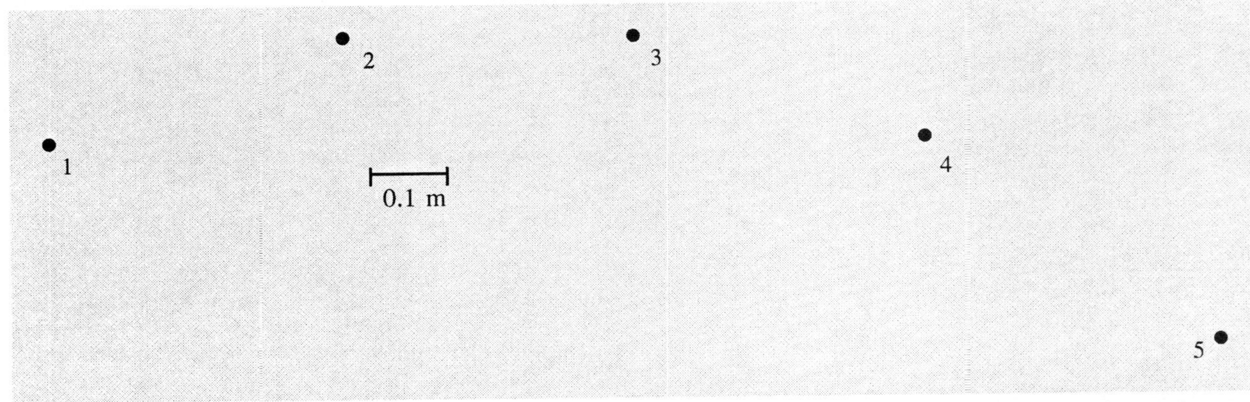

Figure N2.12 (for problem N2S.11)

N2S.11 Figure N2.12 shows a set of dots indicating a certain object's position every 0.10 s (note that 1 cm on this diagram corresponds to 0.1 m of actual distance). After tracing these dots to your own sheet of paper, draw a complete motion diagram for this object and compute the magnitude of its acceleration at points 2, 3 and 4.

RICH-CONTEXT

N2R.1 Galileo was one of the first people to actually measure the motion of a ball rolling down an incline. He discovered that if a ball rolls 1 unit of distance in an interval with a given duration Δt, it rolls 3 units in the next interval of the same duration, 5 units in the next, 7 units in the next, and so on. Argue that this means the acceleration of a ball rolling down an incline is *constant*.

ADVANCED

N2A.1 Argue that the arrow representing the velocity difference $\Delta\vec{v}$ in Figure N2.9b is not exactly *twice* as large as the arrow representing the same in Figure N2.9a, but rather $\sin\theta/\sin(\theta/2)$ times as large, where θ is the angle between the velocity vectors in Figure N2.9a. Using the mathematical fact that if θ is measured in radians,

$$\sin\theta = \theta - \frac{\theta^3}{6} + \frac{\theta^5}{120} - \dots \qquad \text{(N2.11)}$$

then show that the limit of this ratio as $\theta \to 0$ is indeed 2. (If you prefer, you can use a geometrical argument instead of equation N2.11.)

ANSWERS TO EXERCISES

N2X.1 $a_x = dv_x/dt = 0$, $a_y = dv_y/dt = 0$, $a_z = dv_z/dt = b$. So $a_x = a_y = 0$, $a_z = b = 10\ \text{m/s}^2$ at both $t = 0$ and $t = 2$ s.
N2X.2 24 m/s.
N2X.3 The ball's change in velocity during this interval is 30 m/s downward, so the ball's average acceleration in this case must be 10 m/s^2 downward.
N2X.4 The motion diagram will look something like:

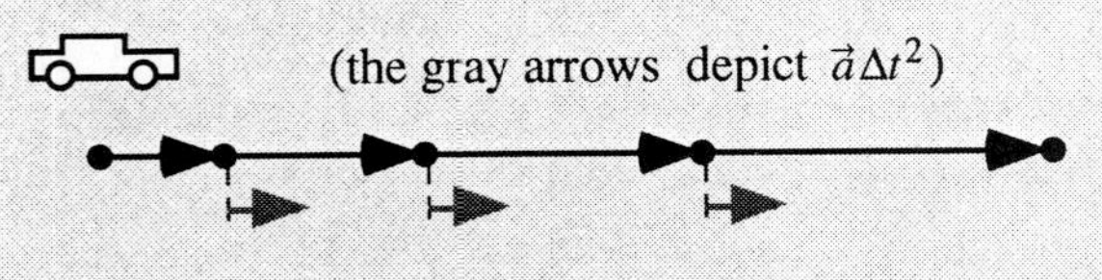

N2X.5 As in case of the ball rolling up the incline, we have to separate the downward and upward portions of the motion. The ball falls with increasing speed toward the ground, suddenly and violently reverses direction, and then moves upward with *decreasing* speed. The motion diagram for the ball is shown at right.

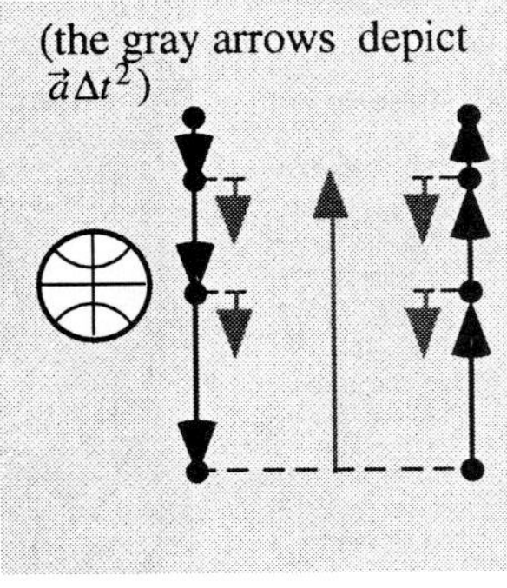

N2X.6 The motion diagram will look something like:

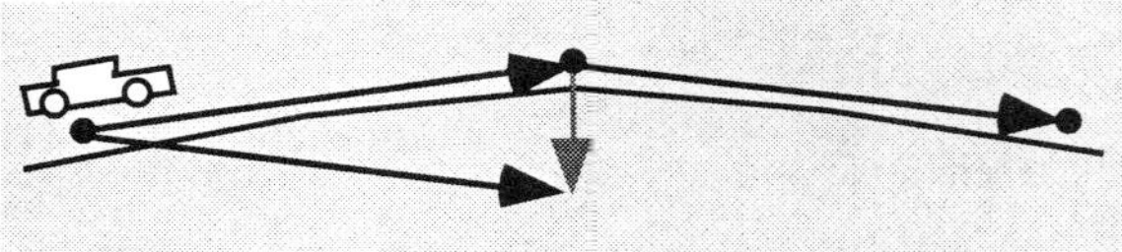

I have only drawn three position points here: five would probably be better (but the diagram would have to be bigger). The direction of the car's velocity is changing, so the car is indeed accelerating, at least at the crest of the hill.
N2X.7 I measure the distance between points 2 and 3 to be about 2.85 cm, so $v_{23} \approx (2.85\ \text{cm})/(0.05\ \text{s}) = 57\ \text{cm/s}$. I measure the distance between points 3 and 4 to be 2.15 cm, so the *difference* in length between the arrows representing $\vec{v}_{23}\Delta t$ and $\vec{v}_{34}\Delta t$ is 0.7 cm. This would be the length of the acceleration arrow $\vec{a}_3\Delta t$ that we would draw for point 3, so $a_3 = (0.7\ \text{cm})/(0.05\ \text{s})^2 = 280\ \text{cm/s}^2$.
N2X.8 I find the length of the arrow between points 2 and 3 to be 2.2 cm, which scales up to (2.2 cm)(5 m/cm) = 11 m in real space. Therefore, $v_{23} = (11\ \text{m})/(0.50\ \text{s}) = 22\ \text{m/s}$. The difference between the lengths of the arrows for $\vec{v}_{12}\Delta t$ and $\vec{v}_{23}\Delta t$ is 14.5 m − 11 m = 3.5 m in real space, so the arrow representing the *change* in the object's velocity (and thus the acceleration) will have this length in real space. Thus $a_2 = (3.5\ \text{m})/(0.50\ \text{s})^2 = 14\ \text{m/s}^2$.
N2X.9 The acceleration arrow should indeed point toward the circle's center when it is moved back to point 3.
N2X.10 about 0.67 m/s^2 west.

N3

NEWTON'S LAWS

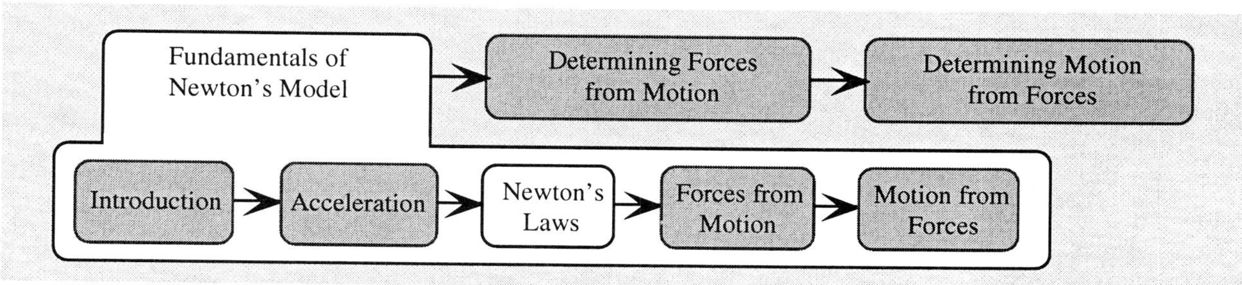

N3.1 OVERVIEW

In the last two chapters we have discussed how we can use calculus to define what we mean by an object's velocity and acceleration at an instant of time. In this chapter, we will use these concepts to talk about the core principles of Newton's model of mechanics.

Newton based his model on three fundamental ideas, which he called *laws*. In this chapter, we will examine each of these laws from a modern perspective, connecting these laws with the deeper ideas of energy and momentum developed in Unit *C*. Here is an overview of the sections in this chapter:

N3.2 *NEWTON'S FIRST LAW* discusses Newton's first law, relating it to the basic principle of conservation of momentum.

N3.3 *NEWTON'S THIRD LAW* explores Newton's third law, again showing it to be a consequence of conservation of momentum.

N3.4 *NEWTON'S SECOND LAW* describes Newton's second law and examines its relationship to the definition of momentum and force presented in Unit *C*. Our primary focus in this unit will be on understanding and describing the consequences of *this* law.

N3.5 *ENERGY AND FORCE* shows how Newton's second law is connected to conservation of energy. This section also shows us how force laws for long-range interactions can be determined from the interaction's potential energy.

N3.6 *CLASSIFICATION OF FORCES* reviews and extends the classification scheme for macroscopic forces that we discussed briefly in Unit *C*.

N3.7 *FREE-BODY DIAGRAMS* presents a simple and general graphical tool that helps us recognize and display the forces that act on an object.

This chapter is in a certain sense the crucial chapter in the unit, since it lays the foundations of the newtonian model that we will be exploring for the remainder of this unit and in units to follow.

N3.2 NEWTON'S FIRST LAW

An important implication of the definition of the CM

In chapter C4, we saw that the total momentum of any system of particles is equal to the system's total mass times the velocity of its center of mass

$$M\vec{v}_{CM} = \vec{p}_1 + \vec{p}_2 + \ldots + \vec{p}_N \equiv \vec{p}_{tot} \tag{N3.1}$$

This tells us that *the total momentum of an extended object* (or even a system of objects!) *can be calculated by treating it as if it were a point particle of equal total mass located at the center of mass.*

Conservation of momentum implies that *the total momentum of an isolated system is conserved.* Since equation N3.1 implies that $\vec{v}_{CM} = \vec{p}_{tot}/M$, and M and $\vec{p}_{tot}$ have fixed values for an isolated object, this means that

Newton's first law

> In the absence of external interactions, an object's (or system's) center of mass moves at a constant velocity.

This is **Newton's first law** as it applies to everyday macroscopic objects.

Should we take this law or conservation of momentum as being fundamental?

From a modern perspective, we consider this law to be a consequence of the law of conservation of momentum, which we now think of as being more fundamental. Newton, however, used this law as one of the basic assumptions of his model, and essentially derived conservation of momentum as a *consequence* of this and his other two laws of motion, rather than taking conservation of momentum as a basic assumption. Newton's approach is not wrong. *Every* physical model is based on assumptions, and it is possible to build the same basic model on *different* sets of assumptions. Given two different sets of assumptions that logically lead to the same physical model, the choice between the different sets becomes more a matter of taste and/or rhetorical aims than anything else.

Newton had a very important pedagogical reason for stating the first law as he did: the first law clearly and vividly contradicts the previously accepted aristotelian assumptions about both terrestrial and celestial motion. Newton wanted to make it clear that the *natural* state of a terrestrial object is *not* to be at rest as close to the center of the earth as possible, but rather constant, unending motion in a straight line. Similarly, the *natural* state of a celestial object is not unending motion in a *circle*, but rather unending motion in a *straight line*. Given the context of the time, making this point clearly and vividly was essential.

In the context of modern physics, in addition to expressing an important and useful consequence of the law of conservation of momentum, Newton's first law also plays an important role in the definition of what we call an *inertial reference frame*. We will explore this issue in more depth in Chapter N10.

N3.3 NEWTON'S THIRD LAW

According to the basic model of interactions presented in chapter C4, an interaction between two objects A and B *transfers* a bit of momentum $d\vec{p}$ out of one object and into the other during a short interval of time dt (rather than creating or destroying momentum). Now, a bit of momentum $d\vec{p}$ flowing out of object A is the same as $-d\vec{p}$ flowing into A, so we could just as easily say that the interaction puts momentum $d\vec{p}$ into object B and puts momentum $-d\vec{p}$ into A (conserving the objects' total momentum). Since the force $\vec{F}$ exerted by an interaction on an object is defined to be the *rate* at which the interaction transfers momentum into the object, we have

Mathematical statement of Newton's third law

$$\vec{F}_B \equiv \frac{d\vec{p}}{dt} = -\left(\frac{-d\vec{p}}{dt}\right) \equiv -\vec{F}_A \tag{N3.2}$$

that is, the rate at which a given interaction transfers momentum into object B is negative the rate at which it transfers momentum into object A.

We might express the implication of equation N3.2 in words as follows:

When objects A and B interact, the force that the interaction exerts on A is equal in magnitude and opposite in direction to the force that it exerts on B.

A verbal statement of Newton's third law

This is a modern statement of **Newton's third law.**

The presentation makes it clear that this law is a direct and necessary consequence of the idea that interactions *transfer* momentum. Newton, on the other hand, again took this law as being fundamental and derived conservation of momentum and the momentum-transfer model from it. Since either approach yields the same final physical model, the argument over which is better is again primarily a matter of rhetorical utility in a given context.

The third law is often counterintuitive

While this law seems logical and straightforward in the context of the momentum-transfer model, when we apply this law to realistic situations, we often get counterintuitive results. For example, consider a 10,000-kg truck traveling at 65 mi/h (30 m/s). Imagine that it hits a parked 500-kg Volkswagen Beetle. Which vehicle exerts the stronger force on the other during the collision? According to the third law, the magnitudes of the forces that the contact interaction exerts on either vehicle *must be the same*: any momentum that the interaction transfers out of the truck must go into the Beetle at the same rate!

But this seems absurd! We know that the Beetle will be totally destroyed in this interaction (probably simultaneously crushed flat and thrown many meters), while the truck will hardly be fazed. How can we possibly *imagine* that the force the Beetle exerts on the truck will be the same as the force that the truck exerts on the Beetle?

The answer is that just because the *forces* are the same on each doesn't mean that each has to *respond* the same way. Because the Beetle is 20 times less massive than the truck, a given amount of momentum flowing into the Beetle will lead to a change in its velocity whose magnitude is 20 times larger than the change in the truck's velocity from the same momentum flow. Thus forces of the same magnitude can have entirely different effects.

Even so, the third law is true and really works!

Part of the trick in successfully understanding and using Newton's third law is to recognize that (1) it does indeed lead to superficially counterintuitive results sometimes, but (2) it is a *necessary* consequence of the momentum flow model, and (3) it really does *correctly* describe interactions between objects (even if they have very different masses) when the situation is carefully analyzed, as in the example above.*

We will discuss consequences and implications of Newton's third law in more detail later (primarily in chapters N4 and N8).

Exercise N3X.1: Imagine that the truck is gently pushing the Beetle at a constant speed of 10 m/s. Which exerts the larger force on the other now?

Exercise N3X.2: Imagine now a small car pushing a large disabled truck in such a way that the pair slowly gains speed. Which exerts the larger force on the other in this circumstance?

Exercise N3X.3: Imagine that in a schoolyard fight, 75-lb Sal shoves a passive 90-lb Terry to the ground. Which exerts the larger force on the other?

---

*The only qualification is that in long range interactions, it can take a finite time for momentum to flow through the interaction "conduit", meaning that the flow rate at the two widely separated ends of the conduit is not necessarily exactly equal and opposite at a given instant of time. In such circumstances, we have to consider the interaction itself as an object in the system, an object that has its own momentum. This is one of the important reasons why we need a field model for long-range interactions, as we will discuss more fully in Unit *E*. Under normal circumstances, however, this is rarely noticeable.

N3.4 NEWTON'S SECOND LAW

Derivation of Newton's second law for point particle

Consider now a single point particle of mass m participating in a number of interactions with external objects. During a given short interval of time dt, the first interaction transfers a certain amount of momentum $d\vec{p}_1$, the second transfers a certain amount of momentum $d\vec{p}_2$, and so on. The net change in the particle's momentum due to all of these interactions is

$$d\vec{p} = d\vec{p}_1 + d\vec{p}_2 + \ldots \tag{N3.3}$$

Dividing both sides of this equation by dt, we get

$$\frac{d\vec{p}}{dt} = \frac{d\vec{p}_1}{dt} + \frac{d\vec{p}_2}{dt} + \ldots \equiv \vec{F}_1 + \vec{F}_2 + \ldots \equiv \vec{F}_{\text{net}} \tag{N3.4}$$

where $\vec{F}_1 \equiv d\vec{p}_1/dt$ is the force that the first interaction exerts on the particle, $\vec{F}_2 \equiv d\vec{p}_2/dt$ is the force that the second exerts, and so on.

We see that the rate at which the particle's *total* momentum changes is equal to the *vector sum* $\vec{F}_{\text{net}}$ of the forces that its interactions exert on it. Now, remembering that $\vec{p} = m\vec{v}$ and that m is a constant, we get

$$\frac{d\vec{p}}{dt} \equiv \frac{d}{dt}(m\vec{v}) = m\frac{d\vec{v}}{dt} \equiv m\vec{a} \tag{N3.5}$$

Therefore

Newton's second law for a point particle

$$\vec{F}_{\text{net}} = \frac{d\vec{p}}{dt} = m\vec{a} \tag{N3.6}$$

Equation N3.6 says that a particle's acceleration in response to the forces that its interactions exert on it is equal to the **net force** $\vec{F}_{\text{net}}$ (that is, the *vector sum* of all the forces) acting on it divided by its mass. Equation N3.6 is **Newton's second law** for a point particle: it links an aspect of the particle's *motion* (its acceleration) with the *forces* acting on it.

Consider now a *system* of particles (such as an extended object) that interacts with its surroundings (it is thus *not* isolated). Each particle in the system may participate in *internal* interactions with other particles *inside* the system and/or *external* interactions with things *outside* the system. However (as we discussed in chapter C4) internal interactions only transfer momentum from particle to particle *inside* the system, so they will not change the system's *total* momentum. Therefore, the rate at which the system's *total* momentum changes is the vector sum of the rates at which *external* interactions with particles in the system supply momentum to that system. This implies that

$$\frac{d\vec{p}_{\text{tot}}}{dt} = \vec{F}_{\text{net,ext}} \tag{N3.7}$$

where $\vec{F}_{\text{net,ext}}$ is the vector sum of all *external* forces acting on all the particles in the system. Using equation N3.1 and recognizing that the total mass M of the system is a constant (assuming that it doesn't lose or gain particles), we have

$$\frac{d\vec{p}_{\text{tot}}}{dt} = \frac{d}{dt}(M\vec{v}_{\text{CM}}) = M\vec{a}_{\text{CM}} \tag{N3.8}$$

Therefore

Newton's second law for extended objects

$$\vec{F}_{\text{net,ext}} = \frac{d\vec{p}_{\text{tot}}}{dt} = M\vec{a}_{\text{CM}} \tag{N3.9}$$

This equation says that the acceleration of the center of mass of a system (or extended object) is equal to the vector sum of all of the external forces acting on the system (or object) divided by its total mass M. Equation N3.9 is **Newton's second law** for *systems* of particles (including extended objects).

An extended object's CM acts like point particle!

Equation N3.6 is really a special case of equation N3.9: all interactions are external to a point particle, and such a particle's position is automatically its center of mass. However, comparing these equations reminds us of a very important point we first encountered in chapter C4: *an extended object's center of mass responds to external forces acting on an object exactly as if it were a point particle responding to those forces.* Therefore, we can always *model* an extended object by treating it as if it were a point particle, completely ignoring all complexities associated with the object's *internal* interactions. Therefore, we will usually express Newton's second law as in equation N3.6, even when we are applying it to extended objects: we *model* the extended object as a point particle.

Let me emphasize again how wonderful this result is! When one object pushes or pulls on another, we do not need to understand in *detail* why and how each atom of the one object interacts with the atoms of the other, nor how the atoms inside each object are interacting. To understand how an object's center of mass moves, we only need to be able to quantify the *net external force* acting on that object (however that force is applied). Everything else is irrelevant. This makes physics *much* simpler than it would otherwise be!

The second law makes three important statements

Newton's second law contradicts common-sense notions about force and motion in some important ways that are worth making explicit. First, it asserts that *forces cause acceleration*. At the literal level, equation N3.9 expresses a quantitative equality between two sides of a mathematical equation. At the more important conceptual level, though, this equation expresses a causal relationship. The net external force is not in any sense equivalent to a mass accelerating (these are completely distinct concepts); rather, the net external force on an object *causes* its center of mass to accelerate. You should read the second law this way.

Moreover (contrary to aristotelian world-view), forces do not cause *motion*, they cause *acceleration*. Newton's first and second laws address this issue from two directions, the first law declaring that rectilinear motion is *natural* and needs no explanatory cause and the second proclaiming that forces cause an object to accelerate away from its natural rectilinear motion.

Finally, Newton's second law tells us that it is the *net* force (the *vector sum* of *all* external forces) on the object that causes this acceleration, not the strongest force or the most recently applied force. A force does not *overcome* another force to cause an object to move in a certain way: rather all forces acting on the object act in concert to direct the object's acceleration.

EXAMPLE N3.1

Problem: The forward speed of a coasting bicycle and rider with a total mass of 62 kg is observed to decrease from 12 m/s to 9.0 m/s during a 6.0-s interval. What must be the average net external force acting on the bike and rider?

Solution Assuming that the bike is moving in a straight line, and assuming that the speeds given refer to the motion of the system's center of mass, its change in velocity is 3.0 m/s rearward in 6.0 s, implying that its average acceleration is (3.0 m/s)/(6.0 s) = 0.50 m/s^2 rearward. Therefore, the average external force on the system must be $\vec{F}_{\text{net,ext}} = M\vec{a}_{\text{CM}}$ = (62 kg)(0.50 m/s^2 rearward) = 31 $\text{kg} \cdot \text{m/s}^2$ rearward = 31 N rearward (since $1\ \text{N} \equiv 1\ \text{kg} \cdot \text{m/s}^2$: see chapter C8).

Exercise N3X.4: A car having a mass of 2200 kg is traveling due west at a constant speed of 28 m/s. What is the magnitude of the net force acting on this car? Explain your response.

Exercise N3X.5: A small jet whose mass is 14,000 kg accelerates for take-off, going from rest to a speed 120 m/s in 45 s. What is the magnitude of the average net external force that must be exerted on the jet?

N3.5 ENERGY AND FORCE

Our goal: linking Newton's second law with energy conservation

We saw in the last section that Newton's second law emerges naturally out of the momentum-transfer model. In this section, we will explore the connection between Newton's second law and the principle of conservation of *energy*. An important result of this exploration will be an understanding of how we can generate the **force law** of an interaction (that is, a statement of how the force exerted by a long-range interaction depends on the separation of interacting objects) knowing the interaction's potential energy function.

We will consider a one-dimensional situation for the sake of simplicity

Consider an isolated system of two interacting objects (A and B). For the sake of simplicity, let us assume that both objects are constrained to move in only one dimension, which we will take to be the x axis. (Dealing with more dimensions requires more difficult and sophisticated mathematics, but does not really offer much new insight.) Let us also assume that the single interaction between the objects can be described by a potential energy function $V(r)$, where r is the separation between the objects:

$$r \equiv x_A - x_B \tag{N3.10}$$

(we will assume that object A is to the right of object B so that this difference is positive). Conservation of energy for this system implies that:

$$E = \tfrac{1}{2}m_A v_A^2 + \tfrac{1}{2}m_B v_B^2 + V(r) = \text{constant} \tag{N3.11}$$

since for one-dimensional motion along the x axis, $v_A^2 = v_{Ax}^2$ and so on.

Taking the time derivative of the law of conservation of energy

Now let us take the time derivative of both sides of this expression. Since E is a constant, its time derivative is zero. According to the product rule, the time derivative of the kinetic energy of object A is:

$$\begin{aligned}\frac{d}{dt}\left(\tfrac{1}{2}m_A v_{Ax}^2\right) &= \tfrac{1}{2}m_A\frac{d}{dt}\left(v_{Ax}v_{Ax}\right) = \tfrac{1}{2}m_A\left[v_{Ax}\frac{dv_{Ax}}{dt} + \frac{dv_{Ax}}{dt}v_{Ax}\right]\\ &= \tfrac{1}{2}m_A(2v_{Ax})a_{Ax} = ma_{Ax}v_{Ax}\end{aligned} \tag{N3.12}$$

Similarly, the time-derivative of object B's kinetic energy is $ma_{Bx}v_{Bx}$. The system's potential energy depends on time because the objects' separation r depends on time. We can thus write the time-derivative of $V(r)$ as follows:

$$\frac{d}{dt}V(r) = \frac{dV}{dr}\frac{dr}{dt} = \frac{dV}{dr}\frac{d}{dt}(x_A - x_B) = \frac{dV}{dr}(v_{Ax} - v_{Bx}) \tag{N3.13}$$

where in the second step I have simply multiplied top and bottom by dr (technically, I am using the chain rule of calculus, but there is no need to worry about this right now), and in the third step I used $r \equiv x_A - x_B$. By gathering together the results discussed since equation N3.11 and reorganizing things somewhat, *you* can show that the time derivative of equation N3.11 is therefore:

$$0 = v_{Ax}\left(m_A a_{Ax} + \frac{dV}{dr}\right) + v_{Bx}\left(m_B a_{Bx} - \frac{dV}{dr}\right) \tag{N3.14}$$

Exercise N3X.6: Verify equation N3.14.

Now, conservation of energy is *always* true, no matter what the values of v_{Ax} and v_{Bx} might be, so equation N3.14 must also be true no matter what these values might be. This equation will only be automatically true if both of the quantities in parentheses are always and separately equal to zero, so

$$m_A a_{Ax} + \frac{dV}{dr} = 0 \quad \text{and} \quad m_B a_{Bx} - \frac{dV}{dr} = 0 \tag{N3.15}$$

Newton's second law emerges from conservation of energy

Now, since the interaction between these objects is by hypothesis the *only* interaction involving each, the force exerted on each object by the interaction is the net force on that object. In chapter C11, we saw that the force exerted by an interaction is related to −1 times the slope of the potential energy function drawn on a potential energy graph. If we identify $dV/dr = -F_{Ax}$, here, then the first equation in equation N3.15 becomes Newton's second law for object A:

$$m_A a_{Ax} - F_{Ax} = 0 \quad \Rightarrow \quad F_{Ax} = m_A a_{Ax} \tag{N3.16}$$

Newton's third law (or the momentum-flow model, either one) tells us that the x-force exerted by the interaction on object B is equal in magnitude and opposite in direction, so $F_{Bx} = -F_{Ax} = +dV/dr$. If you plug this into the second equation in N3.15, you can see it becomes Newton's second law for object B.

Exercise N3X.7: Verify this last statement.

Newton's 2nd law is thus the time derivative of the law of conservation of energy!

So, what do we learn from this? First of all, we see that *Newton's second law is essentially the same thing as time derivative of the law of conservation of energy* or, alternatively, the law of conservation of energy is the antiderivative (integral) of Newton's second law with respect to time. This is a different way of looking at the relationship between Newton's laws and the conservation laws than we saw in in the previous sections of this chapter (where we focused on Newton's laws as consequences of the momentum-flow model).

The link between force and potential energy

Secondly, we see if an interaction between two objects described by a potential energy function $V(r)$ exerts a force on each object whose component F_r along the direction that r increases (if the other object is held still) is given by

$$F_r = -\frac{dV}{dr} \tag{N3.17}$$

A more sophisticated (and difficult) three-dimensional analysis shows that if V depends on the objects' *separation* r alone (that is, on no other aspect of either object's position), then the force points along the line connecting the objects, meaning that the component given by equation N3.17 is the *only* nonzero component of the interaction's force.

Exercise N3X.8: Argue that equation N3.17 correctly describes both the force on object A *and* the force on object B in our one-dimensional example.

How to find a force law from a potential energy

This equation makes it possible for us to find the **force law** for an interaction (that is, the law that specifies how the interaction between objects varies with their separation) if we know the interaction's potential energy function. For example, we know that the gravitational interaction between two particles or spheres is given by $V(r) = -Gm_1m_2/r$, where r is the separation between their centers. Since this depends on r alone, this the gravitational force on either object points along the line connecting them and has an r component

Newton's law of universal gravitation

$$F_r = -\frac{dV}{dr} = +Gm_1m_2\frac{d}{dr}\left(\frac{1}{r}\right) = +Gm_1m_2\left(-\frac{1}{r^2}\right) = -\frac{Gm_1m_2}{r^2} \tag{N3.18}$$

The negative sign here means that the force on each object points in the direction of decreasing r, that is, toward the other object, so the force is *attractive*. Equation N3.18 is **Newton's law of universal gravitation**, which displays the classic inverse-square dependence on the separation of the particles.

Exercise N3X.9: Find F_r for the spring potential energy function given by $V(r) = \frac{1}{2}k_s(r - r_0)^2$.

N3.6 CLASSIFICATION OF FORCES

In the remainder of this unit we will be using Newton's second law to link an object's motion to the forces that act on it. It will help us as we do this to have a clear and agreed-upon scheme for describing and naming these forces.

Basic categories: *contact* and *long-range* forces

Forces ultimately reflect *interactions*. As discussed in chapter C2, it is helpful at the macroscopic level to divide interactions into **long-range interactions** (gravitational and electromagnetic interactions) that can act between objects even when they are physically separated, and **contact interactions** that arise from the microscopic electromagnetic interactions between the atoms of the objects when the objects are in physical contact.

Comments about scheme illustrated in Figure N3.1

Figure N3.1 shows a useful scheme for further subdividing these categories. Let me make a few brief comments about this scheme. First of all, it is important to understand that *all* known long-range interactions are either gravitational or electromagnetic: there are no other possibilities. While *every* object exerts a gravitational force on every other object, unless at least one object in the interacting pair has a mass of planetary proportions, these gravitational forces will be utterly negligible. Loosely speaking, an electromagnetic interaction is *electrostatic* if it involves at least one electrically charged object, and *magnetic* if it involves at least one magnet or flowing electrical current. We will explore the distinction between these categories more fully in Unit *E*.

All forces that are not long-range gravitational or electromagnetic forces arise from contact interactions. The subcategories of contact forces, however, are neither completely exhaustive nor even completely logical: these are simply categories that people have found *useful* in dealing with common situations. Broadly speaking, contact forces can be divided into two subcategories: *compression* forces that arise when the atoms in the contacting surfaces oppose further intermingling or penetration, and *tension* forces, where the atoms (in a string, cable or chain, for example) resist being pulled apart. I have set aside the force exerted by a spring as a third subcategory because it can be either a compression force (if the spring is squeezed) *or* a tension force (if it is stretched).

The primary distinction between *normal* and *frictional* compression forces between two solid objects is the direction that they act. In a certain sense, the basic compression force exerted by the contact interaction on a solid is divided somewhat artificially into a part that is *perpendicular* to the interface between the solids (the normal force) and a part that is *parallel* to this interface (the frictional force), as discussed in chapter C8. (Note that *normal* here does not here mean "typical" or "standard" but "perpendicular": the original Latin root *norma* means "carpenter's square".) A similar distinction distinguishes *lift* forces from *drag* forces on an object moving relative to a fluid: a lift force is the part of the force on the object that acts *perpendicular* to its motion relative to the fluid, and a drag force is the part that acts *parallel* (against) the object's motion.

It turns out to be convenient to divide the frictional forces acting on solid objects in contact into two further categories depending on whether the interacting surfaces are actually sliding relative to each other (kinetic friction) or not (static friction). A *static* friction force acts to *prevent* the surfaces from starting to slide. For example, imagine that you push horizontally on a crate but it doesn't move. An object at rest is not accelerating, so according to Newton's second law it must have zero net force on it, and therefore *some* force must be opposing your push. What is happening is that the crate and floor atoms are intertwined and locked together, and until you supply enough force to break these bonds, they will oppose any sideward push that you might apply.

If it bothers you to call such a force a friction force (because *friction* to you connotes "rubbing" and thus relative motion of the surfaces), you can think of the subscripts SF as meaning *sticking force* instead of *static friction*.

The distinctions between the varieties of fluid forces can be somewhat artificial. (For example, should we think of the force exerted by the air on a helicopter

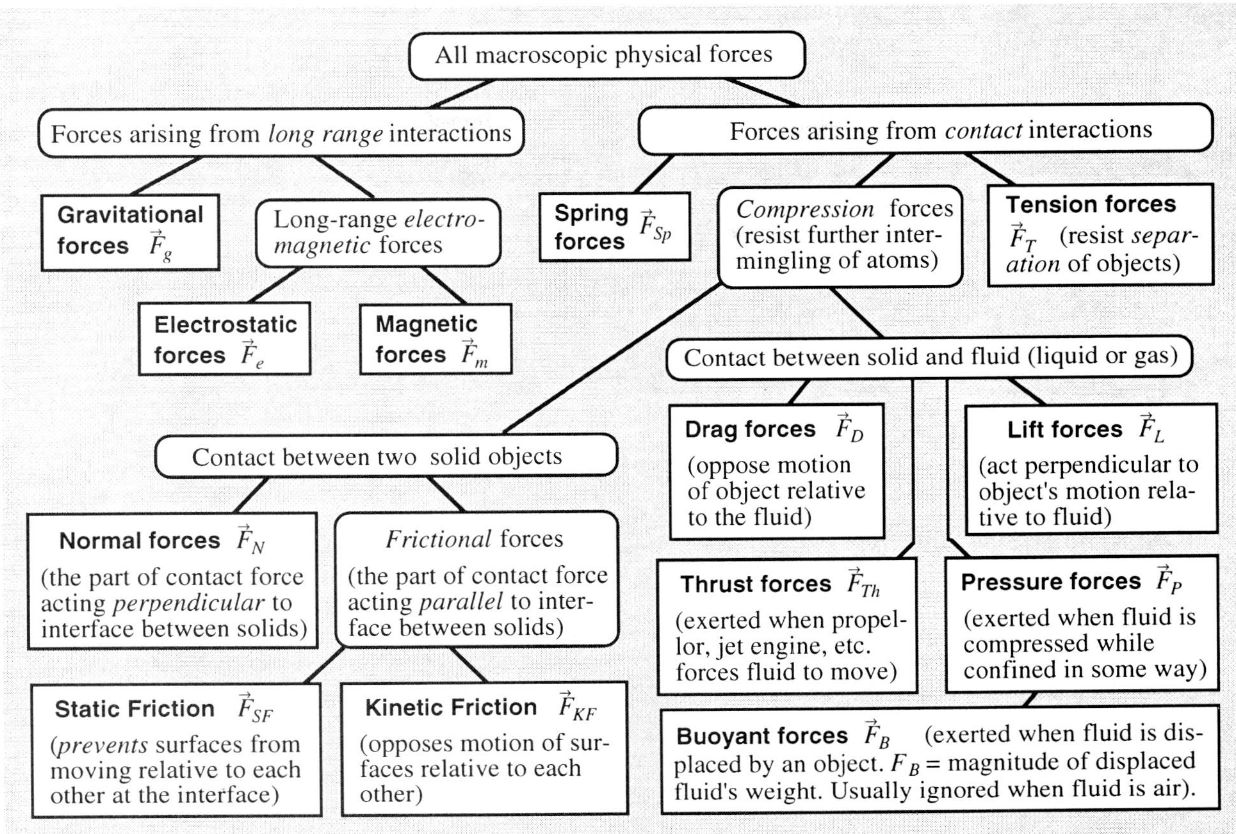

Figure N3.1: A scheme for categorizing forces. The names of the force categories are shown in boldface adjacent to the symbol that we will use for such a force.

rotor a *lift* force or a *thrust* force? A case could be made for either!) Even so, these categories are reasonably helpful in most common contexts.

We looked at *buoyancy* briefly in chapter C12. The most important things to know about buoyancy are that (1) buoyant forces arise from height-dependent pressure *differences* in a fluid body (e.g. the ocean, the atmosphere) confined by gravity, (2) the buoyant force on an object generally acts opposite to its weight and is equal in magnitude to the weight of the fluid it displaces, and (3) buoyant forces on an object in air are generally very small compared to its weight and are by convention ignored unless *explicitly* stated.

Note that in this text, I will *always* use the symbol $\vec{F}$ for a force (this sharply distinguishes forces from other vector quantities). I indicate the various categories of force by attaching subscripts, as shown in Figure N3.1.

Since contact interactions (with the exception of the spring interaction) are not described by potential energy functions, we cannot find nice force laws for contact interactions. The strength of such interactions is usually determined in other ways, as we will discover in the next chapter.

Exercise N3X.10: A motorboat moves at a constant speed across a lake. Classify the forces acting on the boat according to the categories in Figure N3.1.

N3.7 FREE-BODY DIAGRAMS

What is the purpose of a free-body diagram?

Newton's second law says that an object's acceleration is determined by the the external forces acting on it. A **free-body diagram** is a graphical tool that helps us recognize and display these external forces. Drawing a free-body diagram is a very useful first step in almost any problem involving Newton's second law: it helps us clarify for ourselves exactly what forces act on the object and how they are oriented relative to each other.

Steps to follow in drawing a free-body diagram

To draw a free-body diagram of an object, we:

1. ***Imagine the object in its context***, thinking of the things that interact with it. Be especially careful to look along the object's boundary in your mental image for places where it touches things outside itself.
2. ***Draw a sketch of the object alone***, as if it were floating in space. Part of the point of a free-body diagram is to direct our attention to the object itself, reducing its surroundings merely to abstract forces that act on it.
3. For each force that an external interaction exerts on the object, ***draw a dot*** that indicates roughly where the force is applied, and then ***draw an arrow*** that depicts (at least qualitatively) the magnitude and direction of that force, attaching to the dot either the arrow's tail end or tip (preferably the tail end, but whichever makes the diagram clearer). ***Label each arrow*** with an appropriate symbol (in some cases you may also need to provide some explanation to help the reader interpret your symbol).

In the last step, you can depict long-range forces as if they were applied to the object's center of mass, and a contact force acting on a surface as if it were applied to the surface's center. To make a diagram clearer, you may also displace force arrows slightly if they would otherwise overlap.

General do's and don'ts

Here are some general "do's and don'ts" about drawing free-body diagrams:

1. ***Do not draw any arrows on the diagram depicting quantities that are NOT forces***. The main point of a free-body diagram is to display the *forces* that act on an object. Do not confuse the issue by also drawing arrows displaying the object's velocity or acceleration.
2. ***Draw arrows for ONLY those forces that act directly on the object in question***. As stated above, part of the point of a free-body diagram is to direct our attention to a *single* object and display the forces acting on it. Do not confuse the issue by drawing forces that act on other objects.
3. ***Make sure that every force arrow you draw reflects an interaction*** with some other object. A common error is to draw arrows on a free-body diagram for alleged forces associated with an object's velocity and/or acceleration that have no place in the newtonian model. If an alleged force is not obviously associated with an interaction involving an external object, it is not a real force in the newtonian model

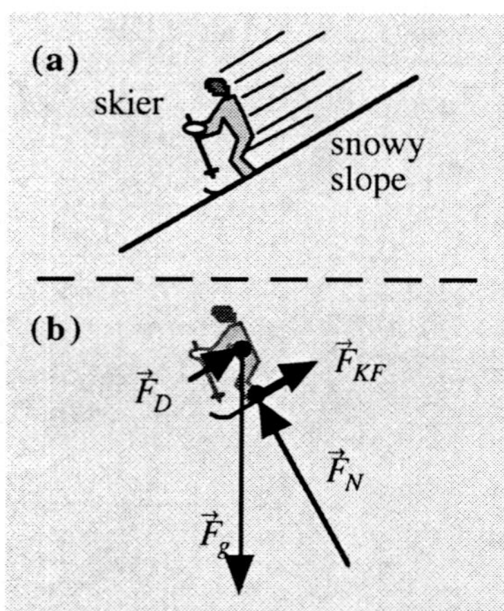

Figure N3.2: (a) A skier sliding down a hill in context. **(b)** A free-body diagram for the skier.

In this chapter, we will use such diagrams only as a device for qualitatively recognizing and displaying the forces that act on an object, and so will not pay much attention the lengths of the forces arrows. (In the next chapter, we will be much more interested in displaying the force magnitudes correctly.)

EXAMPLE N3.2

Problem: Draw a free-body diagram of a skier sliding down a hill.

Solution The skier participates in a gravitational interaction with the earth, which exerts a downward force on the skier's center of mass. The skier also participates in contact interactions with the snow and the air. The force due to the interaction with the snow can be divided into normal and frictional parts. The interaction with the air exerts buoyant and drag forces (we will ignore the former). There is nothing else that the skier is in contact with, so these will be the only forces that act. Figure N3.2b shows the finished free-body diagram.

Exercise N3X.11: Someone takes a rock at the end of string and then whirls it around their head in a horizontal circle. Draw a free-body diagram for the rock ignoring drag. (*Hint*: If you think there is an outward force on the rock, can you describe the interaction that gives rise to it. If not, is it real?)

SUMMARY

I. NEWTON'S LAWS

A. Newton's *first law*: an isolated system's $\vec{v}_{CM}$ is constant

1. this follows from our momentum-transfer model of interactions
2. Newton, on the other hand, took this as a *fundamental* assumption for pedagogical reasons: at the time it was important to emphasize that uniform motion is *natural* and needs no explanation

B. Newton's *third law*: An interaction between two objects A and B exerts forces on them that are opposite but equal in magnitude: $\vec{F}_A = -\vec{F}_B$

1. this is a consequence of our momentum-transfer model
2. but it can be counterintuitive!

C. Newton's *second law* (which will be the main focus of this unit)

1. For a point particle of mass m: $\vec{F}_{\text{net}} = d\vec{p}/dt = m\vec{a}$
   a) $\vec{F}_{\text{net}}$ is the vector sum of forces on the particle
   b) this follows from the definition of force as $\vec{F} \equiv d\vec{p}/dt$ and the way that momentum-transfers add: $d\vec{p} = d\vec{p}_1 + d\vec{p}_2 + \ldots$
2. For an extended object of mass m: $\vec{F}_{\text{net,ext}} = d\vec{p}_{\text{tot}}/dt = m\vec{a}_{CM}$
   a) $\vec{F}_{\text{net,ext}}$ is the vector sum of *external* forces on the object
   b) $d\vec{p}_{\text{tot}}/dt = m\vec{a}_{\text{CM}}$ follows from the definition of the CM
   c) this means that the CM of an extended object behaves like a point particle (and we can model it as such)
3. Important implications:
   a) the net force on an object *causes* it to accelerate (read the equals sign in $\vec{F}_{\text{net}} = m\vec{a}$ this way!)
   b) the net force causes *acceleration* (not motion!)
   c) the *net force* (not the biggest or latest) causes the acceleration

II. FORCE AND ENERGY

A. Newton's second law emerges from the time-derivative of the law of conservation of energy for a fundamental interaction *if* $F_r = -dV/dr$, where F_r is the component of the force that the interaction exerts on either particle in the direction that r increases if the other is held fixed (a three-dimensional analysis implies that all other components are zero)

B. $F_r = -dV/dr$ allows us to find a *force law* for an interaction (an equation telling how F_r depends on r) if we know $V(r)$

III. CLASSIFICATION OF FORCES (See Figure N3.1 for details)

A. All *long-range forces* are either gravitational or electromagnetic

B. All other macroscopic forces are *contact forces.* General categories are:

1. *Compression* forces
   a) between two solids: *normal, static friction, kinetic friction*
   b) between a solid and a fluid (liquid or gas) *lift, drag, thrust, buoyant* and *pressure* forces.
2. *Tension* forces
3. *Spring* forces

IV. FREE-BODY DIAGRAMS

A. A free-body diagram helps us *recognize* and *display* forces on an object

B. To draw a free-body diagram,

1. think about what touches the object
2. sketch the object in isolation
3. for each external interaction:
   a) draw a dot where that interaction's force is applied
   b) draw a force arrow showing the direction and magnitude of the force (with either its tail or head attached to the dot)
   c) label each arrow with an $\vec{F}$ with the appropriate subscript

C. Do's and don'ts:

1. draw only force arrows on a diagram
2. draw only forces that act on that particular object
3. draw only forces arising from *interactions* (only these are real!)

GLOSSARY

Newton's first law: asserts that *the center of mass of an isolated object moves with a constant velocity.* (From our modern perspective, we would say that this law is a consequence of the more fundamental law of conservation of momentum.)

Newton's third law: asserts that an interaction between two objects A and B exerts forces on each object that have equal magnitudes but opposite directions: $\vec{F}_A = -\vec{F}_B$. This law is a consequences of the idea that an interaction transfers momentum from one object to the other.

net force $\vec{F}_{\text{net}}$: the vector sum of all external forces acting on a particle or extended object.

Newton's second law: asserts (in its most general form) that the net *external* force on an object causes its center of mass to accelerate in inverse proportion to the object's total mass m:

$$\vec{F}_{\text{net,ext}} = \frac{d\vec{p}_{\text{tot}}}{dt} = m\vec{a}_{\text{CM}} \tag{N3.9}$$

This law is a consequence of the definition of force as the rate at which a given interaction transfers momentum, the idea that the net momentum an object gains in time dt is the vector sum of the momentum transfers it gains from each interaction, the idea that internal interactions do not affect an object's total momentum, and the definition of the center of mass.

long-range interactions: interactions that act between particles or objects with macroscopic separations. Gravitational and electromagnetic interactions are the only long-range interactions currently known.

contact interactions: interactions that arise when two objects come into physical contact. All known macroscopic interactions are either long-range interactions or contact interactions.

electrostatic force $\vec{F}_e$ **:** a force arising from an interaction between an electrically charged object and something else (a subcategory of an *electromagnetic* force).

magnetic force $\vec{F}_m$ **:** a force arising from an interaction between a magnet (or object carrying an electric current) and something else (a subcategory of an *electromagnetic* force).

compression force: a force arising from a contact interaction that seeks to keep the interacting objects from becoming closer.

tension force $\vec{F}_T$ **:** a force arising from a contact interaction that seeks to keep the interacting objects from separating.

spring force $\vec{F}_{Sp}$ **:** the force exerted by a spring on another object (this may be a compression force if the spring is compressed or a tension force if the spring is stretched).

normal force $\vec{F}_N$ **:** the component of the force arising from a contact interaction between two solid objects that acts perpendicular to the interface between the objects. (*Normal* means *perpendicular* here, not *typical* or *usual.*)

friction force: the component of the force arising from a contact interaction between two solid objects that is parallel to the interface between the objects.

static friction force $\vec{F}_{SF}$ **:** a friction force that *prevents* two solid objects from sliding relative to each other along their interface. (You can think of this as being a "sticking force.")

kinetic friction force $\vec{F}_{KF}$ **:** a friction force that opposes an already existing sliding motion along the interface between two solid objects.

drag force $\vec{F}_D$ **:** a force arising from the contact interaction between a fluid and an object moving relative to the fluid that *opposes* the object's motion.

lift force $\vec{F}_L$ **:** a force arising from the contact interaction between a fluid and an object (like a wing) moving relative to the fluid that acts *perpendicular* to the object's motion.

thrust force $\vec{F}_{Th}$ **:** a force arising from the contact interaction between a fluid and an object (like a propeller) that pushes the fluid backward.

pressure force $\vec{F}_P$ **:** a force arising when something solid presses on a fluid that is somehow confined.

buoyant force $\vec{F}_B$ **:** an upward force exerted on an object at least partially immersed in a fluid. F_B = the magnitude of the weight of the fluid displaced by the object. (This force is usually ignored if fluid is air.)

force law: a mathematical law that expresses how the force exerted on an object by its interaction with another object depends on their separation.

free-body diagram: a diagram of an object that displays all of the external forces acting on it (as labeled arrows) and the points on the object where these forces act.

TWO-MINUTE PROBLEMS

N3T.1 A 6-kg bowling ball moving at 3 m/s collides with a 1.4-kg bowling pin at rest. How do the magnitudes of the forces exerted by the collision on each object compare?
A. the force on the pin is larger than the force on the ball
B. the force on the ball is larger than the force on the pin
C. both objects experience the same magnitude of force
D. the collision exerts no force on the ball
E. there is no way to tell

N3T.2 A large car drags a small trailer in such a way that their common speed increases rapidly. Which tugs harder on the other?
A. the car tugs harder on the trailer than vice versa
B. the trailer tugs harder on the car than vice versa
C. both tug equally on each other
D. the trailer exerts no force on the car at all
E. there is no way to tell

N3T.3 A parent pushes a small child on a swing so that the child moves rapidly away while the parent remains at rest. How does the magnitude of the force that the child exerts on the parent compare to the magnitude of the force that the parent exerts on the child?
A. the force on the child is larger in magnitude
B. the force on the parent is larger in magnitude
C. these forces have equal magnitudes
D. the child exerts zero force on the parent
E. there is no way to tell

N3T.4 A spaceship with a mass of 24,000 kg is traveling in a straight line at a constant speed of 320 km/s in deep space. What is the magnitude of the net thrust force acting on this spaceship?

A. 7.7 MN
B. 7.7 GN
C. 75 N
D. 0.075 N
E. zero
F. other (specify)

N3T.5 Arrows representing four forces having equal magnitudes are shown below. What combinations of these forces, acting together on the same object, will allow that object to move with a constant velocity?

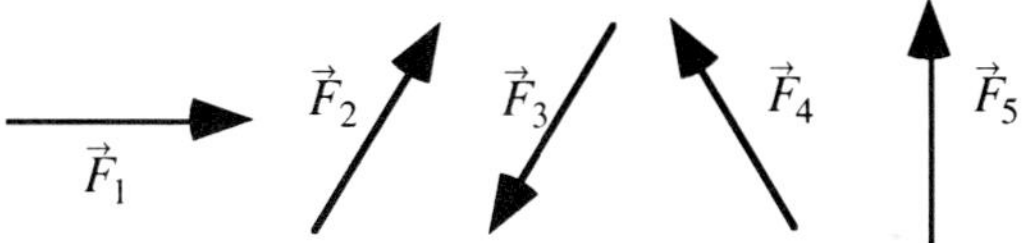

A. $\vec{F}_2$ and $\vec{F}_3$
B. $\vec{F}_3$ and $\vec{F}_4$
C. $\vec{F}_1$, $\vec{F}_2$, and $\vec{F}_4$
D. $\vec{F}_1$, $\vec{F}_3$, and $\vec{F}_4$
E. *A* and *D*
F. none of the above

N3T.6 The free-body diagram at the right is supposed to represent a box sliding at a constant speed toward the right along a tabletop as it is pulled by a string. How should we label the leftward force?

A. $\vec{F}_D$
B. $\vec{F}_{KF}$
C. $\vec{F}_{SF}$
D. The diagram is wrong: the left force should not be there at all
E. other (specify)

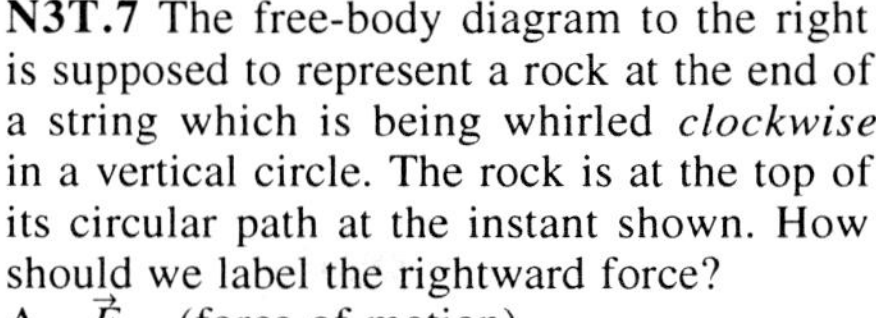

N3T.7 The free-body diagram to the right is supposed to represent a rock at the end of a string which is being whirled *clockwise* in a vertical circle. The rock is at the top of its circular path at the instant shown. How should we label the rightward force?

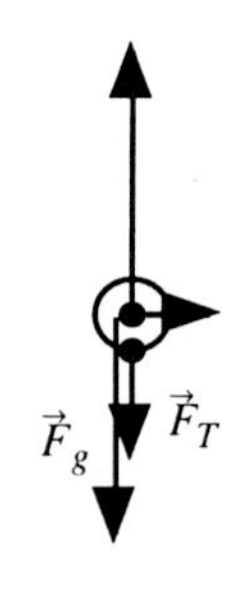

A. $\vec{F}_M$ (force of motion)
B. $\vec{F}_I$ (force of inertia)
C. $\vec{F}_D$
D. The diagram is wrong: the right force should not be there at all
E. the diagram is wrong: there should be a leftward force instead of a rightward force
F. other (specify)

N3T.8 Consider again the free-body diagram in the previous problem. How should we label the upward force?

A. $\vec{F}_C$ (centrifugal force)
B. $\vec{F}_C$ (centripetal force)
C. $m\vec{a}$
D. $\vec{F}_I$ (force of inertia)
E. the diagram is wrong: there is no upward force

N3T.9 Imagine that you serve a tennis ball directly forward and parallel to the ground. *Ignoring* air resistance, what is the approximate direction of the *net* force acting on the ball after it leaves your racquet?

A. directly forward
B. directly downward
C. zero
D. forward and a bit downward
E. downward and a bit backward
F. other (specify)

HOMEWORK PROBLEMS

BASIC SKILLS

N3B.1 Imagine a cannon that is free to roll on wheels. According to Newton's third law, the interaction between the cannon and a cannonball as the latter is fired should exert equal forces on both. If this is so, why does the cannonball fly away at a very high speed and the cannon only recoil modestly? If the cannon is bolted to the ground, it does not recoil at all. How is this possible if the interaction with the cannonball exerts a force on the cannon? Please explain carefully.

N3B.2 The moon is about 1/81 times that of the earth. How does the gravitational force that the moon exerts on the earth compare to the gravitational force the earth exerts on the moon? Explain your answer.

N3B.3 A car with a mass of 1300 kg decreases its forward speed from 22 m/s to 14 m/s in 4.0 s. What is the magnitude and direction of the net force on the car? Show how you arrived at your result.

N3B.4 The magnitude of the net external force on a flying 0.25-kg ball is 2.5 N. What is the magnitude of the ball's acceleration? Explain how you arrived at your result.

N3B.5 A batter hits a baseball vertically in the air. Ignoring air resistance, draw a free-body diagram of the ball after it has left the bat but is still moving upward.

N3B.6 A crane lifts a crate at a constant velocity upward. Draw a free-body diagram of the crate.

N3B.7 An airplane is flying horizontally at a constant velocity. Draw a free-body diagram for this airplane.

N3B.8 A baseball player slides into second base. Draw a free-body diagram of the player.

SYNTHETIC

N3S.1 Two friends push on a 2200-kg car. The car is initially at rest, but after pushing on it for about 2.0 s, they manage to get it to a final speed of 0.80 m/s. What is the *minimum* average magnitude of the total force the friends exert on the car? Show your work, and explain why this is the *minimum* average force.

N3S.2 A 52-kg skater sliding due north at an initial speed of 2.5 m/s runs into a snow bank and comes to rest in a time interval of 0.25 s. What is the magnitude and direction of the average force exerted on the skater by the snowbank?

N3S.3 You push on a 32-kg box with a force of 75 lbs forward. The sliding friction between the box and the floor is 60 lbs backward. If the box starts from rest, how long will it take you to get it moving at a speed of 5 mi/h? (Remember that 1 lb = 4.45 N, 1 m/s = 2.2 mi/h.)

N3S.4 Two people, one with a mass of 70 kg and one with a mass of 45 kg sit on 5-kg frictionless carts that are initially at rest on a flat surface. The lighter person pushes on the more massive person, causing the latter accelerate at a rate of 1.0 m/s^2 toward the east. What is the lighter person's acceleration?

N3S.5 Use the potential energy formula for the electrostatic attraction to find the force law (called **Coulomb's law**) for the attraction or repulsion between two charged particles. Check that your force law is repulsive if the charges have like sign and attractive otherwise.

N3S.6 The potential energy function associated with the magnetic interaction between two small magnets has the form $V(r) = C/r^3$, where C is a constant that depends on the magnets' relative orientation and strength. What is the force law for the interaction between these magnets?

N3S.7 A runner accelerates forward after the starting gun sounds in a race. What kind of force causes the runner's forward acceleration? Is there any force opposing the runner's forward motion? If so, what kind of force is it? Explain your response.

N3S.8 Draw free-body diagrams for a box that **(a)** is sliding down an incline at a constant speed, and **(b)** is being pulled up an incline at a constant speed by a rope.

N3S.9 **(a)** Draw a free-body diagram for a motorboat at rest on a still lake. How should the vertically upward force on the boat be classified in this case? **(b)** Now draw a free-body diagram for the boat when it is moving at a constant velocity. There is an additional upward force in this case (that causes the boat to rise somewhat in the water). How should this force be classified? Explain in both cases.

RICH-CONTEXT

N3R.1 Imagine that you have discovered a way to create a repulsive force field and you are trying to use it to make a device to protect against car crashes. Your device is mounted in a 1000-kg car and is capable of exerting a constant repulsive force of 200,000 N on another car until the cars come to rest with respect to each other. Generating the force field creates a lot of electrical power, though, and you are trying to decide how big a battery you need to use. You want the battery to be able to supply power long enough to bring the cars to rest with respect to each other (not necessarily the road) when the protected car is involved in a head-on collision with a 2000-kg car when both are traveling at 30 m/s. How long must your battery be able to supply power to the force field?

ADVANCED

N3A.1 (Do this problem only if you are familiar with partial derivatives.) One can show that Newton's second law follows from the time-derivative of the law of conservation of energy. Assume that two particles participate in a long-range interaction that can be described by some potential energy function $V(r) = V(x_1\ y_1, z_1, x_2, y_2, z_2)$, where x_1, y_1, and z_1 are the coordinates of the first particle and x_2, y_2, and z_2 are the same for the second particle.**(a)** Show that the time derivative of the kinetic energies of the particles can be written

$$\frac{dK_1}{dt} = m_1\vec{a}_1 \cdot \vec{v}_1 \quad \text{and} \quad \frac{dK_2}{dt} = m_2\vec{a}_2 \cdot \vec{v}_2 \qquad \text{(N3.19)}$$

(*Hint:* The dot product also obeys the product rule, and the dot product is commutative.) **(b)** Show that the time derivative of the potential energy can be written

$$\frac{dV}{dt} = -\vec{F}_1 \cdot \vec{v}_1 + -\vec{F}_2 \cdot \vec{v}_1 \qquad \text{(N3.20)}$$

if we define

$$\vec{F}_1 \equiv \left[-\frac{\partial V}{\partial x_1}, -\frac{\partial V}{\partial y_1}, -\frac{\partial V}{\partial z_1}\right] \qquad \text{(N3.21)}$$

and $\vec{F}_2$ analogously. **(c)** Use these results to compute the time derivative of the law of conservation of energy, and argue (in a manner similar to what we did in section N3.5) that this equation will be satisfied for all possible particle velocities if and only if

$$\vec{F}_1 = m_1\vec{a}_1 \quad \text{and} \quad \vec{F}_2 = m_2\vec{a}_2 \qquad \text{(N3.22)}$$

ANSWERS TO EXERCISES

N3X.1 They both exert the same force on each other.
N3X.2 They both exert the same force on each other. (The car/truck system accelerates *not* because the car exerts more force on the truck than the truck exerts on the car. But because the forward force exerted on the small car by its wheels exceeds the friction forces acting on both the car and the truck. These are the only *external* horizontal forces acting on the system, and for the system to gain momentum, it needs *external* forces.)
N3X.3 Each exerts the same force on the other, even though Terry does nothing to actively exert a force on Sal. Terry falls to the ground and Sal does not probably because Terry was surprised and did not brace against the push, but Sal was braced to exert the push (and thus receive a passive push in return).
N3X.4 The car is not accelerating, so the net force on it must be *zero*.
N3X.5 The plane's change in velocity during is 120 m/s during the 45 s run, so the plane's average acceleration is (120 m/s)/(45 s) = 2.7 m/s^2 forward. The average thrust force required to produce this average acceleration will be (14,000 kg)(2.7 m/s^2) = 37,000 kg·m/s^2 = 37,000 N.
N3X.6 Adding everything together, we get

$$0 = m_A v_{Ax} a_{Ax} + m_B v_{Bx} a_{Bx} + \frac{dV}{dr}(v_{Ax} - v_{Bx}) \qquad \text{(N3.23)}$$

Gathering terms involving v_{Ax} and v_{Bx} yields N3.14.
N3X.7 When you plug $F_{Bx} = +dV/dr$ into the second equation in N3.15, we get $m_B a_{Bx} - F_{Bx} = 0$, which implies that $F_{Bx} = m_B a_{Bx}$.
N3X.8 For object A, the r direction is the $+x$ direction, since as A moves in that direction, it moves away from B. Thus $F_{Ar} = +F_{Ax}$. But $F_{Ax} = -dV/dr$ according to our statement just above equation N3.16, so $F_{Ar} = -dV/dr$ as claimed. On the other hand, the r direction is the $-x$ direction for B, since that is the direction it moves to go away from A. Therefore, $F_{Br} = -F_{Bx}$. But $F_{Bx} = +dV/dr$ according to what we said just above exercise N3X.7. Therefore, $F_{Br} = -dV/dr$ in this case also.
N3X.9 The radial component of the force in this case is

$$F_r = -\frac{d}{dr}\left[\tfrac{1}{2}k_s(r^2 - 2r_0 r + r_0^2)\right] = -\left[\tfrac{1}{2}k_s(2r - 2r_0)\right]$$
$$= -k_s(r - r_0) \qquad \text{(N3.24)}$$

Note that this force acts opposite to the direction of increasing r (and thus opposes its increase) if $r > r_0$, and in the direction of increasing r if $r < r_0$.
N3X.10 The boat experiences a gravitational force downward, a buoyant force upward, a thrust force forward and a drag force backward.
N3X.11 Ignoring drag from the air, the string is the only thing touching the rock, and it exerts an inward tension force as shown. Otherwise, the only force that can act on the rock is gravity. An outward force is neither possible nor necessary (as we will see in chapter N10).

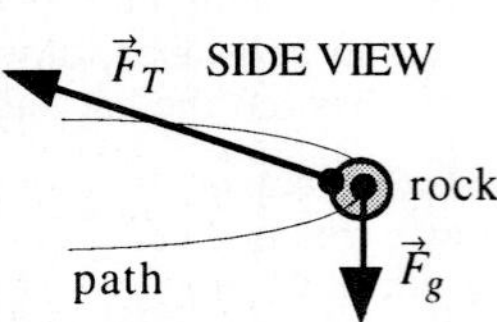

N4

FORCES FROM MOTION

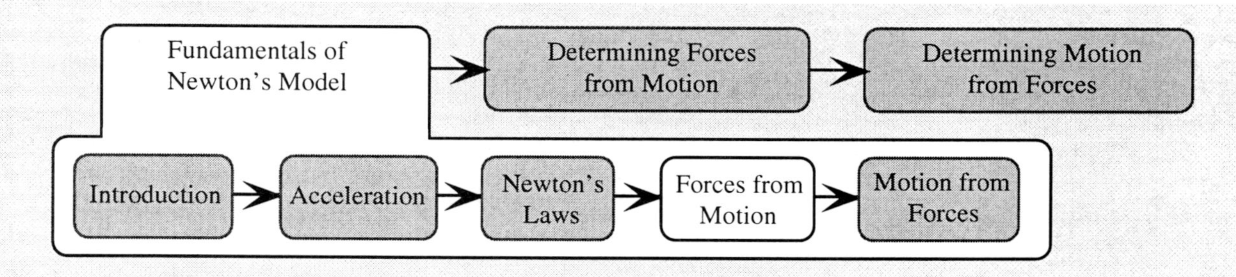

N4.1 OVERVIEW

In the last chapter, we discussed Newton's laws of motion with particular emphasis on the second law, which links the net force acting on an object to that object's acceleration. We also learned how to classify forces and describe them using a force diagram.

In the remainder of this unit, we will use Newton's second law of motion either (1) to determine the forces acting on an object when its motion is known, or (2) to determine an object's motion when the forces acting on it are known. In this chapter and the next, we will briefly examine these two applications of Newton's second law in order to lay sturdy foundations for the deeper study of each of these approaches in chapters N6–N10 and N11–N14 respectively.

In this chapter, we will focus on how we can use Newton's second law to infer the magnitudes (or even the existence) of forces by observing how an object moves. Here is a summary of this chapter's sections.

N4.2 *THE KINEMATIC CHAIN* summarizes the mathematical relationships between position, velocity, and acceleration and the link between an object's acceleration and the forces acting on it.

N4.3 *NET-FORCE DIAGRAMS* discusses a kind of diagram that serves as a useful supplement to a free-body when we want to display more exactly the relationships between the forces that act on an object.

N4.4 *QUALITATIVE EXAMPLES* shows how we can use motion diagrams, force diagrams, and free-body diagrams in conjunction with Newton's second law to learn things about the forces acting on an object.

N4.5 *THIRD-LAW AND SECOND-LAW PAIRS* examines pairs of forces that are equal in magnitude but opposite in direction and distinguishes between pairs that are equal and opposite because of Newton's *third* law and those that are equal and opposite because of Newton's *second* law.

N4.6 *COUPLING FORCES* shows how we can use Newton's second and third laws together to find the forces that coupled objects exert on each other.

N4.7 *GRAPHS OF ONE-DIMENSIONAL MOTION* explores how we can use graphs to analyze and visually describe the motion of an object and determine the forces acting on an object as a function of time.

N4.8 *A FEW QUANTITATIVE EXAMPLES* glances at a few simple quantitative calculations as a foretaste of what we will do in chapters N6–N10.

**** ***MATH SKILLS:** DERIVATIVES AND SLOPES* examines the relationship between the derivative of a function and the *slope* of that function on a graph.

N4.2 THE KINEMATIC CHAIN

In chapters N1 and N2, we saw that an object's velocity is the time-derivative of its position ($\vec{v}(t) \equiv d\vec{r}/dt$), and its acceleration is the time-derivative of its velocity ($\vec{a}(t) \equiv d\vec{v}/dt$). This means that an object's position, velocity, and acceleration as a function of time are connected by the following chain of derivative relationships:

The kinematic chain

$$\vec{r}(t) \xrightarrow{\text{time derivative}} \vec{v}(t) \xrightarrow{\text{time derivative}} \vec{a}(t) \qquad \text{(N4.1)}$$

We will call this chain of relationships the **kinematic chain** (**kinematics** is the mathematical study of motion).

At the component level this chain means that

$$x(t) \xrightarrow{\text{time derivative}} v_x(t) = \frac{dx}{dt} \xrightarrow{\text{time derivative}} a_x(t) = \frac{dv_x}{dt} \qquad \text{(N4.2a)}$$

$$y(t) \xrightarrow{\text{time derivative}} v_y(t) = \frac{dy}{dt} \xrightarrow{\text{time derivative}} a_y(t) = \frac{dv_y}{dt} \qquad \text{(N4.2b)}$$

$$z(t) \xrightarrow{\text{time derivative}} v_z(t) = \frac{dz}{dt} \xrightarrow{\text{time derivative}} a_z(t) = \frac{dv_z}{dt} \qquad \text{(N4.2c)}$$

We see that the chain of vector derivatives in equation N4.1 is equivalent to the three separate chains of component derivatives listed in equations N4.2.

Finding forces from motion: an overview

In many situations, there are one or more forces acting on an object whose magnitude and/or direction we do not know. The kinematic chain implies that if we know how an object *moves* (that is, we know either its position or its velocity as a function of time), then we can compute the object's *acceleration* as a function of time. Newton's second law in turn links an object's acceleration to the *net force* acting on the object, and knowing the net force usually is sufficient information to determine the unknown forces. This is how we determine *forces from motion*. We will talk about this in fairly general terms in the remainder of this chapter and in much more detail in chapters N6 through N10.

Exercise N4X.1: Say that we know that $x(t) = \frac{1}{2}bt^2 + c$, where b and c are constants. What are the units of b and c? What are $v_x(t)$ and $a_x(t)$ in this case? Do the units come out right for these quantities?

N4.3 NET-FORCE DIAGRAMS

In the last chapter, we learned how to draw a *free-body diagram* of an object. Free-body diagrams are very useful for helping us to recognize what forces might act on an object and to display these forces in a vivid way. Such diagrams are not so good, however, at displaying the *quantitative* relationships between these forces in an accessible way. In situations where we are trying to determine the magnitudes of forces more precisely, it helps *supplement* a free-body diagram with a different kind of diagram that we call a **net-force diagram**.

Newton's second law tells us that the acceleration of an object's center of mass is determined by the *net external force* acting on the object, that is, the vector sum of the external forces acting on the object. The most important purpose of a net-force diagram is to display the vector sum of the forces acting on the object.

How to draw a net-force diagram

To draw a net-force diagram, we do the following:

1. ***Draw a free-body diagram first.*** A net-force diagram is essentially a rearrangement of a free-body diagram to make the quantitative relationships between forces clearer. A net-force diagram makes little sense without an accompanying free-body diagram.

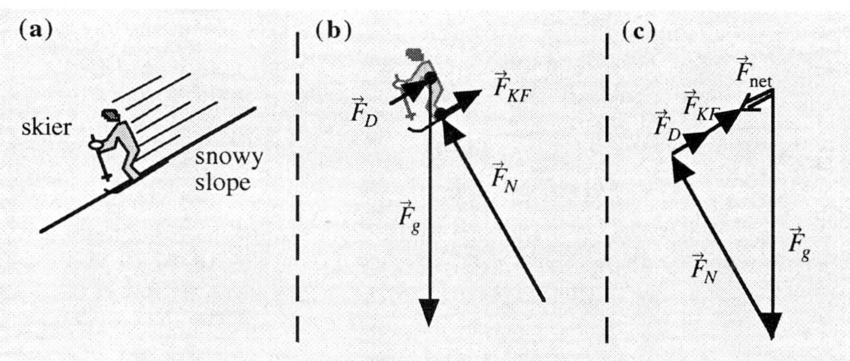

Figure N4.1: **(a)** A skier sliding down a snowy slope. **(b)** A free-body diagram of the same skier. **(c)** A net-force diagram showing the vector sum of the force acting on the skier.

2. ***Copy the force arrows*** (and labels) from the free-body diagram to the net-force diagram, ***arranging them in sequence*** (that is, drawing each new force arrow with its tail end starting at the tip of the previously drawn arrow). If two force arrows point in opposite directions, draw them right next to each other (displaced by a small distance).

3. ***Draw the arrow that represents the vector sum*** of these force arrows. To distinguish it visually from the other arrows, make its body a double line instead of a single line. If the force arrows add up to zero, you can simply write "$\vec{F}_{net} = 0$."

Figure N4.1 displays both a free-body diagram and a net-force diagram for a skier sliding down a hill. Note that if the forces on the skier really have the magnitudes shown, the net force on the skier is a vector pointing down the incline and thus the skier will be *accelerating* down the incline.

N4.4 QUALITATIVE EXAMPLES

In this section, we will do a number of qualitative example problems that illustrate how we can use motion diagrams, free-body diagrams, net-force diagrams and Newton's second law to determine the magnitude, direction, and/or even the very existence of certain forces acting on an object. The basic process for doing such qualitative problems is as follows:

An outline of the steps that we will follow in the following examples

1. ***Determine the direction of the object's acceleration (or determine that it is zero)*** using information from the problem statement. You often will find a motion diagram useful in establishing the direction of the acceleration when the acceleration is nonzero.

2. ***Draw a free-body diagram*** of the object, displaying the directions and possible magnitudes of the forces that you know act on the object.

3. ***Draw a net-force diagram*** using these force arrows. If the net force arrow that you construct does not have the same direction as the acceleration found in the first step (as required by Newton's second law), adjust the magnitudes of the force arrows and/or add new forces until the net force and acceleration are consistent. Be sure to update your free-body diagram to keep it consistent with the net-force diagram.

In this section, we will be applying this process explicitly and in detail to make some very basic qualitative statements about the forces that must acting on an object to make it move in the manner stated. When we do more quantitative problems of this type in chapters N6-N9, we will have to follow this same process (in more compressed form) before we can do the calculations. What we are doing here is thus necessary background for doing those problems correctly.

EXAMPLE N4.1

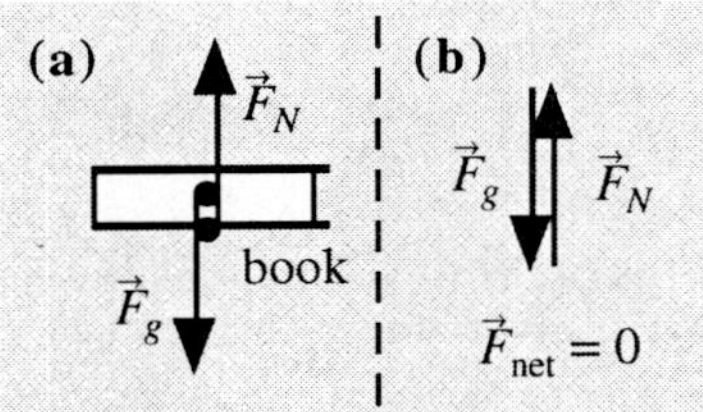

Figure N4.2: **(a)** A free-body diagram of the book. **(b)** A net-force diagram for the book.

Problem: A book sits at rest on a table. Does the contact interaction between the book and the table exert a force on the book? If so, what is the magnitude of this force compared to the gravitational force $\vec{F}_g$ acting on the book?

Solution A book sitting at rest has a constant velocity of zero, so it has no acceleration. Newton's second law then implies that the net force exerted on the book must be zero. Since gravity clearly exerts a downward gravitational force $\vec{F}_g$ on the book, *something* must be exerting an upward force on the book to cancel the gravitational force. This force must be exerted by the *table*, since it is the only substantial thing in contact with the book. (The surrounding air does not significantly support the book, as you can vividly illustrate by removing the table suddenly.) This force must be equal and opposite to $\vec{F}_g$ to cancel it. This means that the force exerted by the contact interaction must be perpendicular to the book/table interface, so it is a *normal* force $\vec{F}_N$ (see Figure N4.2).

EXAMPLE N4.2

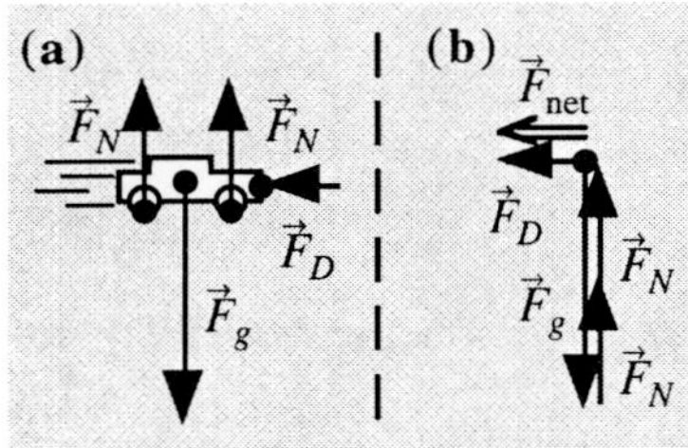

Figure N4.3: **(a)** A first draft for a free-body diagram for the car. **(b)** The net-force diagram shows that if the free-body diagram is correct, there will a non-zero net force on the car.

Problem: A car travels at a constant velocity along a level, straight highway. Draw an accurate free-body diagram for the car. (Do not ignore air drag.)

Solution As usual, the car experiences a gravitational force $\vec{F}_g$ directed toward the center of the earth. The car only touches the road and the air that rushes by. The air exerts a rearward drag force $\vec{F}_D$ on the car. The road plausibly also exerts upward normal force $\vec{F}_N$ on each of the car's tires for reasons similar to those discussed in the previous example. A plausible first guess for a free-body diagram might look as shown in Figure N4.3a.

However, we are told that the car is moving at a constant velocity, which means that its acceleration is zero. Therefore, by Newton's second law, the net force on the car must also be zero. However, the net-force diagram of Figure N4.3b makes it clear that the forces drawn in Figure N4.3a do not add up to zero. Therefore, some other force must be acting on the car in the forward direction to cancel the rearward drag force. What is this force?

A natural first guess would be an "engine force." The engine *must* be involved somehow, since without the engine, the car would not be able to maintain its forward velocity. However, an "engine force" does not appear in our list of categories in the previous chapter. Moreover, the force that moves the car forward must be an *external* force, and there is no way that an engine that is *inside* the car could exert an external force on the car.

A second guess might be that the force is a *thrust* force, but the engine (in a normal car at least) does not propel the car forward by pushing a fluid backward.

The only thing that the car touches that might be able to exert such a force is the *road*. Also, think about this: if the road were very slippery (icy, perhaps), the engine would not be effective in causing the car to move forward. So the *road* must somehow exert a forward force on the car's tires that pushes the car forward. Since this force must be parallel to the interface between the road and the tire, and since (under normal conditions) the tire and road do not slide relative to each other, the force fits the description of a *static friction* force.

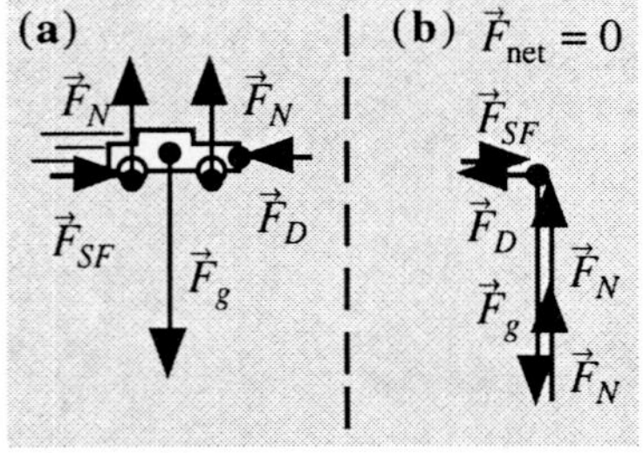

Figure N4.4: **(a)** The final draft for the free-body diagram for the car. **(b)** The net-force diagram shows now that the net force on the car will be zero, consistent with its observed motion.

What has this static friction force to do with the engine? The car's engine turns the wheels against the road surface. This causes the atoms of the tire to interact with the road atoms in such a way that the latter get pushed backward. Newton's *third* law, however, tells us that this contact interaction must then push the atoms in the tire forward with the same force. So the engine, by compelling the tires to rotate against the road, sets up an interaction with the road that exerts a forward static friction force on the tires that pushes the car forward.

Complete and correct free-body and net-force diagrams for the car are shown in Figures N4.4. Note that the net-force diagram clearly indicates that the normal forces exerted on the tires must add up to a vector that is equal in magnitude to the gravitational force and the drag and static friction forces must also have equal magnitudes if the net force is to be zero.

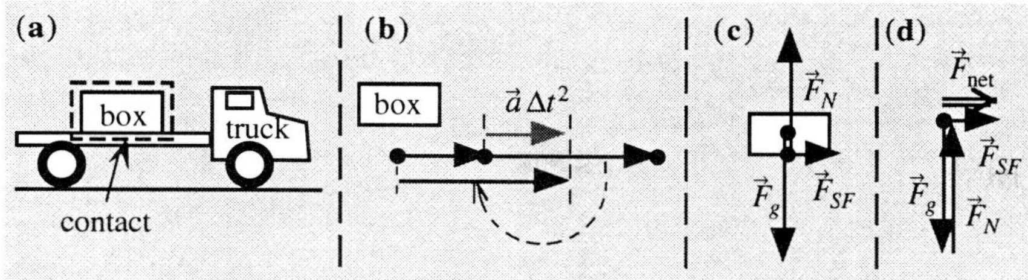

Figure N4.5: **(a)** The box in its context. **(b)** A motion diagram for the box, showing that its acceleration is forward. **(c)** A free-body diagram for the box. **(d)** A net-force diagram of the box, showing that the forces drawn are consistent with the box's forward acceleration.

EXAMPLE N4.3

Problem: A box sits in the back of a pickup truck at rest relative to the truck. The truck is speeding up after a stop sign. Draw a free-body diagram for the box. (Ignore the box's interaction with the air in this case.)

Solution The earth exerts a downward gravitational force $\vec{F}_g$ on the box. If we ignore air resistance, the only thing touching the box is the bed of the truck, which will exert a normal force $\vec{F}_N$ on it that opposes the force of gravity. However, the box (along with the truck) is speeding up, *so it must be accelerating to the right* (forward), as we see from the motion map in Figure N4.5b. No matter what the relative magnitudes of the vertical forces $\vec{F}_N$ and $\vec{F}_g$ might be, they cannot add up to a forward force. Therefore there must be some other force acting on the box. What could this possibly be?

A clue is that if the truck bed were very slippery, we know intuitively that the box would slide backward (relative to the truck) as the truck speeds up. Therefore, the contact interaction between the box and truck bed must be exerting a horizontal forward force on the box to prevent it from sliding relative to the truck. This can *only* be a static friction force (sticking force) $\vec{F}_{SF}$.

The free-body diagram appears in Figure N4.4c. If $\vec{F}_N$ and $\vec{F}_g$ have equal magnitudes, they cancel, and the net force on the box is then equal to the forward force $\vec{F}_{SF}$ (as shown in the net-force diagram in Figure N4.4d) and therefore is consistent with the box's described forward acceleration.

EXAMPLE N4.4

Problem: A car goes over the crest of a hill at a constant speed. Draw a free-body diagram of the car at the instant that it goes over the top of the hill.

Solution As usual, we have the opposed forces of gravity $\vec{F}_g$ and the normal forces $\vec{F}_N$ exerted by the tires' contact interaction with the road. We also have a rearward drag force $\vec{F}_D$ due to the car's motion through the air, and (as discussed in example N4.2) a forward static friction force $\vec{F}_{SF}$ exerted on the tires by contact interaction between the tires and road. These are the only possible forces acting on the car, since it is only in contact with the air and the road. The car's free-body diagram thus must look at least qualitatively as shown in Figure N4.6c.

However, as the motion diagram in Figure N4.6b shows, the car is accelerating *downward* as it goes over the crest of the hill. What can possibly make the car accelerate downward? We already have listed all of the forces that can act on the car, so the only way to get the net force on the car to be downward is to adjust their magnitudes. The horizontal forces must be equal in magnitude if the net force is to be downward, and the gravitational force on the car has a fixed magnitude (the car's weight), so the only way to make the net force on the car be downward is for $2\,F_N$ to be less than $F_g = mg$ (as shown in Figure N4.6d).

This serves as an important lesson: the magnitude of the total normal force on an object is not *automatically* equal to the magnitude of the object's weight $\vec{F}_g$. These magnitudes are *often* equal, because an object's vertical acceleration is commonly zero. But if it is not, then these forces *cannot* cancel!

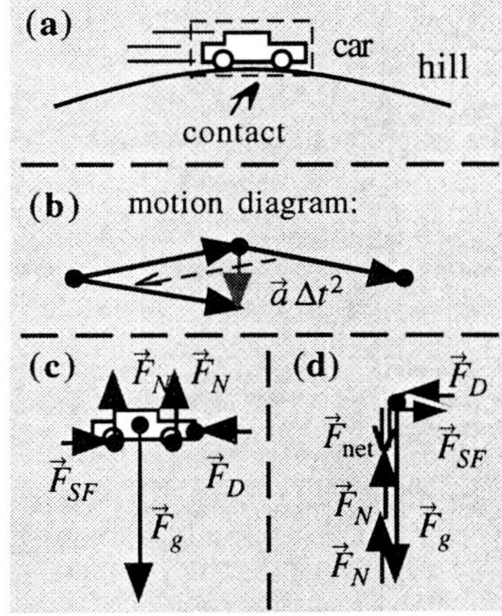

Figure N4.6: **(a)** The car in context. **(b)** A motion diagram of the car. **(c)** A free-body diagram for the car. **(d)** A net-force diagram for the car.

How does the normal force adjust itself to the correct value in the previous example? Think of it this way. If the total normal force were equal in magnitude to the gravitational force, the net force on the car would be zero, and the car

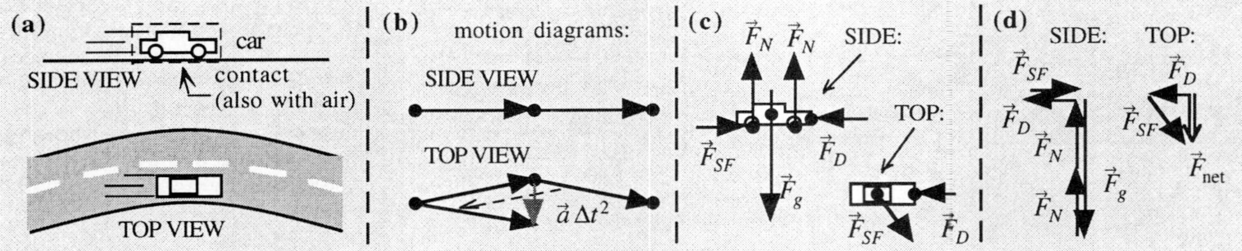

Figure N4.7: **(a)** Side and top views of the car in its context. **(b)** Side and top views of the car's motion diagram. **(c)** Side and top views of a free-body diagram for the car. **(d)** Side and top views of the net-force diagram for the car. Note how the interaction with the road must exert an angled total static friction force to keep the car in its circular path.

would move in a straight line at a constant velocity. But the road does *not* go in a straight line. Therefore, if the car were to follow a straight line, it would begin to move *away* from the road. As the car begins to pull away, its atoms withdraw from the atoms of the road, and the normal force arising from their interpenetration decreases. If the total normal force decreases too far, though, then gravity will pull the car back down toward the road, which increases the total normal force again. Therefore, the total normal force will adjust to whatever value is required to keep the car on the road: no more, no less.

EXAMPLE N4.5

Problem: A car travels around a bend in the road at a constant speed. Draw a free-body diagram of the car. (Do not ignore air resistance.)

Solution The situation is almost exactly the same as the previous example, except that the car is going around a horizontal bend instead of over a hill. The car thus stays in the horizontal plane (and thus does not accelerate vertically), but as the motion diagram in Figure N4.7b shows, the car *is* accelerating *horizontally* (toward the center of the circle). According to Newton's second law, this means that the net force on the car has to be toward the center of the circle, so there has to be *some* force on the car that has a component toward the circle's center (perpendicular to the car's motion!). Where does this force come from?

The clues are that (1) the road is the only solid thing that touches the car, and (2) if the road were very slippery, the car could not negotiate the curve. Thus the *road* must exert the force on the car (through its tires). Moreover, since this force acts parallel to the tire/road interface, and the tires are not sliding relative to the road, it must be a *static friction force*.

As shown in Figure N4.7c, the *total* static friction force exerted by the road on the tires must therefore be tipped at an angle, so that it it has both a forward component and a component toward the center of the curve. (Note that in the top view of this diagram, I have added up the static friction force exerted on each tire and simply presented the *total* static friction force as if it acted on the car's center.) The net-force diagram in Figure N4.7d shows that the vector sum of all the forces on the car will then (and only then) yield a net force directed horizontally toward the center of the road's curve, which is what is necessary to be consistent with the car's observed motion around the curve.

Note that in this case, we need two views to completely illustrate this intrinsically three-dimensional free-body diagram!

Exercise N4X.2: Draw free-body and net-force diagrams for a box sitting at rest on an incline.

Exercise N4X.3: Draw motion, free-body and net-force diagrams for a car skidding toward a stop (with its wheels locked).

Exercise N4X.4: Draw motion, free-body and net-force diagrams for an upward-bound elevator just as it begins to pull away from the ground floor.

N4.5 THIRD-LAW AND SECOND-LAW PAIRS

We have seen now a number of situations where we have determined that certain pairs of opposing forces acting on an object (for example, the gravitational and normal force) must have equal magnitudes. Newton's third law talks about forces that are equal in magnitude but opposite in direction. How are the cases we have considered so far related to Newton's third law?

The short answer is that the cases we have considered so far of equal and opposite pairs acting on an object have NOTHING to do with Newton's third law (and in fact has much more to do with Newton's *second* law). The purpose of this section is to clarify when pairs of forces are related by Newton's third law and when they are related by Newton's second law.

Newton's *third* law asserts that:

Newton's third law

> When objects A and B interact, the force that the interaction exerts on A is equal in magnitude and opposite in direction to the force that it exerts on B.

This law is *always* true because all known interactions simply transfer momentum from one object to the other (see the discussion in chapter N3).

Recognizing third-law partners

This statement implies that if a given pair of forces are connected by Newton's third law, they will have the following characteristics:

1. Each force in the pair must act on a *different* object.
2. Both forces must reflect the *same interaction* between the objects.

We will call any pair of forces fitting these conditions **third-law partners.** An example of third-law partners are the gravitational forces that two objects interacting gravitationally exert on each other.

Since third-law partners must act on *different* objects, it follows that two forces that act on the same object (and thus appear on the same free-body diagram) *cannot* be third-law partners. For example, the normal and gravitational forces on the book in Figure N4.2 are indeed equal in magnitude and opposite in direction, but they both act on the *same* object and they do not reflect the same interaction (one reflects a gravitational interaction and the other a contact interaction). So they *cannot* possibly be third-law partners.

Recognizing pairs of forces that are linked by Newton's *second* law

The particular forces shown in Figure N4.2 are in fact equal in magnitude because of Newton's *second* law: since the book does not accelerate vertically, the net vertical force on the book must be zero, and since these two forces are the only ones acting in the vertical direction, they must be equal and opposite so that they cancel. In general, a pair of opposing forces whose magnitudes are constrained to be equal by Newton's *second* law have the following characteristics:

1. Both forces act on the *same* object.
2. The forces oppose each other along a certain axis.
3. No other forces have a component along that axis.
4. The component of the object's acceleration along that axis is *zero*.

If these things are true, then Newton's second law implies that the net force along that axis must be zero, which implies that the two forces must be equal in magnitude (so that they cancel).

Exercise N4X.5: Which of the following pairs of opposing forces are third-law partners and which are instead linked by Newton's second law?
(a) The drag force and static friction force in Figure N4.6.
(b) A helicopter's rotor exerts a downward force on the air; the air exerts an upward opposing force on the helicopter.
(c) A car's tires push backward on the road; the road pushes forward on the tires.
(d) You exert an upward force on a ball sitting on your hand; gravity exerts an opposing downward force on the ball.

N4.6 COUPLING FORCES

We can use Newton's second and third laws *together* to determine the magnitude and direction of the forces that two objects that are coupled together exert on each other. The example below illustrates how this can be done.

EXAMPLE N4.6

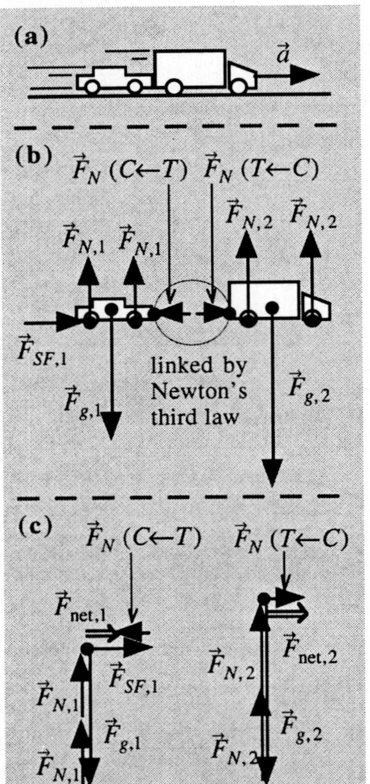

Figure N4.8: (a) The car and truck in context. **(b)** Free-body diagrams for the car and truck. **(c)** Net-force diagrams for the car and truck.

Problem: A small car of mass m_1 is pushing a disabled truck m_2 along a level road in such a way that both accelerate forward with a certain acceleration $\vec{a}$. Assuming that both the car and the truck are moving slowly enough so that frictional forces are negligible, draw free-body diagrams of both the car and truck, describe which pair (or pairs) of forces on these diagrams are linked by Newton's third law, and determine the magnitude of the force the truck exerts on the car.

Solution Figure N4.8a shows the situation described in the problem. Figure N4.8b shows free-body diagrams for both the car and the truck. The car must exert a forward contact force on the truck to get the latter accelerating. Since this force acts perpendicular to the bumper of the car that is exerting it, it is a normal force, so I have labeled this force $\vec{F}_N\,(T\leftarrow C)$ (meaning "the normal force exerted on the truck by the car") to distinguish it from the normal forces that the road exerts on the tires of each. Newton's *third* law then implies that the truck must exert an opposing force of equal magnitude on the car, which I have given the label $\vec{F}_N\,(C\leftarrow T)$. It is these forces that are equal by Newton's third law.

The net force on the car must be rightward as well if it is to accelerate in that direction, so the contact interaction between the car and the road must exert a forward static friction force $\vec{F}_{SF,1}$ on the car that exceeds the rearward force applied by the truck. This is shown in both the free-body and net-force diagrams for the car in Figures N4.8b and N4.8c.

The net-force diagram for the truck in Figure N4.8c shows that the net force on the truck is simply equal to the force $\vec{F}_N\,(T\leftarrow C)$ exerted on it by the car, so by Newton's *second* law, we must have

$$\vec{F}_N\,(T\leftarrow C) = \vec{F}_{\text{net},2} = m_2\vec{a} \qquad \text{(N4.3)}$$

So if we know the mass of the truck and its acceleration, we can find the force the car exerts on the truck $\vec{F}_N\,(T\leftarrow C)$. But Newton's *third* law implies that the the forces the car and truck exert on each other must be equal in magnitude, so

$$F_N(C\leftarrow T) = F_N(T\leftarrow C) = m_2 a \qquad \text{(N4.4)}$$

(where I have taken the magnitude of equation N4.3 to arrive at the second equality). This gives us the magnitude of the force the truck exerts on the car.

Note that the net-force diagram for the car implies that net force on the car has the magnitude $F_{SF,1} - F_N(C\leftarrow T)$. From this and equation N4.4, you should be able to use Newton's second law to show that

$$F_{SF,1} = (m_1 + m_2)a \qquad \text{(N4.5)}$$

just as if we were to consider the truck and car to be a single unit acted on by the single horizontal external force $\vec{F}_{SF,1}$.

Exercise N4X.6: Verify equation N4.5.

N4.7 GRAPHS OF ONE-DIMENSIONAL MOTION

Free-body diagrams help us see how force and motion are related at an instant of time: they are like snapshots of the force on an object. To understand more about how position, velocity, acceleration, and force are related in time-*dependent* situations, it helps to draw graphs of these quantities versus time.

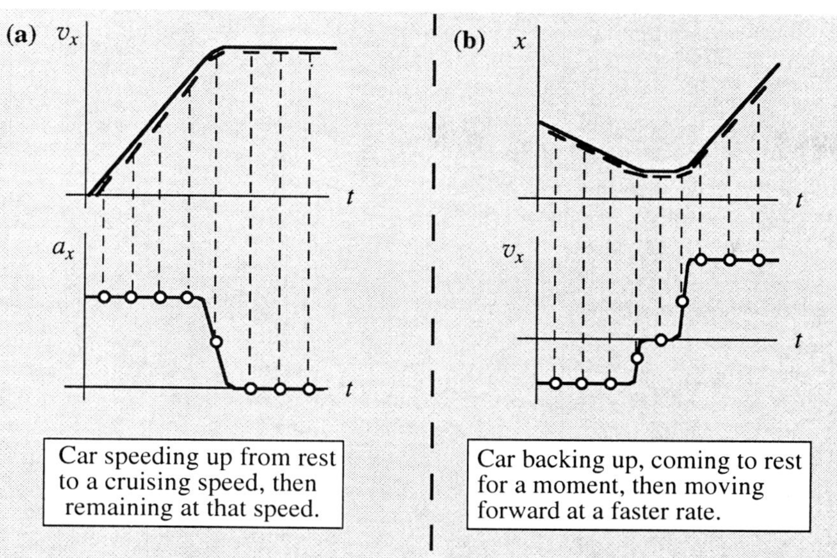

Figure N4.9: (a) How to construct a graph of $a_x(t)$ from a graph of $v_x(t)$ for a car speeding up to and then holding at a certain cruising speed. Each little line segment on the top graph indicates the *slope* of that graph of $v_x(t)$ at a certain instant, while the circle on the bottom graph indicates the value of a_x at that instant. Note that at first, the car's velocity is steadily increasing, so its slope is a constant positive number. When the car reaches its cruising speed, the slope quickly falls to zero and remains zero. **(b)** How to construct a graph of $v_x(t)$ from a graph of $x(t)$ using the same technique.

In the case of 1D motion, vectors are represented by signed numbers

In cases where an object's motion is confined to a line, we can generally choose to orient our reference frame so that the line of motion coincides with the x axis (or we might choose the z axis if the motion is vertical). If we do this, then only the x component of an object's position vector $\vec{r}(t)$ is nonzero, and this means that only the x-component of its velocity vector $\vec{v}(t)$ is nonzero:

$$\vec{r}(t) = \begin{bmatrix} x(t) \\ 0 \\ 0 \end{bmatrix} \quad \Rightarrow \quad \vec{v}(t) \equiv \frac{d\vec{r}}{dt} = \begin{bmatrix} dx/dt \\ d(0)/dt \\ d(0)/dt \end{bmatrix} = \begin{bmatrix} dx/dt \\ 0 \\ 0 \end{bmatrix} \qquad \text{(N4.6)}$$

Similarly, the object's acceleration $\vec{a}(t)$ only has one nonzero component

$$\vec{a}(t) \equiv \frac{d\vec{v}}{dt} = \begin{bmatrix} dv_x/dt \\ 0 \\ 0 \end{bmatrix} \qquad \text{(N4.7)}$$

Therefore, in this very special case of one-dimensional motion (and *only* in this case), we can represent an object's position, velocity, and acceleration vectors each by a *single signed number* considered to be a function of time (these numbers are $x(t)$, $v_x(t)$, and $a_x(t)$, respectively). These signed numbers provide all that we need to know about the vectors $\vec{r}(t)$, $\vec{v}(t)$, and $\vec{a}(t)$ in this case: the *absolute value* of the number is the same as the *magnitude* of the vector (for example, $\text{mag}(\vec{a}) = |a_x|$) and the *sign* of the number tells us the *direction* of the corresponding vector (a positive sign indicates that the vector points in the $+x$ direction, whereas a negative sign indicates that it points in the $-x$ direction).

Constructing a graph of $a_x(t)$ from one of $v_x(t)$

Now, according to equation N4.7, $a_x(t) = dv_x/dt$. This means that (as discussed in the ***Math Skills*** section on *Derivatives and Slopes* in this chapter), the *value* of the object's x-acceleration at every instant of time will correspond to the *slope* of $v_x(t)$ when the latter is plotted versus t on a graph. Therefore, given a graph of $v_x(t)$, we can in principle construct a graph of $a_x(t)$. Figure N4.9a illustrates the process of constructing such a graph.

Constructing a graph of $v_x(t)$ from one of $x(t)$

Similarly, since $v_x = dx/dt$, the *value* of v_x at every instant of time will correspond to the *slope* of a graph of $x(t)$ versus t (see Figure N4.9b). It is conventional when drawing pairs of graphs like these to put the graph of $x(t)$ above the paired graph of $v_x(t)$ and a graph of $v_x(t)$ above a paired graph of $a_x(t)$ so that the slope of the upper graph always corresponds to the value on the lower graph (always remember: "*slope* above equals *value* below").

Keep components and magnitudes distinct!

In this text, in order to keep the distinction between the components and the magnitude of a vector clear, I will *always* use the symbols x, v_x, and a_x to refer to these signed numbers (because they really are just components of the corresponding vectors) and *always* use r, v, and a to refer to the (non-negative) magnitudes of the corresponding vectors. I ask you to follow the same convention. (Note, however, that many standard texts blur this distinction, using v in some places to mean mag($\vec{v}$) and in other places to represent a velocity component. This can be very confusing, as you might guess!)

Inferring forces

According to Newton's second law an object's acceleration is proportional to the net force acting on the object, so a graph of $a_x(t)$ essentially tells us how the $F_{\text{net},x}$ depends on time. This can give us insight into the forces acting on the object, as the next example illustrates.

EXAMPLE N4.7

Problem: Imagine that you hold a basketball some distance above the floor, and then release it from rest. It falls toward the floor, hits the floor, and then rebounds upward at about the same speed. Draw graphs of the ball's vertical velocity and acceleration components as functions of time.

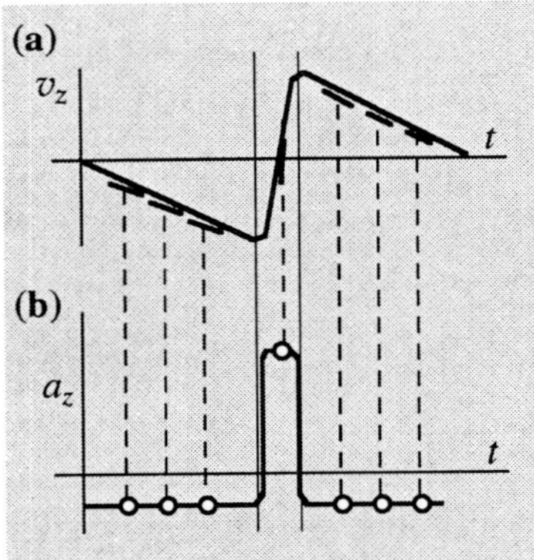

Figure N4.10: (a) A graph of $v_z(t)$, and **(b)** a graph of $a_z(t)$ for a basketball dropped on the floor. The ball is in contact with the floor during the time interval bracketed by the thin vertical lines.

Solution The first thing that we have to do is define our reference frame. Let us define our z axis to be vertically upward, and take $z = 0$ to correspond to the floor. (In the case of vertical motion, it is more natural to align the z axis with the object's motions than to use the x axis.) Let us also define $t = 0$ to be the instant when the ball is dropped.

According to the description of the situation, the ball is released from rest ($v_z = 0$). As the ball falls, its unopposed gravitational interaction with the earth transfers downward momentum to the ball at a constant rate, and thus the ball's z-velocity should steadily become more negative as time decreases. When the ball hits the floor, its z-velocity rapidly changes from being negative to positive (downward to upward). After the ball leaves the floor, its z-velocity again becomes steadily more negative.

From this information, we can construct the graph of v_z versus t shown in Figure N4.10a. Then we can use the methods discussed in this section to construct the graphs of $a_z(t)$ below this graph (Figure N4.10b). Note how in each case the slope of the upper graph at a given t is equal to the value of the graph below it (as indicated at selected places).

We see from Figure N4.10b that the net force on the ball when it is *not* in contact with the floor must be constant and downward: this is the constant force of gravity. When the ball is in contact with the floor, however, the graph indicates that the net force on the ball must become very large and upward. The gravitational force on the ball always has a constant (downward) value, so we must have a new upward force acting on the ball. This must be a normal force arising from the ball's interaction with the ground. The graph clearly indicates that this force must be larger than the gravitational force, and indeed, the more brief the bounce is, the larger this normal force has to be.

Drawing such sets of graphs is an excellent way to visualize (and thus think more carefully about) the time dependence of these motion functions. *Drawing such graphs is a skill that you should practice and master:* becoming adept at both constructing and interpreting these graphs is one of the surest ways to gain a firm understanding of the mathematical relationship between position, velocity and acceleration. Note that it is not critical that your graphs be *quantitatively* accurate: the important thing is that they be *qualitatively* accurate.

Exercise N4X.7: How does the net x-force on the car described in Figure N4.9a change as time passes? What force provides the forward force on the car, and why does it change with time? (Ignore air resistance.)

Exercise N4X.8: A car traveling along a straight road at a constant speed suddenly brakes to a stop to avoid an animal and remains at rest thereafter. Draw graphs of $v_x(t)$ and $a_x(t)$ for this situation.

N4.8 A FEW QUANTITATIVE EXAMPLES

So far in this chapter, we have focused exclusively on determining *qualitatively* the existence and/or direction of forces from an analysis of an object's motion. These techniques of qualitative analysis are the bedrock on which we can construct more detailed quantitative analyses: without the guidance that these qualitative tools provide, one can quickly become lost in complicated problems. We will examine a number of progressively more complex problems in chapters N6 through N10, but I would like to close this chapter with a few simple examples of quantitative analysis based on the work we have already done to give you a foretaste of what we can do.

A simple way to get a quick *estimate* of the magnitude of the net force on an object that is accelerating along a straight line during a certain time interval Δt is to assume that the object's instantaneous acceleration during the interval is approximately equal to its average acceleration over that interval. If the motion is along the x direction, for example, this assumption means that:

$$a \approx a_{\Delta t} = |a_{\Delta t,x}| = \left|\frac{\Delta v_x}{\Delta t}\right| \tag{N4.8}$$

where $\Delta v_x \equiv v_{f,x} - v_{i,x}$ is the change in the object's x-velocity during the interval. This is a useful (if often crude) first model for a variety of situations. Once we know the magnitude of the acceleration, we can use the magnitude of Newton's second law to determine the magnitude of the net force. The example below illustrates this process.

EXAMPLE N4.8

Problem: Imagine that in example N4.3, the box being carried by the pickup truck has a mass $m = 25$ kg. If the truck goes from 0 to 60 mi/h in 10 s, what is the magnitude of the average static friction force exerted on the box? How does this compare to magnitude of the object's weight?

Solution We can see from our previous qualitative analysis (see particularly the net-force diagram in Figure N4.5c) that the net force on the box in this case is equal to the static friction force. Let's define the $+x$ direction to be the direction of the truck's motion. The truck is initially at rest ($v_{i,x} = 0$) and is moving with x-velocity $v_{f,x} = +60$ mi/h: if the box remains at rest with respect to the truck, this will be its initial and final x-velocities as well. The magnitude of Newton's second law applied to the box then implies that

$$\begin{aligned} F_{SF} &= F_{\text{net}} = ma = m|a_x| \approx m\frac{|\Delta v_x|}{\Delta t} \\ &= \frac{10\ \cancel{\text{kg}}}{10\ \cancel{\text{s}}}|+60\ \cancel{\text{mi/h}}|\left(\frac{1\ \cancel{\text{m/s}}}{2.24\ \cancel{\text{mi/h}}}\right)\left(\frac{1\ \text{N}}{1\ \cancel{\text{kg}}\cdot\cancel{\text{m/s}^2}}\right) = 27\ \text{N} \end{aligned} \tag{N4.9}$$

We saw in chapter C8 that if an object has mass m the magnitude of its weight is $F_g = mg$, which in this case is $F_g = (10\text{ kg})(9.8\text{ m/s}^2) = 98\text{ kg}\cdot\text{m/s}^2 = 98\text{ N}$. So the magnitude of the static friction force acting on the box in this case is a bit less than a third of the magnitude of its weight.

Exercise N4X.9: If the box's interaction with the truck bed can exert no more than 45 N on the box before the box begins to slip, what is the minimum time that the driver should allow to go from 0 to 60 mi/h?

Exercise N4X.10: If the basketball in example N4.7 is traveling at 3.0 m/s when it hits the ground, and the bounce lasts 0.10 s, how does the magnitude of the normal force exerted by the floor compare to the basketball's weight?

When an object is moving in a circular path of radius R at a constant speed v, we saw in chapter N2 that the magnitude of its acceleration is $a = v^2/R$. We can use this in conjunction with Newton's second law to determine the magnitude of forces acting on the object. The next example illustrates this process.

EXAMPLE N4.9

Problem: Imagine that in Example N4.4, the car going over the hill is traveling at a constant speed of 30 m/s and (near the top of the hill, anyway) the road curves vertically as if it were a part of a circle whose radius is 450 m. What is the ratio of the magnitudes of the total normal force and gravitational forces acting on the car in this situation?

Solution The net-force diagram in Figure N4.6d makes it clear that the magnitude of the net force acting on the car is $F_{\text{net}} = F_g - F_{N,\text{tot}} = mg - F_{N,\text{tot}}$. This means that the total normal force acting on the car is $F_{N,\text{tot}} = mg - F_{\text{net}}$. The magnitude of Newton's second law implies that $F_{\text{net}} = ma$. The car's acceleration in this case is given by $a = v^2/R$. Putting all this together, we find that the requested ratio is equal to

$$\frac{F_{N,\text{tot}}}{F_g} = \frac{mg - ma}{mg} = 1 - \frac{a}{g} = 1 - \frac{v^2}{Rg} = 1 - \frac{(30\ \cancel{\text{m/s}})^2}{(450\ \cancel{\text{m}})(9.8\ \cancel{\text{m/s}^2})} = 0.80 \quad \text{(N4.10)}$$

So the total normal force acting on the in this case is only 80% as strong as the gravitational force on the car as it goes over the top of the hill.

MATH SKILLS: Derivatives and Slopes

The meaning of the *derivative* of a function $f(t)$ can be more fully understood in terms of graphs of $f(t)$. This section reviews the graphical interpretation of the derivative.

Recall from chapter N1 that the derivative of the function $f(t)$ with respect to its variable t is defined to be:

$$\frac{df}{dt} \equiv \lim_{\Delta t \to 0} \frac{\Delta f}{\Delta t} \equiv \lim_{\Delta t \to 0} \frac{f(t+\Delta t) - f(t)}{\Delta t} \quad \text{(N4.11)}$$

Figure N4.10 illustrates that on a graph of $f(t)$ plotted versus t, the ratio $\Delta f/\Delta t$ for nonzero Δt is equal to the slope of a straight line drawn between the points on the curve at t and $t + \Delta t$. As Δt approaches zero, the value of this slope approaches a specific value that equals the slope of a line drawn tangent to the graph at t. We call this the **slope of $f(t)$ at that point**.

Since an object's x-velocity $v_x(t)$ is defined to be dx/dt, this means that the value of $v_x(t)$ at any time t is equal to the slope of a graph of $x(t)$ at that time t. Similarly, since an object's x-acceleration $a_x(t)$ is defined to be dv_x/dt, the value of $a_x(t)$ at any time t is equal to the slope of a graph of $v_x(t)$ at t.

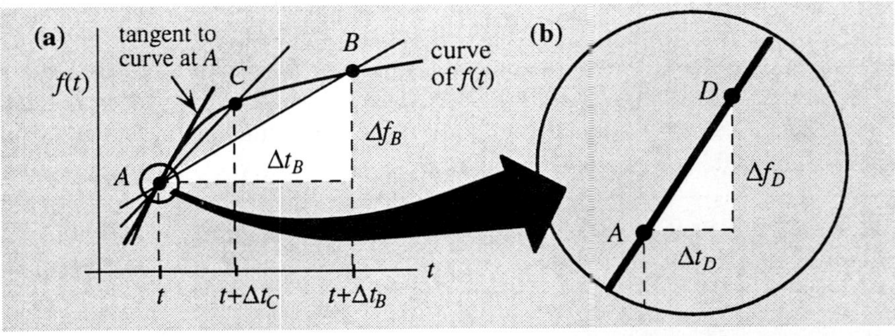

Figure N4.10: (a) The ratio $\Delta f_B/\Delta t_B$ specifies the slope of the line going through points *A* and *B*. The ratio $\Delta f_C/\Delta t_C$ specifies the slope of the line going through points *A* and *C*. The latter is closer to the slope of the line tangent to the curve of $f(t)$ at point *A* (dark straight line). **(b)** As Δt gets *very* small, the curve and the tangent line become indistinguishable, and $\Delta f/\Delta t$ specifies the slope of *both* in the neighborhood of *A*.

SUMMARY

I. THE KINEMATIC CHAIN

A. A statement of the *kinematic chain* [which follows from the definitions of $\vec{v}(t)$ and $\vec{a}(t)$]: $\vec{r}(t) \dashv\!\left[\text{time derivative}\right]\!\mapsto \vec{v}(t) \dashv\!\left[\text{time derivative}\right]\!\mapsto \vec{a}(t)$

B. This implies that if we know about the object's motion, we can:

1. use the chain to determine its acceleration
2. use Newton's second law to determine the net force on the object

C. Knowing the net force, we can usually infer the existence, directions, and/or magnitudes of other forces acting on the object.

II. NET-FORCE DIAGRAMS

A. A net-force diagram clearly displays the net force acting on object

B. Steps in constructing a net-force diagram:

1. construct a free-body diagram (if you haven't already)
2. copy the force arrows from the free-body diagram, placing them in sequence (i.e. the tail of next arrow at the tip of the previous one)
3. draw a double-line arrow to represent vector sum $\equiv \vec{F}_{\text{net}}$

III. QUALITATIVE EXAMPLES

A. The basic process:

1. find the object's acceleration (maybe using a motion diagram)
2. draw a free-body diagram and a net-force diagram
3. adjust the forces in these two diagrams until the net force points in the direction of the acceleration determined in the first step

B. Some implications (from the examples considered)

1. for an object that is sitting at rest (or moving with a constant velocity) on a level surface, $\text{mag}(\vec{F}_{N,\text{tot}}) = \text{mag}(\vec{F}_g)$
2. but $\text{mag}(\vec{F}_{N,\text{tot}}) \neq \text{mag}(\vec{F}_g)$ in general (see Example N4.4)
3. static friction forces make it possible for the car to maintain its speed against drag, slow down, speed up, and round corners

IV. THIRD-LAW AND SECOND-LAW PAIRS

A. A statement of Newton's third law:
The interaction between two objects exerts on each a force that is opposite but equal in magnitude to that it exerts on the other.

B. This statement implies that a pair of forces (*third-law partners*) that are connected by Newton's *third* law have the following characteristics

1. each force in the pair must act on a different object
2. each must reflect the same interaction between those objects

C. In contrast, a pair of opposing forces that are equal in magnitude because of Newton's *second* law generally act on the *same* object (which is not accelerating in their direction) and reflect *different* interactions

D. We can use Newton's 2nd and 3rd laws together to determine the magnitude of the forces due to an interaction that couples two objects

V. GRAPHING ONE DIMENSIONAL MOTION

A. When an object moves along a straight line (which we can define to be the x axis), $\vec{r}(t)$, $\vec{v}(t)$, and $\vec{a}(t)$ are completely described by their x components $x(t)$, $v_x(t)$, and $a_x(t)$ (which are simple signed numbers)

B. Graphs of these functions allow one to find $a_x(t)$ (and thus $F_{\text{net},x}$) from $x(t)$ or $v_x(t)$. The conventional process is as follows:

1. place a graph of $x(t)$ above a graph of $v_x(t)$, $v_x(t)$ above $a_x(t)$
2. then the *slope above* corresponds to the *value below*

VI. QUANTITATIVE EXAMPLES

A. The qualitative tools we have explored make quantitative analysis easier

B. Ways to estimate the quantitative magnitude of an object's acceleration:

1. if the object moves a line, make the line the x axis and approximate $a = a_x$ by the average x-acceleration $a_{\Delta t,x} = \Delta v_x/\Delta t$
2. when the object moves in a circular path of radius R at a constant speed v, use $a = v^2/R$

GLOSSARY

kinematics: the mathematical study of motion.

kinematic chain: the chain of relationships linking an object's position to its velocity to its acceleration:

$$\vec{r}(t) \xrightarrow{\text{time derivative}} \vec{v}(t) \xrightarrow{\text{time derivative}} \vec{a}(t) \qquad (N4.1)$$

net-force diagram: a diagram used as a supplement to a free-body diagram that displays the vector sum of all the forces acting on a body (that is, the net force on that object) as a double-line arrow.

third-law partners: a pair of forces that we know are opposite in direction and equal in magnitude because of Newton's third law. Such a pair of forces represent the two ends of an interaction between two objects: therefore they must (1) act on different objects and (2) reflect the same basic interaction.

slope [of a function $f(t)$] **at a point:** defined to be the slope of a line that is tangent to the graph of the function at that point, that is, a line that touches the graph only at that point. The slope of this line has the value df/dt.

TWO-MINUTE PROBLEMS

N4T.1 A car passes a dip in the road, first going down, then up. At the very bottom of the dip, when the car's instantaneous velocity is passing through horizontal, how does the magnitude of the total normal force on the car compare to the magnitude of the car's weight?

A. $F_{N,\text{tot}} < F_g$
B. $F_{N,\text{tot}} = F_g$
C. $F_{N,\text{tot}} > F_g$
D. $F_{N,\text{tot}} = 0$
E. Not enough information for a reasonable answer

N4T.2 A crate sits on the ground. You push as hard as you can on it, but you cannot move it. At any given time when you are pushing, what is the magnitude of the static friction force exerted on the crate by its contact interaction with the ground compared to the magnitude of your push (which is a *normal* force)?

A. $F_{SF} < F_N$
B. $F_{SF} = F_N$
C. $F_{SF} > F_N$
D. $F_{SF} = 0$!
E. Not enough information for a reasonable answer

N4T.3 A box sits at rest on an inclined plank. How do the magnitudes of the normal force and the gravitational force exerted on the box compare? (*Hint:* Draw a picture!)

A. $F_N < F_g$
B. $F_N = F_g$
C. $F_N > F_g$
D. $F_N = 0$!
E. Not enough information for a reasonable answer

N4T.4 The drawing below is supposed to be a free-body diagram of a box that sits without slipping on the back of a truck that is moving to the right but is slowing down. Is the diagram correct?

A. Yes.
B. No: $\vec{F}_{SF}$ should point leftward
C. No: the $\vec{F}_{SF}$ label should be $\vec{F}_{KF}$
D. No: there should be a leftward drag force
E. No: F_N should not be equal to F_g
F. No: some other problem (specify)

N4T.5 The drawing below is supposed to be a free-body diagram of a crate that is being lowered by a crane and is speeding up as it is being lowered. Is the diagram correct? (Ignore air resistance.)

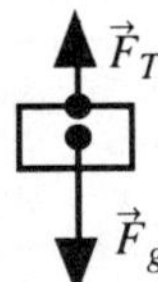

A. Yes.
B. No: $\vec{F}_T$ should be labeled $\vec{F}_N$
C. No: $\vec{F}_T$ should be equal to F_g
D. No: $\vec{F}_T$ should be greater than F_g
E. No: There should be an upward $\vec{F}_D$
F. No: some other problem (specify)

N4T.6 Which of the following are third-law partners? Answer T if the pair of forces described are third-law partners, F if they are not.

(a) A thrust force from its propeller pulls a plane forward; a drag force pushes it backward.
(b) A car exerts a forward force on a trailer; the trailer tugs backward on the car.
(c) a motorboat propeller pushes backward on the water; the water pushes forward on the propeller
(d) gravity pulls down on a person sitting in a chair; the chair pushes back up on the person.

N4T.7 A large truck with mass m_1 pushes on a disabled small car with mass $m_2 < m_1$, giving it a forward acceleration a. Each vehicle exerts a force on the other as a result of their contact interaction. Which vehicle exerts the *greater* force on the other?

A. the car
B. the truck
C. both forces have the same magnitude
D. the car doesn't exert any force on the truck
E. not enough information for a meaningful answer

N4T.8 In the situation described in the previous problem, what is the *magnitude* of the force $\vec{F}$ that the car exerts on the truck?

A. $F = m_1 a$
B. $F = m_2 a$
C. $F = (m_1 + m_2)a$
D. $F = 0$
E. $F = -m_2 a$
F. other (specify)

N4T.9 An object's x-position $x(t)$ is shown in the boxed graph at the top left. Which of the other graphs in the set below most correctly describes its x-velocity?

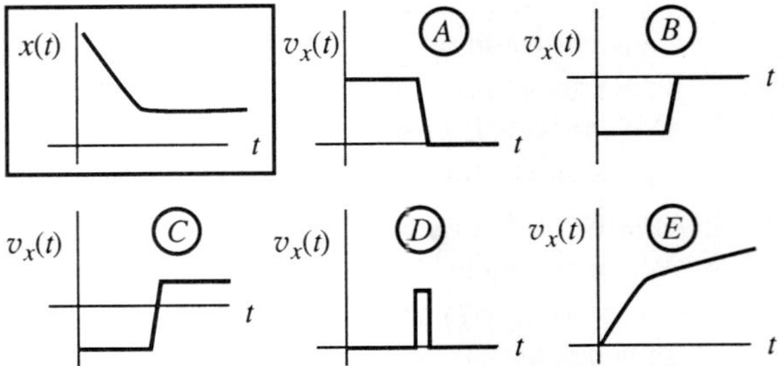

N4T.10 Which graph below best describes the x-acceleration of the object described in problem N4T.9?

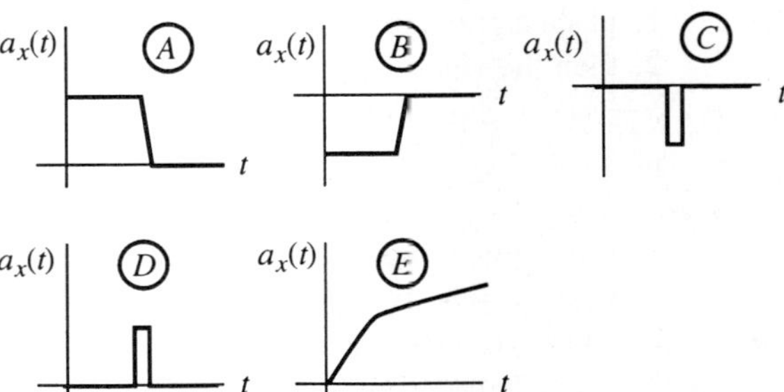

HOMEWORK PROBLEMS

BASIC SKILLS

N4B.1 Draw a motion diagram, a free-body diagram, and a net-force diagram for a moving motorboat whose motor has just run out of gas.

N4B.2 Draw a motion diagram, a free-body diagram, and a net-force for a child bouncing on a trampoline (at the instant that the child is at the lowest part of the bounce).

N4B.3 Draw a motion diagram, a free-body diagram, and a net-force diagram for a box sitting in an elevator whose downward speed is increasing as it begins its descent from the top floor.

N4B.4 Draw a motion diagram, a free-body diagram, and a net-force diagram for a child holding on for dear life (with feet in the air!) to a rapidly spinning merry-go-round.

N4B.5 A car travels at a constant speed through a dip in the road that takes the car first down and then up. Draw a motion diagram, a free-body diagram, and a net-force diagram for the car as it passes the *bottom* of the dip, with special attention to correctly indicating the relative magnitudes of the vertical forces on the car.

N4B.6 Imagine a person hanging on to the landing gear of a helicopter as it accelerates upward. Draw free-body diagrams for both the person and the helicopter. Indicate which pairs of forces (if any) are third-law partners and explain your reasoning.

N4B.7 Imagine that a person jumps off the floor. Draw free-body diagrams of the earth and the person at some instant while the person is beginning the jump (but has not yet left the floor). Indicate on your diagram which pairs of forces (if any) are third-law partners and explain your reasoning.

N4B.8 Construct a graph of x-acceleration as a function of time for an object whose x-velocity is as shown below.

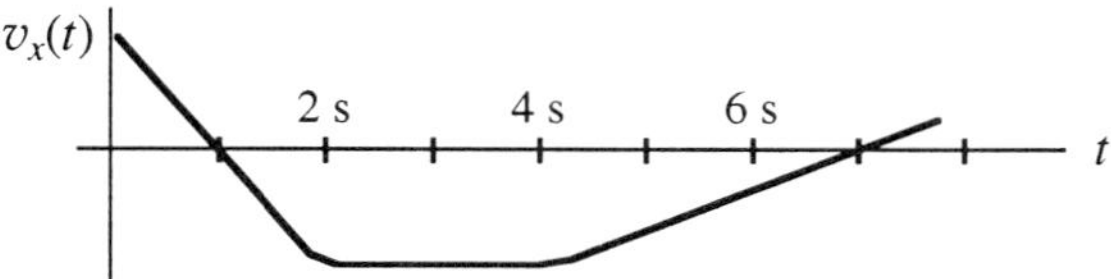

N4B.9 Construct a graph of x-velocity as a function of time for an object whose x-position is as shown below.

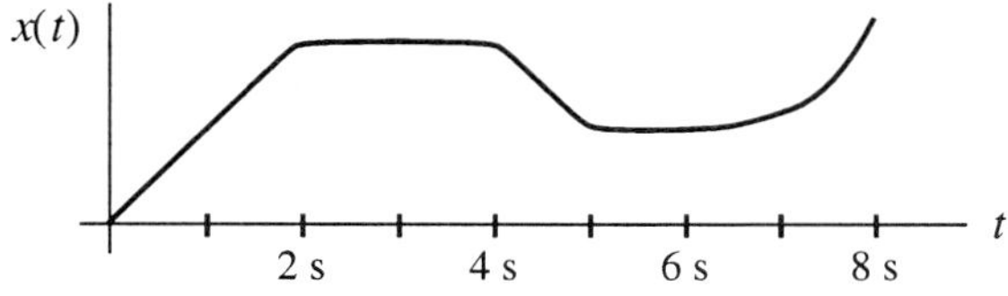

SYNTHETIC

N4S.1 An airplane travels at a constant speed in a horizontal circular path around an airport. Draw a *top-view* motion diagram and a *rear-view* free-body diagram of the plane. (*Hint*: A plane has to "bank", that is lower the inner wing and raise the outer wing, when it flies in a circular turn to orient the lift force exerted on the plane by the wings away from vertical. Why is this?)

N4S.3 A box sitting on the floor of a van slides toward the front of the van when the van suddenly brakes to a stop. Draw a motion diagram (as viewed from the *ground*, not the van), a free-body diagram, and a net-force diagram of the box. Explain qualitatively why the box moves forward relative to the van.

N4S.3 Draw a motion diagram and a free-body diagram for a roller-coaster car at the top of a "loop-de-loop" (that is, when the car is upside down). Do not ignore air resistance, and assume that the car is moving rapidly enough to remain in firm contact with the rails. In particular, explain why there is no *outward* force on the car.

N4S.4 Imagine that in a movie chase scene, the director wants a 1000-kg car traveling at 15 m/s to run head-on into a brick wall. If the wall brings the car to rest in about 0.2 s, roughly what force will the car exert on the wall during the collision? (This calculation would give the director's assistant some advice about how strong the wall has to be.)

N4S.5 Imagine that unpowered go-cart (with a child rider) traveling at an initial speed of 5 m/s coasts on a level road for about 50 m before coming to rest. If the cart and rider have a mass of 40 kg, about what force would you have to apply to keep the cart moving at a constant speed. Assume that the friction forces acting on the cart are essentially independent of speed. (*Hint:* Use energy concepts to compute the magnitude of the total friction/drag force acting on the cart, then link this force to your force.)

N4S.6 Imagine that the car in Example N4.5 must convert chemical energy to other forms at a rate of 10,000 W to keep moving at a constant speed of 25 m/s. If the car's mass is m = 1100 kg, its constant speed is v = 25 m/s, and the radius of the curve is R = 120 m, what angle does the static friction force on the car make with the car's direction of travel as it rounds the curve? Be sure to describe your reasoning (*Hint:* See chapter C12 for a discussion of how to relate the power a car uses to the forces acting on it.)

N4S.7 During a certain interval of time, the x-velocity of a car with a mass m = 1200 kg is given by the expression $v_x(t) = v_0 - bt^2$, where v_0 = 20 m/s and b = 5 m/s^3. Sketch graphs of $v_x(t)$ and $a_x(t)$ for this situation. Find an expression (in terms of m, b, t and whatever else you need) for the x component of the combined static friction and drag forces acting on the car during that time interval, and calculate the magnitude of this force at time t = 0.5 s.

RICH-CONTEXT

N4R.1 The top of a small hill in a certain highway has a circular (vertical) cross-section with an approximate radius of 57 m. A car going over this hill too fast might leave the ground and thus lose control. What speed limit should be posted? (*Hint*: The magnitude of force of gravity on a car of mass m is mg, where g = 9.8 m/s^2.)

N4R.2 You are designing an ejector seat for an automobile. Your seat contains two rocket engines that will burn for no more than 0.5 s (so as not to fry passers-by) and yet must throw the seat and occupant at least 150 m in the air after the engines shut off (to allow the parachute to deploy). Estimate the combined thrust that your rockets will have to exert, and check whether the seat's acceleration will exceed the safe limit of about 10g.

ADVANCED

[There are no advanced problems for this chapter.]

ANSWERS TO EXERCISES

N4X.1 The units of b are m/s^2, units of c are m. $v_x = bt$ and $a_x = b$. The units of these expressions are correct: (m/s and m/s^2, respectively).

N4X.2 A drawing of the situation (a), a free-body diagram of the box (b), and a net force diagram (c) are shown below:

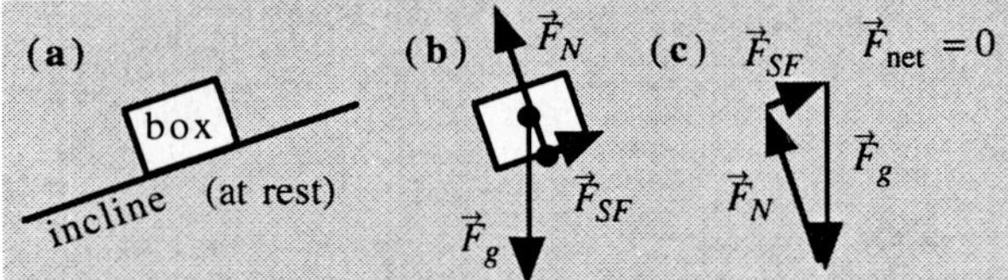

N4X.3 A drawing of the situation (a), a motion diagram (b), a free-body diagram (c), and a net force diagram for the car (d) are shown below:

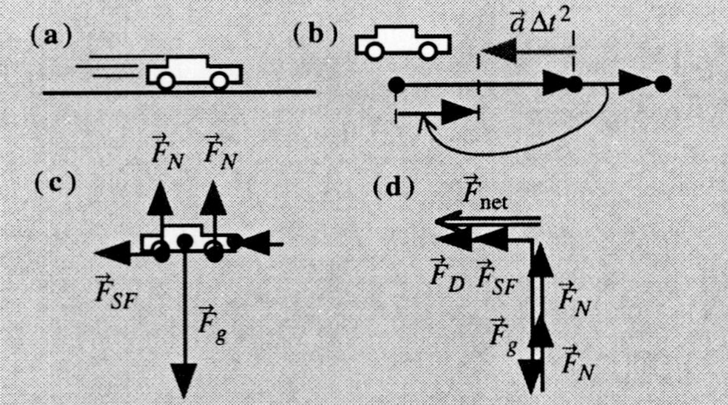

N4X.4 A drawing of the situation and motion diagram (a), a free-body diagram (b), and a net force diagram (c) for the elevator are shown below (ignoring friction forces):

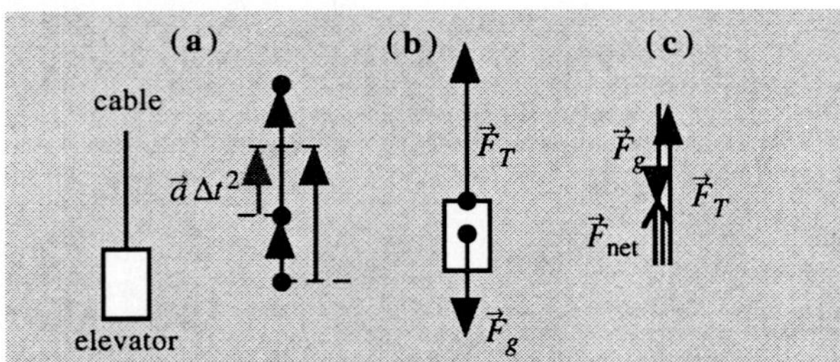

N4X.5 (a) Linked by the second law. (b) Linked by the third law. (c) Linked by the third law. (d) Linked by the second law.

N4X.6 According to Newton's second law, the magnitude of the net force on the car is linked to the magnitude of its acceleration as follows:

$$F_{SF,1} - F_N(C \leftarrow T) = F_{net} = m_1 a \tag{N4.12}$$

But according to equation N4.4, $F_N(C \leftarrow T) = m_2 a$. Plugging this into the equation above, we find that

$$F_{SF,1} - m_2 a = m_1 a \quad \Rightarrow \quad F_{SF,1} = (m_1 + m_2)a \tag{N4.13}$$

N4X.7 Initially the net x-force is positive and it remains relatively constant, but then drops to zero. This force is a static friction force exerted by the road on the wheels (as a result of the engine rotating the wheels against the road). Presumably, this force drops to whatever value is equal in magnitude to the drag force when the driver decides the car has reached cruising speed and lets up on the gas pedal.

N4X.8 Graphs of $v_x(t)$ and $a_x(t)$ look as follows:

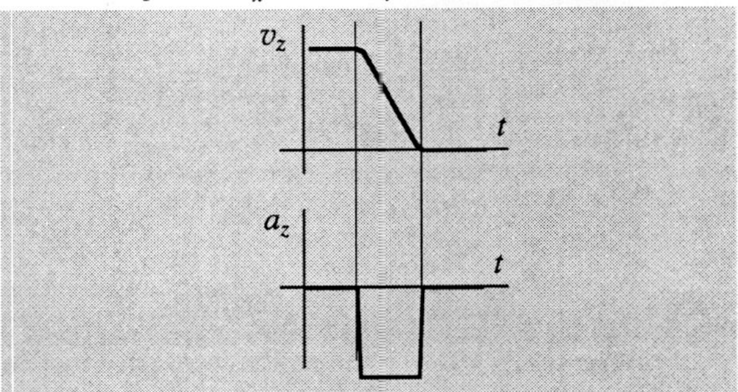

N4X.9 In this case the maximum possible deceleration that the box can have before it begins to slip is given by $a_{max} = F_{net,max}/m$. If we approximate this acceleration by the average acceleration, then we have

$$\frac{F_{net,max}}{m} = a_{max} \geq a \approx \frac{|\Delta v_x|}{\Delta t} \tag{N4.14}$$

Since $|\Delta v_x|$ is given, the acceleration will get larger when Δt is smaller, so this equation puts a lower limit on Δt:

$$\Delta t \geq \frac{m|\Delta v_x|}{F_{net,max}} = \frac{(10\text{ kg})(60\text{ mi/h})}{45\text{ N}}\left(\frac{1\text{ m/s}}{2.24\text{ mi/h}}\right)$$

$$= 6.0\,\frac{\text{kg}\cdot\text{m/s}}{\text{N}}\left(\frac{1\text{ N}}{1\text{ kg}\cdot\text{m/s}^2}\right) = 6.0\text{ s} \tag{N4.15}$$

N4X.10 If we define the x direction to be positive upward, the change in the basketball's x-velocity is given by $\Delta v_x = v_{f,x} - v_{i,x} = +3.0$ m/s $-$ (-3.0 m/s) $= 6.0$ m/s). If this takes place in 0.10 s, and we approximate the ball's acceleration by its average acceleration during this interval, we find that the magnitude of the ball's acceleration is $a = |\Delta v_x|/\Delta t = 60$ m/s^2. The net upward force on the ball while it touches the floor has a magnitude of

$$F_{net} = F_N - F_g = F_N - mg \tag{N4.16}$$

since the ball's weight opposes the normal force that is driving the ball upward. The magnitude of Newton's second law then implies that:

$$F_N - mg = F_{net} = ma \quad \Rightarrow \quad F_N = m(a+g)$$

$$\Rightarrow \quad \frac{F_N}{mg} = \frac{m(a+g)}{mg} = \frac{a}{g} + 1 = \frac{60\text{ m/s}^2}{10\text{ m/s}^2} + 1 = 7 \tag{N4.17}$$

So the magnitude of the normal force is about 7 times larger than the ball's weight.

N5

MOTION FROM FORCES

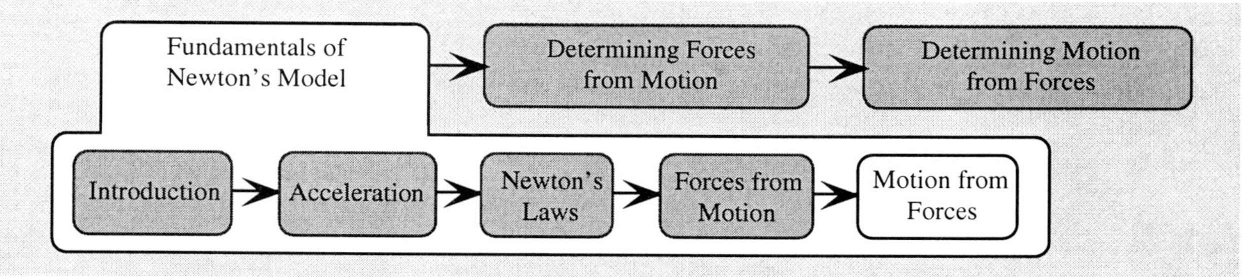

N5.1 OVERVIEW

In the last chapter we saw that with the help of Newton's second law and some graphical aids, we can determine a number of useful things about the forces acting on an object if we know how it moves. The quintessential application of newtonian mechanics, however, is to do the reverse: that is, predict an object's motion knowing the forces that act upon it. This is the core of Newtonian synthesis discussed in chapter N1: Newton's great triumph was to demonstrate that the characteristics of planetary orbits could be *predicted* using his three laws of motion and the law of universal gravitation.

Predicting an object's motion knowing the forces determining that motion is actually much more difficult than doing the reverse. We will study a number of important cases in detail in chapters N11 through N14. The purpose of this chapter is to lay some foundations for that study (as well as generate some results that we will find useful in chapters N6-N10). Here is an overview of the sections in this chapter:

N5.2 *THE REVERSE KINEMATIC CHAIN* reviews the kinematic chain and discusses (in general terms) how we can follow it in reverse to determine an object's motion knowing the forces acting on it.

N5.3 *GRAPHICAL ANTIDERIVATIVES* shows how (in the case of one-dimensional motion) we can work our way from graphs of an object's x-acceleration as a function of time to the same for an object's x-position.

N5.4 *INTEGRALS IN ONE DIMENSION* demonstrates that we can do the same thing (for one-dimensional motion) using integral calculus. This section also makes clear the importance of *initial conditions* in determining an object's motion.

N5.5 *FREE-FALL IN ONE DIMENSION* explores the important special case of an object that is freely falling vertically.

N5.6 *INTEGRALS IN THREE DIMENSIONS* shows how we can (in principle at least!) extend the mathematics that we developed for one dimensional motion in section N5.4 to situations where an object moves in three dimensions.

N5.7 *CONSTRUCTING TRAJECTORIES* explores how we can qualitatively construct an an approximation to object's trajectory in three dimensions by reversing the process that we used to construct motion diagrams.

**** ***MATH SKILLS:*** *ANTIDERIVATIVES AND INTEGRALS* reviews the definition of a function's *antiderivative* and how it is linked to the function's *integral.*

N5.2 THE REVERSE KINEMATIC CHAIN

The kinematic chain

In the last chapter, we discussed the *kinematic chain* of relationships that link an object's position to its velocity and its acceleration. This chain of relationships can be summarized as follows:

$$\vec{r}(t) \quad \dashv\!\left[\begin{smallmatrix}\text{time}\\ \text{derivative}\end{smallmatrix}\right]\!\mapsto \quad \vec{v}(t) \quad \dashv\!\left[\begin{smallmatrix}\text{time}\\ \text{derivative}\end{smallmatrix}\right]\!\mapsto \quad \vec{a}(t) \tag{N5.1}$$

This chain of relationships means that given an object's position or velocity as a function of time, we can determine its acceleration as a function of time. As we demonstrated in the last chapter, we can then use Newton's second law to link the object's acceleration to the net force on the object, and use that information to learn things about the forces that go into that net force.

We can reverse the chain by taking *antiderivatives*. The **antiderivative** of a function $f(t)$ is simply that function $F(t)$ such that $dF/dt = f(t)$. For example, $\vec{v}(t)$ is the antiderivative of $\vec{a}(t)$, since $d\vec{v}/dt = \vec{a}(t)$. Therefore, the kinematic chain in reverse looks like this

Reversing the kinematic chain using antiderivatives

$$\vec{a}(t) \quad \dashv\!\left[\begin{smallmatrix}\text{time anti-}\\ \text{derivative}\end{smallmatrix}\right]\!\mapsto \quad \vec{v}(t) \quad \dashv\!\left[\begin{smallmatrix}\text{time anti-}\\ \text{derivative}\end{smallmatrix}\right]\!\mapsto \quad \vec{r}(t) \tag{N5.2}$$

Therefore, if we know how to compute the antiderivative of a function, and we know an object's acceleration as a function of time, we can find its velocity and its position as functions of time as well. (For more information about antiderivatives, see the ***Math Skills*** section on *Antiderivatives and Integrals* at the end of this chapter.)

Finding motion from forces

If we know all of the forces that act on an object, we can find the net force on the object. Newton's second law then allows us to find the object's acceleration at all times, and the reversed kinematic chain then allows us to find the object's velocity and position at all times. This is how we determine *motion from forces* (in principle).

The problem here is not the physics but the mathematics. With the help of a finite (and relatively small) number of rules, one can fairly easily take the derivative of almost any useful function. Finding the antiderivative of an arbitrary function is generally not nearly so easy.

In this chapter, we will discuss *four* different approaches to finding the antiderivatives of acceleration: two approaches that work for one-dimensional motion and two for multi-dimensional motion. When an object moves in only one dimension, we can essentially reverse the graphical construction techniques we learned in the last chapter to construct graphs of $v_x(t)$ and $x(t)$ from a graph of $a_x(t)$. Alternatively, we can use the techniques of integral calculus to find these functions mathematically, an approach that is powerful and flexible but abstract.

If the object moves in three dimensions, we can use the mathematical approach as well (we simply have to keep track of three vector components instead of one). We can also reverse the approach we used to construct motion diagrams: where before, we used information about the object's position at equally spaced instants of time to construct arrows representing the object's acceleration, we now use known acceleration arrows to construct the object's trajectory. This reversed motion diagram technique turns out to be especially easy to turn into a computer program that is able to compute trajectories. (It was also essentially the approach that Newton himself used in the *Principia*.)

N5.3 GRAPHICAL ANTIDERIVATIVES

We begin with the graphical approach to one dimensional motion, which is simplest (and yet still illuminating in important ways). In the last chapter, we learned how to use the "slope above equals value below" method to construct a graph of $v_x(t)$ below a graph of $x(t)$ and/or a graph of $a_x(t)$ below a graph of $v_x(t)$. In this section, we will learn how to do the process in reverse, starting

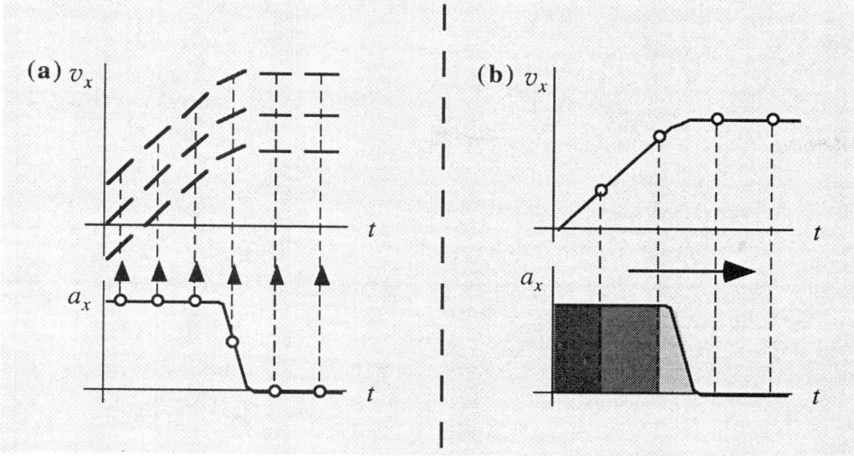

Figure N5.1: Two methods for constructing a graph of $v_x(t)$ from a graph of $a_x(t)$.
(a) The antiderivative method involves drawing short line segments on the upper graph whose *slopes* are consistent with the *value* of the lower graph at the same time. We can only choose between the various possible upper curves we generate this way by actually choosing a value for, say, $v_x(0)$.
(b) The integral method involves linking the value on the upper graph with the area under the lower graph. We still have to choose a value for $v_x(0)$: the single curve I have drawn corresponds to the choice $v_x(0) = 0$.

with a graph of $a_x(t)$, constructing a graph of $v_x(t)$ from it, and then using the graph of $v_x(t)$ in turn to construct a graph of $x(t)$.

The slope method

There are essentially two ways to do this. One technique (which I will call the **slope method**) is a literal reversal of the approach that we used in the previous chapter. For example, imagine we are trying to construct a graph of $v_x(t)$ from a graph of $a_x(t)$. If we were instead constructing the graph of $a_x(t)$ from $v_x(t)$, we would look at the slope of the upper graph and plot its value on the lower graph. To do this in reverse, we should, at a series of instants of time, draw a little line segment on the upper graph whose slope reflects the value that you see at that time on the lower graph. If you draw these line segments so that they nearly touch end-to-end, as shown in Figure N5.1a, then they sketch out the desired upper curve.

Unfortunately, there is a certain ambiguity in this process, because we can draw several different upper graphs that satisfy this criterion, as shown. These different graphs all have the same *shape*, but are offset vertically from each other by different constant values. The only way to resolve this ambiguity is *choose* the value that the upper graph is to have at a certain time (generally at $t = 0$). In some cases, the problem statement may suggest an appropriate value (it might say, for example, that the object starts at *rest*), but often we simply have to *pick* an arbitrary value. Once we have chosen or found a value for $v_x(t)$ at any given time, the rest of the graph is completely determined.

The area method

An alternative approach (which I will call the **area method**) is to recognize that the difference between the *value* on the upper graph at time t and its *value* at $t = 0$ will correspond to the *area* under the lower graph during this interval (see the *Math Skills* section on *Antiderivatives and Integrals*):

$$v_x(t) - v_x(0) = \text{area under curve of } a_x(t) \text{ between 0 and } t \tag{N5.3a}$$

$$x(t) - x(0) = \text{area under curve of } v_x(t) \text{ between 0 and } t \tag{N5.3b}$$

The phrase to remember is "*value* above equals *area* below" when we stack graphs in the conventional way (x above v_x above a_x).

This method is illustrated in Figure N5.1b. Note that for the first half of the graph, the acceleration is constant and positive, so as t increases, the area between t and 0 increases steadily, and thus so does the velocity. However, after the acceleration has fallen to zero, no new area is added as t increases, so the velocity remains constant.

Either way, we must find or choose v_x and/or x at some time

Note that this approach does not eliminate the need to determine or choose a *value* for the upper graph at some time (usually $t = 0$). Indeed, equations N5.3 make it clear that knowing the area does not help us determine $v_x(t)$ if we do not also know $v_x(0)$.

EXAMPLE N5.1

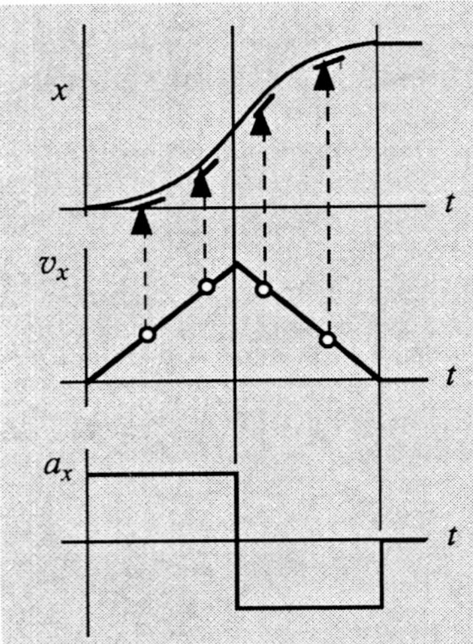

Figure N5.2: Graphs of $a_x(t)$, $v_x(t)$, and $x(t)$ for the rocket car.

Problem: Imagine a car (initially at rest at $x = 0$) that is powered by a rocket engine that pivots so that we can direct its thrust either forward or backward. The rocket engine ignites at $t = 0$, subsequently exerting a constant forward thrust on the car for a certain amount of time. Then the engine is turned around so that it exerts a rearward thrust on the car of the same magnitude for the same amount of time. Draw graphs of $a_x(t)$, $v_x(t)$, and $x(t)$ for this car. (Ignore drag or friction.)

Solution Assuming that there is no significant drag or friction force opposing the thrust on the car, and assuming that the car moves along a level road (so that the vertical normal and gravitational forces on the car cancel out), then the net force on the car will be the same as that applied by the rocket engine. The problem description implies (if we take the $+x$ direction to be forward) that the car's x-acceleration is some positive constant a for a certain Δt, and then $-a$ for Δt. This means the graph of $a_x(t)$ should look as shown in the bottom graph.

Working upwards, we can use either the slope or area method to show that the car's velocity first increases linearly and then decreases linearly. The problem description states that $v_x(0) = 0$, so a graph of the car's $v_x(t)$ must look as shown in the middle graph.

To create the position graph, we start with $x = 0$ at $t = 0$. I find the slope method somewhat easier to use to construct this graph (though you may not). Note that as the velocity increases, the slope of the $x(t)$ graph increases, and as the velocity decreases, the slope of $x(t)$ decreases, as shown in the top graph.

Note that we have (qualitatively at least) completely described the car's motion in terms of the forces applied to it!

To get more practice drawing such graphs, make sure that you could generate the upper graph in Figures N4.10 and N4.9b given the lower graph.

Exercise N5X.1: Plot $z(t)$ for the basketball described in example N4.7.

N5.4 INTEGRALS IN ONE DIMENSION

Using *integrals* to do the same thing mathematically

In the last section, we saw that we can construct graphs of an object's x-velocity and x-position as functions of time if we know the object's x-acceleration (or equivalently, the net force acting on it) as a function of time. In this section, we will see how we can do the same thing mathematically using *integrals*.

According to the definitions of velocity and acceleration, $dx/dt = v_x(t)$ and $dv_x/dt = a_x(t)$. Integrating both sides of these equations from 0 to t (see the ***Math Skills*** section on *Antiderivatives and Integrals*) we get

$$\int_0^t \frac{dv_x}{dt}dt = \int_0^t a_x(t)dt \quad \Rightarrow \quad v_x(t) - v_x(0) = \int_0^t a_x(t)dt \tag{N5.4a}$$

$$\int_0^t \frac{dx}{dt}dt = \int_0^t v_x(t)dt \quad \Rightarrow \quad x(t) - x(0) = \int_0^t v_x(t)dt \tag{N5.4b}$$

where $\int_a^b f(t)dt$ is the **definite integral** of $f(t)$, which we calculate by finding any function $F(t)$ whose derivative is $f(t)$ and evaluating the difference $F(b)-F(a)$ between that function's values at the limits of the integration. Equations N5.4 are essentially the mathematical expression (for one-dimensional motion) of the reverse kinematic chain described in equation N5.2. Since Newton's second law links an object's acceleration to the net force acting on it as follows

$$\vec{F}_{\text{net}} = m\vec{a} \quad \Rightarrow \quad F_{\text{net},x} = ma_x \tag{N5.5}$$

these equations allow us to determine the motion of an object if we know how the net x-force $F_{\text{net},x}$ acting on it varies with time.

EXAMPLE N5.2

Problem: Imagine a car waiting at a stoplight. After the stoplight turns green (call this time $t = 0$), the car accelerates from rest with a constant acceleration of 2.0 m/s^2 forward along a straight stretch of road (which we will take to define our x axis). What are the car's x-velocity and x-position as a function of time as long as this is true? What are the car's x-velocity and x-position after 5.0 s?

Solution We are told that the car's acceleration has a *constant* magnitude and is directed forward (that is, in the $+x$ direction), so $a_x(t) = +a$, where a is the constant magnitude of the acceleration = 2.0 m/s^2.

$$v_x(t) - v_x(0) = \int_0^t a_x(t)dt = \int_0^t a\,dt = at - a \cdot 0 = at \quad \text{(N5.6a)}$$

since $d(at)/dt = a$. Since we are told that the car starts "at rest" at $t = 0$, this means that $v_x(0) = 0$ in this case, so

$$v_x(t) = at \quad \text{(N5.6b)}$$

Plugging this into equation N5.4b, we find that

$$x(t) - x(0) = \int_0^t v_x(t)\,dt = \int_0^t at\,dt = \tfrac{1}{2}at^2 - \tfrac{1}{2}a \cdot 0^2 = \tfrac{1}{2}at^2 \quad \text{(N5.7a)}$$

We are not given what the car's initial x-position is, but if we define our reference frame so that $x(0) \equiv 0$, then equation N5.7a becomes simply

$$x(t) = \tfrac{1}{2}at^2 \quad \text{(N5.7b)}$$

Evaluating at $t = 5.0$ s, we find that $v_x(5.0\text{ s}) = at = (2.0\text{ m/s}^2)(5.0\text{ s}) = 10$ m/s and $x(t) = \frac{1}{2}at^2 = \frac{1}{2}(2.0\text{ m/s}^2)(5.0\text{ s})^2 = 25$ m.

In general, when an object's x-acceleration is some *constant* a_x (which may be either positive or negative), we have

$$v_x(t) - v_x(0) = \int_0^t a_x(t)dt = \int_0^t a_x\,dt = a_x t - a_x \cdot 0 = a_x t$$

$$\Rightarrow \quad v_x(t) = v_{0x} + a_x t \quad \text{(N5.8a)}$$

$$x(t) - x(0) = \int_0^t v_x(t)\,dt = \int_0^t [a_x t + v_x(0)]dt = \tfrac{1}{2}a_x t^2 + v_x(0)t - 0 - 0$$

$$\Rightarrow \quad x(t) = \tfrac{1}{2}a_x t^2 + v_{0x}t + x_0 \quad \text{(N5.8b)}$$

where $v_{0x} \equiv v_x(0)$ and $x_0 \equiv x(0)$. These results are important and useful for at least two reasons. First of all, this is a case that is relatively simple mathematically but illustrates the integral approach nicely. Note in particular that the object's x-position $x(t)$ is not specified until we specify its initial x-position x_0 *and* its initial x-velocity v_{0x}, as we have previously discussed.

Secondly, this result is useful because the assumption that a_x = constant is a reasonably good *model* for a variety of situations. Physicists often use this assumption to make quick estimates even when it is clear that a_x is *not* particularly constant. These equations are therefore part of a physicist's basic repertoire, and are therefore good to have close at hand.

On the other hand, I think that that these equations are overused. One of the most common physics errors that I have seen students make is to think that these equations are more general then they actually are. These equations really apply only to the very *specific* special case where a_x happens to be *constant.* While this is occasionally an excellent approximation to a realistic acceleration and more often an adequate first approximation, there are many more situations where these formulas are not even close to being correct.

Moreover, the fact that both integrals evaluated at the lower integration limit $t = 0$ happen to yield zero in this case often misleads people when they encounter slightly more complicated situations. Note that in the next example, evaluating the integrals at their lower limits does *not* yield zero in either case; instead, we have to keep track of these non-zero lower-limit terms very carefully in order to get meaningful results. So it is not wise take the constant-acceleration case too literally as a paradigm for all cases.

EXAMPLE N5.3

Problem: Imagine a similar situation as discussed in the last example: a car is sitting at a stoplight and then accelerates from rest when the light turns green. But in this case, let us assume that a drag force that increases with speed opposes the constant forward force applied to the car, so that the car's x-acceleration ends up being $a_x(t) = b/(t+T)^3$, where b = 2,000 m·s and T = 10 s (both are constants). What are the car's x-velocity and x-position as a function of time in this case? What are the car's x-velocity and x-position at t = 5.0 s?

Solution Note that the opposing drag force makes $a_x(t)$ *decrease* as t increases: the given $a_x(t)$ decreases to 1/8 its original value by time $t = T$. Note also that at $t = 0$, $a_x = (2000 \text{ m}\cdot\text{s})/(10 \text{ s})^3 = 2.0 \text{ m/s}^2$, which is the same as the *constant* acceleration value in the previous example. Since $v_x(0) = 0$ in this case, equation N5.4a implies that

$$v_x(t) = \int_0^t a_x(t)\,dt = \int_0^t \frac{b}{(t+T)^3}\,dt = b\int_0^t \frac{dt}{(t+T)^3}$$
$$= \frac{b}{-2}\left[\frac{1}{(t+T)^2} - \frac{1}{(0+T)^2}\right] = \frac{b}{2T^2} - \frac{b}{2(t+T)^2} \tag{N5.9}$$

(where I have used equation N5.30d in the ***Math Skills*** section to evaluate this integral). Plugging this into equation N5.4b, again defining the car's initial position to be $x(0) \equiv 0$, and using the product and sum rules, we get

$$x(t) = \int_0^t v_x(t)\,dt = \int_0^t \left[\frac{b}{2T^2} - \frac{b}{(t+T)^2}\right]dt = \frac{b}{2T^2}\int_0^t dt - \frac{b}{2}\int_0^t \frac{dt}{(t+T)^2}$$
$$= \frac{b}{2T^2}[t-0] - \frac{b}{2}\left[\frac{-1}{(t+T)} - \frac{-1}{T}\right] = \frac{b}{2}\left[\frac{t}{T^2} + \frac{1}{(t+T)} - \frac{1}{T}\right] \tag{N5.10}$$

Evaluating $v_x(t)$ and $x(t)$ at time t = 5.0 s, we find that:

$$v_x(5.0 \text{ s}) = \frac{(2000 \text{ m}\cdot\text{s})}{2(10 \text{ s})^2} - \frac{(2000 \text{ m}\cdot\text{s})}{2(15 \text{ s})^2} = 5.6 \text{ m/s} \tag{N5.11a}$$

$$x(5.0 \text{ s}) = \frac{2000 \text{ m}\cdot\text{s}}{2}\left[\frac{5.0 \text{ s}}{(10 \text{ s})^2} + \frac{1}{15 \text{ s}} - \frac{1}{10 \text{ s}}\right] = 17 \text{ m} \tag{N5.11b}$$

Note that these values are smaller than the results in the previous example, as we might expect.

Exercise N5X.2: *Why* might we expect the results to be smaller?

Exercise N5X.3: Check that the expressions given in equations N5.9 and N5.10 for for $v_x(t)$ and $x(t)$ yield $v_x(0) = 0$ and $x(0) = 0$ when $t = 0$, consistent with our assumptions about the car's initial position and velocity.

The importance of initial conditions

Let me emphasize again that just knowing $a_x(t)$ is *not* sufficient to determine an object's motion. All of the examples we have encountered so far make it clear that we also need to specify the object's initial x-position $x_0 \equiv x(0)$ and

x-velocity $v_{0x} \equiv v_x(0)$ (which we describe as stating its **initial conditions**) to completely determine its motion. This is *always* the case when we are trying to determine an object's trajectory from its acceleration, so dealing with initial conditions is going to be a recurring theme in chapters N11 through N14.

Exercise N5X.4 A car whose initial x-position and x-velocity are $x(0) = 0$ and $v_x(0) = +12$ m/s experiences forces that give it an x-acceleration of $a_x(t) = -a$, with $a = 2.0$ m/s^2. Find $v_x(t)$ and $x(t)$ and evaluate these quantities at $t = 5.0$ s.

N5.5 FREE-FALL IN ONE DIMENSION

The definition of a freely falling object

An important application of the equations for constant acceleration is the case of a *freely falling* object. If the only significant force acting on an object near the earth is its weight (which, according to chapter C8, is directed downward and has a magnitude mg), then we say that is **freely falling**, and we have (according to Newton's second law):

$$m\vec{a} = \vec{F}_{net} = \vec{F}_g = m\vec{g} \quad \Rightarrow \quad \vec{a} = \vec{g} \qquad \text{(N5.12)}$$

where $\vec{g}$ is a *vector* whose magnitude is $g = 9.8$ m/s^2 and whose direction is downward. Note that this equation implies that *all* objects fall with the same acceleration $\vec{g}$, independent of their mass! This is an interesting and non-obvious result that has important implications that we will explore in chapter N10. ($\vec{g}$ is properly called the **gravitational field vector**, but people often call it the "acceleration of gravity" for obvious reasons.)

If we define the z axis to be vertically upward, then $a_z = -g$. So, if the falling object only moves along the z axis, we adapt equations N5.8 by switching the x subscripts to z subscripts, yielding:

$$v_z(t) = v_{0z} - gt \qquad \text{(N5.13a)}$$

$$z(t) = -\tfrac{1}{2}gt^2 + v_{0z}t + z_0 \qquad \text{(N5.13b)}$$

The following example illustrates how we can apply these equations.

EXAMPLE N5.4

Problem: Imagine that at $t = 0$ we drop a ball from rest at an initial z-position $z_0 = 25$ m above the ground (which we define to have z-position $z = 0$). After 2.2 s, how fast is the ball moving? How far above the ground is it?

Solution Plugging $v_{0z} = 0$ and $t = 2.2$ s into equation N5.13a yields

$$v_z(2.2\text{ s}) = 0 - (9.8\text{ m/s}^2)(2.2\text{ s}) = -21.6\text{ m/s} \qquad \text{(N5.14)}$$

So the ball is moving *downward* with a speed of 21.6 m/s = 48.3 mi/h after falling for 2.2 s. Similarly, plugging numbers into equation N5.13b yields

$$z(2.2\text{ s}) = -\tfrac{1}{2}(9.8\text{ m/s}^2)(2.2\text{ s})^2 + 0 + 25\text{ m} = +1.3\text{ m} \qquad \text{(N5.15)}$$

So the object is 1.3 m above the ground at this time (almost there!)

Exercise N5X.5 When will the ball in example N5.4 hit the ground?

Exercise N5X.6 Rework example N5.4 assuming that at $t = 0$, we throw the ball directly upward with a speed of 12 m/s instead of dropping it from rest.

Exercise N5X.7 Argue that the downward component of a falling object's velocity increases by 22 mi/h every second.

N5.6 INTEGRALS IN THREE DIMENSIONS

We can easily generalize the mathematics developed in the last section to handle the situation where the object in question is *not* confined to move in one dimension. An object's acceleration vector $\vec{a}(t)$ is *defined* to be the time derivative of its velocity vector: $d\vec{v}/dt \equiv \vec{a}(t)$. If we integrate both sides of this expression from 0 to t and treat these vector functions as if they were *scalars*, we get

The integral of the vector acceleration (in abstract)

$$\vec{v}(t) - \vec{v}(0) = \int_0^t \vec{a}(t)\,dt \tag{N5.16}$$

But can we do this when the functions are really *vector* functions? We can show that N5.16 is correct by considering each component separately. The vector equation $d\vec{v}/dt \equiv \vec{a}(t)$ is equivalent to three independent component equations

$$\begin{bmatrix} dv_x/dt \\ dv_y/dt \\ dv_z/dt \end{bmatrix} \equiv \begin{bmatrix} a_x(t) \\ a_y(t) \\ a_z(t) \end{bmatrix} \quad \Rightarrow \quad \begin{matrix} dv_x/dt \equiv a_x(t) \\ dv_y/dt \equiv a_y(t) \\ dv_z/dt \equiv a_z(t) \end{matrix} \tag{N5.17}$$

Each one of these three independent equations only involves scalar functions of time. If we integrate both sides of each of these scalar equations, we get

What this means in terms of components

$$\begin{matrix} v_x(t) - v_x(0) = \int_0^t a_x(t)\,dt \\ v_y(t) - v_y(0) = \int_0^t a_y(t)\,dt \\ v_z(t) - v_z(0) = \int_0^t a_z(t)\,dt \end{matrix} \quad \Rightarrow \quad \begin{bmatrix} v_x(t) \\ v_y(t) \\ v_z(t) \end{bmatrix} = \begin{bmatrix} \int_0^t a_x(t)\,dt \\ \int_0^t a_y(t)\,dt \\ \int_0^t a_z(t)\,dt \end{bmatrix} + \begin{bmatrix} v_{0x} \\ v_{0y} \\ v_{0z} \end{bmatrix} \tag{N5.18}$$

where $v_{0x} \equiv v_x(0)$, $v_{0y} \equiv v_y(0)$ and so on. The leftmost equation shows what equation N5.16 really means at the component level.

Similarly, an object's velocity vector $\vec{v}(t)$ is defined to be the time derivative of its position vector: $d\vec{r}/dt \equiv \vec{v}(t)$. Integrating both sides of *this* expression from 0 to t, we get:

The integral of the vector velocity

$$\vec{r}(t) - \vec{r}(0) = \int_0^t \vec{v}(t)\,dt \quad \Rightarrow \quad \begin{bmatrix} x(t) \\ y(t) \\ z(t) \end{bmatrix} = \begin{bmatrix} \int_0^t v_x(t)\,dt \\ \int_0^t v_y(t)\,dt \\ \int_0^t v_z(t)\,dt \end{bmatrix} + \begin{bmatrix} x_0 \\ y_0 \\ z_0 \end{bmatrix} \tag{N5.19}$$

Equations N5.18 and N5.19 provide a complete and general description of how we can compute an object's velocity components and position components as a function of time if we know the object's initial velocity, its initial position, and its acceleration at all times.

We see that when it comes to integrating and taking derivatives, we *can* treat vector functions (symbolically) as if they were scalar functions.

Exercise N5X.8 What do equations N5.18 and N5.19 become in the particular case of a freely-falling object? (We will thoroughly study the implications of these equations in chapter N11.)

N5.7 CONSTRUCTING TRAJECTORIES

While equations N5.18 and N5.19 provide a means for determining an object's trajectory from its motion that is both straightforward and general, actually computing the integrals in these equations can be *very* difficult in many situations of interest. In this section, we will see how we can reverse the procedure

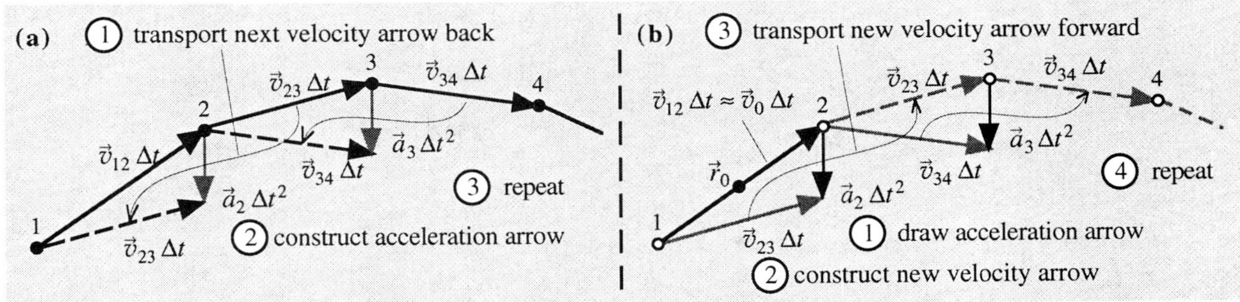

Figure N5.3: (a) How to construct acceleration arrows from a given trajectory. (b) How to construct a trajectory from given acceleration arrows. In both figures, black arrows represent known vectors, gray arrows represent constructed vectors, dashed arrows represent transported vectors, and white dots represent constructed positions.

that we used to construct motion diagrams (which allowed us to determine an object's acceleration from its trajectory) to construct an object's trajectory from its acceleration under *any* circumstances.

Figure N5.3a reviews how we construct a motion diagram. We first plot the object's position at equally-spaced instants of time and draw displacement arrows $\Delta\vec{r}_{12} = \vec{v}_{12}\,\Delta t$, $\Delta\vec{r}_{23} = \vec{v}_{23}\,\Delta t$, and so on between the dots. Since Δt is the same for all these arrows, these arrows depict both the direction and relative magnitudes of the average velocities $\vec{v}_{12}$, $\vec{v}_{23}$, and so on between the dots. To construct an arrow representing the object's acceleration at, say, point 2, we move the velocity arrow $\vec{v}_{23}\,\Delta t$ just after the point back to that its tail end coincides with the tail end of the velocity arrow $\vec{v}_{12}\,\Delta t$ just before the point and construct the vector difference $\Delta\vec{v}\,\Delta t \approx \vec{a}_2\,\Delta t^2$. As long as Δt is reasonably small compared to the time that it takes the object's acceleration to change significantly, this process yields reasonably accurate acceleration arrows $\vec{a}_2\,\Delta t^2$, $\vec{a}_3\,\Delta t^2$, etc.

How to draw a trajectory diagram:

To construct a **trajectory diagram** of the object's motion, we simply do this process backwards, as illustrated in Figure N5.3b. We start by drawing the arrow $\vec{v}_0\,\Delta t$ (where $\vec{v}_0$ is the object's initial velocity) so that it is *centered* on the object's initial position $\vec{r}_0$. If we define $t_1 = -\frac{1}{2}\Delta t$ and $t_2 = +\frac{1}{2}\Delta t$, then the middle of the interval from t_1 to t_2 is $t = 0$, and since the object's average velocity during a given time interval most closely approximates its instantaneous velocity at the center of the interval, we have $\vec{v}_{12}\,\Delta t \approx \vec{v}_0\,\Delta t$. Since $\vec{v}_{12}\,\Delta t$ stretches between the first two position points by definition, the endpoints of the vector $\vec{v}_0\,\Delta t$ approximately locate the object's positions at times t_1 and t_2.

Construct the first two position points using the specified initial conditions

Now, to find the object's position at time $t_3 = \frac{3}{2}\Delta t$, we carefully draw the object's acceleration arrow $\vec{a}_2\,\Delta t^2$ (which we have presumably calculated using Newton's second law) so its tail end coincides with point 2. We then construct the arrow $\vec{v}_{23}\,\Delta t$ so that the difference $\vec{v}_{23}\,\Delta t - \vec{v}_{12}\,\Delta t = \vec{a}_2\,\Delta t$, that is so that

Use the known acceleration vector to construct future position points

$$\vec{v}_{23}\,\Delta t = \vec{a}_2\,\Delta t + \vec{v}_{12}\,\Delta t \tag{N5.20}$$

We then transport the arrow $\vec{v}_{23}\,\Delta t$ so that its tail end coincides with point 2; its tip now indicates the position of the object at time t_3.

We can then repeat the process indefinitely, using the velocity arrow $\vec{v}_{23}\,\Delta t$ and the computed acceleration arrow $\vec{a}_3\,\Delta t^2$ to find the object's position at time t_4, and so on. Figure N5.3b shows how this works. If you carefully compare Figures N5.3a and N5.3b, you will see that the construction processes are simple inverses of each other.

This process works in all cases (if Δt is small)

We can use this construction process *any* time that we know the object's initial velocity and position and can compute the object's acceleration as a function of time. It is, however, based on two approximations: we are assuming in the construction that the object's average velocity during an interval is actually *equal* to the instantaneous velocity during that interval, and the same for the acceleration. Since this is true only in the limit that Δt goes to zero, the constructed trajectory is only an approximation if $\Delta t \neq 0$. Still, as long as Δt is fairly small compared to the time it takes the acceleration to change appreciably, the constructed trajectory will be reasonably accurate.

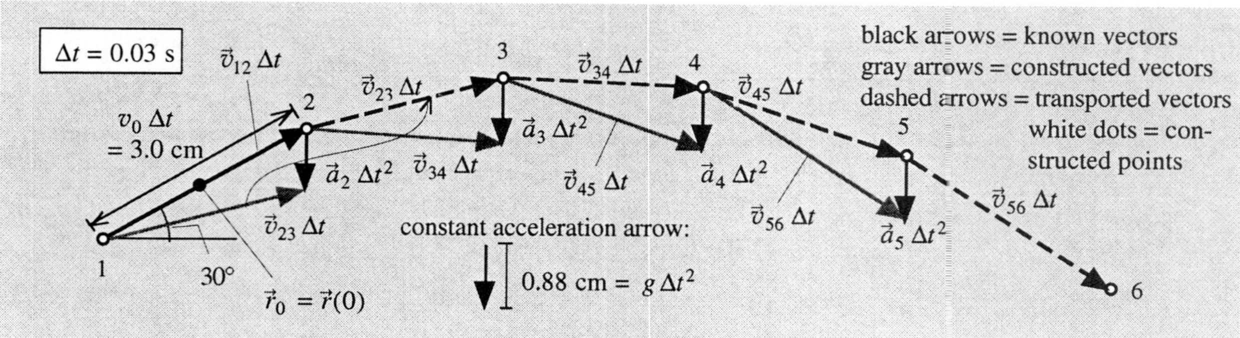

Figure N5.4: An actual-size trajectory diagram for the freely-falling marble discussed in example N5.4.

While we can always predict an object's trajectory using this process, it is especially easy to apply when the acceleration is constant, as the examples below illustrate. (The process also happens to yield the exact trajectory when $\vec{a}$ is constant, no matter how large we choose Δt to be.)

EXAMPLE N5.5

Problem: Imagine that we launch a marble with an initial velocity of 1.0 m/s at an angle of 30° above the horizontal, and that it subsequently falls freely. Use the graphical construction method to predict its future trajectory. (Suggestion: Set Δt = 0.03 s: this yields a diagram with a convenient size.)

Solution With Δt = 0.03 s, the the marble's acceleration arrow at all points on the diagram (according to equation N5.12) should have a magnitude of $g\,\Delta t^2 = (9.8 \text{ m/s})(0.03 \text{ s})^2 = 0.0088 \text{ m} = 0.88 \text{ cm}$ long and always points downward. The length of the marble's initial velocity vector arrow should be $\vec{v}_0\,\Delta t = (1.0 \text{ m/s})(0.03 \text{ s}) = 0.03 \text{ m} = 3.0 \text{ cm}$. Figure N5.4 shows the resulting constructed trajectory. You may already know that a freely falling object generally follows a parabolic path: our constructed trajectory looks plausibly parabolic.

Exercise N5X.9 In this example, if we choose Δt = 0.09 s, how long should we draw the acceleration and initial velocity arrows?

Exercise N5X.10 Construct the first few steps of the trajectory of an marble whose initial velocity is 0.80 m/s and horizontal. Use Δt = 1/30 s = 0.033 s.

EXAMPLE N5.6

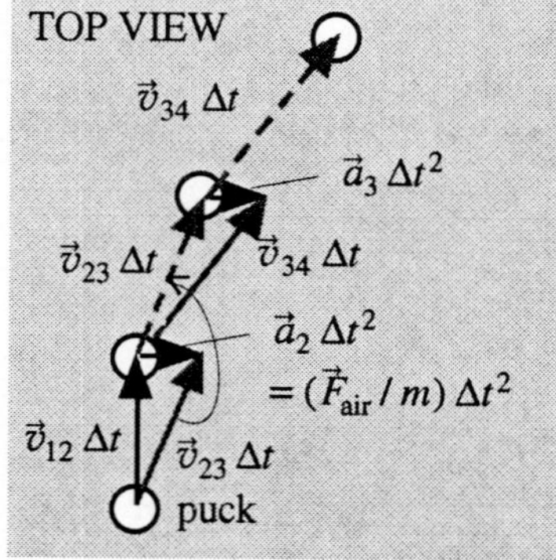

Figure N5.5: A north-bound puck that is being blown east.

Problem: A hockey puck slides due north on a level, frictionless plane of ice. Imagine that we blow due east on the puck, so that the puck experiences a constant eastward force. Qualitatively, what will be the puck's trajectory after we begin to blow on it?

Solution Because the ice is level and frictionless, the gravitational force on the puck and the force exerted on it by its interaction with the ice will cancel, meaning that the net force on the puck will then be the force associated with our blowing on it. Therefore Newton's second law implies that the puck's acceleration will be constant. Therefore, a top view of its trajectory must look qualitatively as shown in Figure N5.5. This trajectory is consistent with what we might expect from a momentum-transfer analysis. The puck should pick up speed in the eastward direction as its interaction with the blown air transfers eastward momentum to it, but its northward momentum should be unchanged. Thus the puck's trajectory should slowly bend toward the east, as shown.

This process provides a good basis for a computer calculation

Doing these constructions becomes quickly tedious, and it is hard to do them accurately. However, because the process is simple, repetitive, and universally applicable, it provides an excellent basis for a computer algorithm to compute trajectories. Your instructor may provide you with a computer program or spreadsheet that automates this procedure.

MATH SKILLS: Antiderivatives and Integrals

The **antiderivative** $F(t)$ of a function $f(t)$ is defined to be any function whose derivative is $f(t)$. For example, one possible antiderivative for $f(t) = at$, (where a is a constant) is $F(t) = \frac{1}{2}at^2$ because

$$\frac{dF}{dt} = \frac{d}{dt}\left[\tfrac{1}{2}at^2\right] = \tfrac{1}{2}a\frac{d}{dt}[t^2] = \tfrac{1}{2}a(2t) = at \qquad \text{(N5.21)}$$

Note that there is not a single *unique* antiderivative for a given function: if $F(t)$ is an antiderivative of $f(t)$, then so is $F(t) + C$ (where C is a constant), since

$$\frac{d}{dt}[F(t)+C] = \frac{dF}{dt} + \frac{dC}{dt} = \frac{dF}{dt} + 0 = \frac{dF}{dt} = f(t) \qquad \text{(N5.22)}$$

Thus the antiderivative $f(t)$ is really a whole *family* of functions $F(t)$ that differ by an additive constant. We usually write the antiderivative of $f(t)$ as $F(t) + C$, where C is an *unspecified* constant called a **constant of integration.**

Now, the **definite integral** of a function $f(t)$ from a point t_A to a point t_B on the horizontal axis is defined as follows:

$$\int_{t_A}^{t_B} f(t)dt \equiv [\text{total area under curve of } f(t) \text{ from } t_A \text{ to } t_B] \qquad \text{(N5.23a)}$$

(with area above the t axis being considered positive and area below the t axis negative). Figure N5.6 shows that we can approximate this area by a set of bars, the ith bar of which (where i is an arbitrary integer between 1 and the number of bars N) has a width Δt (the same for all bars) and a height equal to the value of $f(t_i)$, where $t_i = t_A + (i - \frac{1}{2})\Delta t$ is the value of t halfway across the width of the bar. Perhaps you can convince yourself that as the bar width Δt goes to zero, the area under the curve of the graph can be:

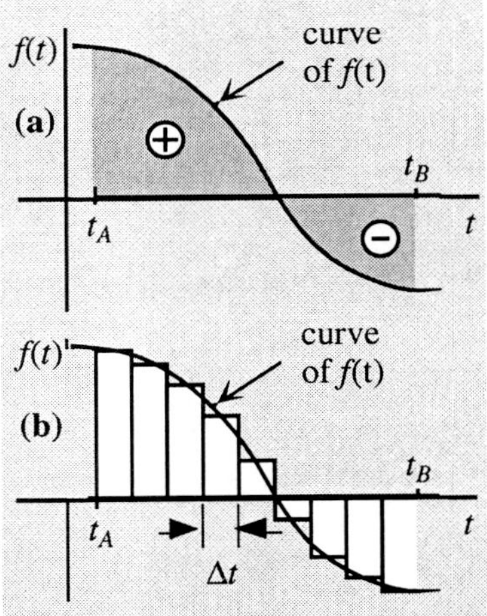

Figure N5.6: (a) The definite integral of a function $f(t)$ from t_A to t_B is defined to be the area under a graph of $f(t)$ between t_A and t_B (that is, the shaded area shown). Area above the horizontal axis is treated as positive and area below as negative. **(b)** The area is approximately equal to the total area of the bars shown, where each bar is Δt wide and as tall as the function's value halfway across the bar.

$$\int_{t_A}^{t_B} f(t)dt \equiv \lim_{\Delta t \to 0} \sum_{i=1}^{N} f(t_i)\Delta t \qquad \text{(N5.23b)}$$

where $N = (t_B - t_A)/\Delta t$ is the number of bars in the set. Indeed, the symbol $\int$ used for the integral is simply a stylized s standing for "sum".

The following properties of the integral follow directly from its definition:

Properties of the definite integral:

The Sum Rule

$$\int_{t_A}^{t_B}[f(t)+g(t)]dt = \int_{t_A}^{t_B} f(t)dt + \int_{t_A}^{t_B} g(t)dt \qquad \text{(N5.24a)}$$

The Constant Rule

$$\int_{t_A}^{t_B} cf(t)dt = c\int_{t_A}^{t_B} f(t)dt \qquad \text{(N5.24b)}$$

$$\int_{t_A}^{t_C} f(t)dt = \int_{t_A}^{t_B} f(t)dt + \int_{t_B}^{t_C} f(t)dt \quad [\text{if } t_C \ge t_B \ge t_A] \qquad \text{(N5.24c)}$$

The last of these equations says that the total area under the curve between t_A and t_C is the sum of the areas between t_A and t_B, and t_B and t_C, (something that is not really very surprising). The first two equations are easiest to prove if we remember that the integral is the limit of a sum (see equation N5.23b).

The fundamental theorem of calculus

At first glance, it may not seem that the *antiderivative* of $f(t)$ [which is a function whose derivative is $f(t)$] should anything to do with the *integral* of $f(t)$ [which is the area under the curve of $f(t)$]. However, the **fundamental theorem of calculus** asserts that if $F(t)$ is *any* antiderivative of $f(t)$, then

$$F(t) - F(t_0) = \int_{t_0}^{t} f(t)dt \qquad \text{(N5.25)}$$

where t_0 is some arbitrary fixed value of the variable t. Note that when we use the variable t as the upper limit of the integration, the value of the integral (that is, the area from t_0 to t) is essentially a function of t.

Proof of the fundamental theorem of calculus

We can prove the fundamental theorem as follows. The definition of the derivative implies that the time derivative of the right side of equation N5.25 is

$$\frac{d}{dt}\int_{t_0}^{t} f(t)dt = \lim_{\Delta t \to 0}\frac{1}{\Delta t}\left[\int_{t_0}^{t+\Delta t} f(t)dt - \int_{t_0}^{t} f(t)dt\right] \qquad \text{(N5.26)}$$

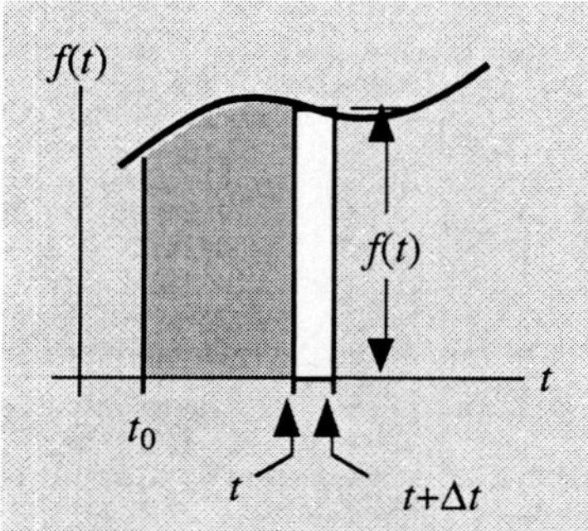

Figure N5.7: The *difference* between the area from t_0 to t and the area from t_0 to $t+\Delta t$ will be *approximately* equal to the area of the white bar, which has width Δt and height $f(t)$. This approximation becomes exact as $\Delta t \to 0$.

However, as Figure N5.7 shows, the difference between the area under the curve between t_0 and t and the area between t_0 and $t+\Delta t$ (as Δt becomes very small) becomes essentially the area of a bar of width Δt and height $f(t)$. Therefore

$$\frac{d}{dt}\int_{t_0}^{t} f(t)dt = \lim_{\Delta t \to 0}\frac{1}{\Delta t}[f(t)\Delta t] = f(t) \qquad \text{(N5.27)}$$

So we see that the integral is indeed an antiderivative of $f(t)$. Therefore, the *general* antiderivative of $f(t)$ can be written as

$$F(t) = \int_{t_0}^{t} f(t)dt + C \quad (C \text{ is an undetermined constant}) \qquad \text{(N5.28)}$$

Therefore, for *any* antiderivative $F(t)$ of the function $f(t)$, we have

$$F(t) - F(t_0) = \int_{t_0}^{t} f(t)dt + C - \left[\int_{t_0}^{t_0} f(t)dt + C\right] = \int_{t_0}^{t} f(t)dt \qquad \text{(N5.29)}$$

since the Cs cancel and the area under the curve from t_0 to t_0 is clearly zero. This equation N5.25, so we have completed the proof.

Indefinite integrals

The **indefinite integral** $\int f(t)\,dt$ is *any* one of the possible antiderivatives of the function $f(t)$ (usually, the antiderivative with the simplest algebraic form). Some useful indefinite integrals to know are:

$$\int \frac{df}{dt}dt = f(t) \quad \text{for any function } f(t) \qquad \text{(N5.30a)}$$

$$\int c\,dt = ct \quad \text{where } c \text{ is any constant} \qquad \text{(N5.30b)}$$

$$\int t^n\,dt = \frac{t^{n+1}}{n+1} \quad \text{for } n \neq -1 \qquad \text{(N5.30c)}$$

The last follows from equation N1.18. One can show in fact that

$$\frac{d}{dt}\left[\frac{(t+c)^{n+1}}{n+1}\right] = (t+c)^n \quad \Rightarrow \quad \int (t+c)^n\,dt = \frac{(t+c)^{n+1}}{n+1} \qquad \text{(N5.30d)}$$

This is a useful extension of equation N5.30c. A **table of integrals** provides a list of indefinite integrals for a variety of functions Feel free to look up indefinite integrals in such a table if you have access to one. Remember that a given indefinite integral is only one of a family of possible antiderivatives: all others can be generated by adding some constant C to the one you have.

Going from indefinite integrals to definite integrals

Once you know an indefinite integral for a function, you can find its definite integral from t_0 to t by simply evaluating the indefinite integral at these two points and subtracting. For example, $\int_{t_0}^{t} c\,dt = ct - ct_0$. This works no matter which member of the family of possible antiderivatives the indefinite integral you have chosen is: equation N5.29 applies to *any* antiderivative of a given function $f(t)$ (note how the arbitrary constant C cancels out).

We generally use *definite* integrals in physics

In physics, it is generally easiest to use *definite* integrals when trying to determine an object's motion from its acceleration. If you use indefinite integrals, you have to explicitly choose the constant of integration C to match the initial conditions (and this can be tricky sometimes). If you use definite integrals, the initial conditions are automatically accounted for: this is both easier and reduces the probability of error.

SUMMARY

I. THE REVERSE KINEMATIC CHAIN

A. The reversed chain: $\vec{a}(t)$ —[time antiderivative]→ $\vec{v}(t)$ —[time antiderivative]→ $\vec{r}(t)$

1. If we know the net force acting on an object, we can use Newton's second law to determine its acceleration
2. We can then use the reversed chain to predict the object's motion

B. The problem is that computing antiderivatives is *hard*. We will look at *four* different approaches to this problem in this chapter

II. GRAPHICAL ANTIDERIVATIVES (IN ONE DIMENSION)

A. Our convention is to stack a graph of $x(t)$ above a graph of $v_x(t)$ above a graph of $a_x(t)$ (and the discussion below assumes this)

B. How to construct an antiderivative graph above from a graph below

1. The *slope method*: at selected instants of time,
   a) Draw a short line segment on the upper graph whose slope reflects the value on lower graph ("slope above = value below")
   b) As you draw successive segments, arrange them vertically so that they sketch out a continuous curve
2. The *area method* ("value above = area below"):
   a) $v_x(t) = dx/dt$ and the fundamental theorem of calculus imply that $x(t)–x(0)$ = the area under a graph of $v_x(t)$ between 0 and t
   b) Similarly, $v_x(t) - v_x(0)$ = area under $a_x(t)$ between 0 and t

C. In either case, we have to choose $x(0)$ or $v_x(0)$ to be consistent with the problem statement (or arbitrarily, if we are given no information)

III. INTEGRALS OF ONE DIMENSIONAL MOTION

A. The definitions of velocity and acceleration imply that

1. $v_x(t) - v_x(0) = \int_0^t a_x(t)dt, \quad x(t) - x(0) = \int_0^t v_x(t)dt$ (N5.4)
2. Note again that we have to know $x(0)$ and $v_x(0)$ (the object's *initial conditions*) to determine $x(t)$ and $v_x(t)$: this is always true

B. An important *special case*: IF a_x = constant, then

$v_x(t) = a_x t + v_{0x}, \quad x(t) = \frac{1}{2}a_x t^2 + v_{0x}t + x_0$ (N5.8)

C. Free-fall in one dimension

1. An object is *freely falling* if the only force on it is its weight
2. Newton's second law then implies that $m\vec{a} = \vec{F}_{\text{net}} = m\vec{g}$, which implies that $\vec{a} = \vec{g}$ (and so all objects fall with the same acceleration!)
3. If the object falls in one dimension, we can adapt equation N5.8 by substituting z subscripts for x and using $a_z = -g$

IV. INTEGRALS IN THREE DIMENSIONS

A. In 3 dimensions, the definitions of velocity and acceleration imply that

1. $\vec{v}(t) - \vec{v}(0) = \int_0^t \vec{a}(t)\,dt, \quad \vec{r}(t) - \vec{r}(0) = \int_0^t \vec{v}(t)\,dt$ (N5.16,19)
2. These compactly express three independent component equations (see equations N5.18 and N5.19 for the details)

B. Note the similarity to equation N5.4

V. CONSTRUCTING TRAJECTORY DIAGRAMS

A. How to construct a trajectory diagram:

1. Draw an arrow $\vec{v}_0\,\Delta t$ centered on the object's initial position $\vec{r}_0$
   a) This arrow expresses the initial conditions
   b) $\vec{v}_0\,\Delta t \approx \vec{v}_{12}\,\Delta t$ if we define $t_1 = -\frac{1}{2}\Delta t$ and $t_2 = +\frac{1}{2}\Delta t$, so the endpoints of this arrow are points 1 and 2
2. Draw the arrow $\vec{a}_2\,\Delta t^2$ with its tail attached to point 2
3. Construct the arrow $\vec{v}_{23}\,\Delta t = \vec{v}_{12}\,\Delta t + \vec{a}_2\,\Delta t^2$
4. Move $\vec{v}_{23}\,\Delta t$ so that its tail is at point 2: its tip then is point 3
5. Repeat steps 2 through 4 to find points 4, 5, 6, ...

B. Comments about the process

1. This process works in *all* cases (as long as Δt is sufficiently small)
2. It is tedious (though well-suited for computers)

GLOSSARY

antiderivative: the antiderivative $F(t)$ of a function $f(t)$ is that function whose derivative is $f(t)$: $dF/dt = f(t)$. $F(t)$ is determined only up to an overall additive constant: if $F(t)$ is an antiderivative of $f(t)$ then so is $F(t) + C$, where C is any constant. Thus the antiderivative of $f(t)$ is usually written $F(t) + C$, where $F(t)$ is any *specific* antiderivative of $f(t)$ and C is an unspecified **constant of integration**.

reverse kinematic chain: the chain of relationships linking an object's acceleration, velocity , and position:

$$\vec{a}(t) \;\text{-}\big[\text{time anti-derivative}\big]\!\!\to\; \vec{v}(t) \;\text{-}\big[\text{time anti-derivative}\big]\!\!\to\; \vec{r}(t) \qquad \text{(N5.1)}$$

slope method: a method of drawing a graph of a function from a graph of its derivative that involves drawing a connected set of short line segments on the former graph at successive instants of time, the *slope* of each of which reflects the corresponding *value* on the derivative graph.

area method: a method of drawing a graph of a function from a graph of its derivative that involves plotting a value on the former graph at a given t that corresponds to the accumulated area under the derivative graph up to that t.

initial conditions: the value of an object's position and velocity at a given instant of time (usually $t = 0$). These values are needed to determine the object's position and velocity at other times from its acceleration.

definite integral: The definite integral from t_0 to t of a function $f(t)$ is defined to be the area under a graph of $f(t)$ between t_0 and t.

the fundamental theorem of calculus: states that if $F(t)$ is any antiderivative of $f(t)$ then the definite integral of $f(t)$ from t_0 to t is

$$\int_{t_0}^{t} f(t)\,dt = F(t) - F(t_0) \qquad \text{(N5.25)}$$

indefinite integral: the indefinite integral $\int f(t)dt$ is defined to be any (convenient) antiderivative of $f(t)$.

table of integrals: a list of indefinite (sometimes also some definite) integrals of various functions.

freely falling: we say that an object is *freely falling* if the *only* (significant) force acting on it is its weight.

gravitational field vector $\vec{g}$: the quantity that links an object's mass to its weight at a given place in a gravitational field, characterizing the strength and direction of the gravitational field at that point. This quantity also specifies the acceleration that all freely-falling objects will have at that point, and thus is sometimes called the **acceleration of gravity**. Near the surface of the earth, g has the magnitude 9.8 m/s^2 = (22 mi/h)/s.

trajectory diagram: A diagram that is very similar to a motion diagram except that its purpose is reversed: instead of allowing us to determine an object's acceleration at various times based on its known motion, a trajectory diagram allows us to construct an object's trajectory based on its initial conditions and its known acceleration.

TWO-MINUTE PROBLEMS

N5T.1 An object's x-velocity $v_x(t)$ is shown in the boxed graph at the top left. Which of the other graphs in the set most correctly describes its x-position?

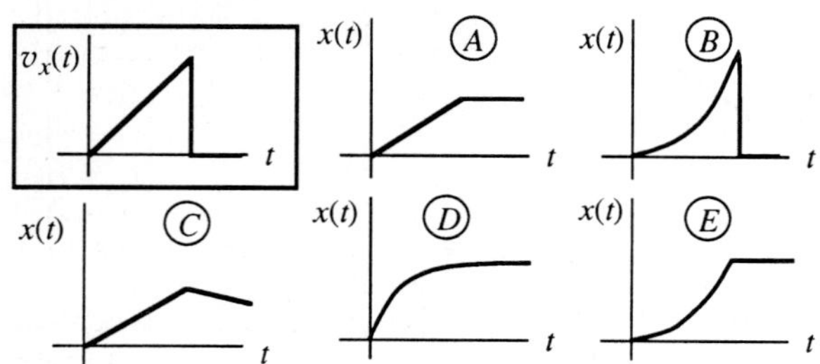

N5T.2 An object's x-acceleration $a_x(t)$ is shown in the boxed graph at the top left. Which of the other graphs in the set most correctly describes its x-velocity?

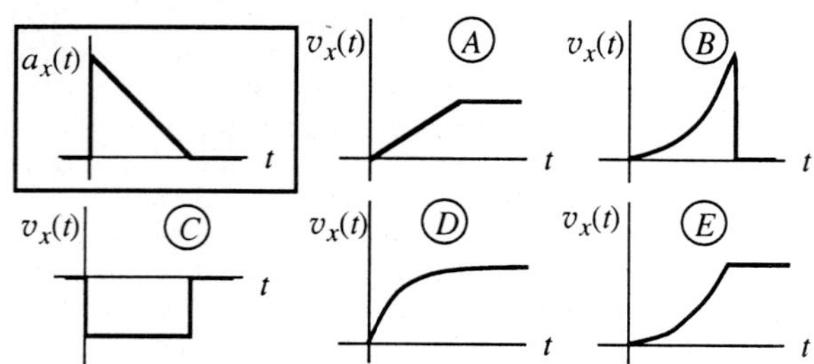

N5T.3 If a car has an x-acceleration of $a_x(t) = -bt + c$, and its initial x-velocity at time $t = 0$ is $v_x(0) = v_0$, which function below best describes $v_x(t)$?

A. $-b$
B. $-b + v_0$
C. $\frac{1}{2}bt^2 + ct + v_0$
D. $-\frac{1}{2}bt^2 + ct + v_0$
E. $-2bt^2 + v_0$
F. $-\frac{1}{2}bt^2 + v_0$

N5T.4 If a car's x-position at time $t = 0$ is $x(0) = 0$, and it has an x-velocity of $v_x(t) = b(t-T)^2$, where b and T are constants, which function below best describes $x(t)$?

A. $x(t) = 2b(t-T)$
B. $x(t) = 3b(t-T)^3$
C. $x(t) = \frac{1}{3}b(t-T)^3$
D. $x(t) = \frac{1}{2}b(t-T)$
E. $x(t) = \frac{1}{3}b[(t-T)^3 + T^3]$
F. other (specify)

N5T.5 Imagine that you are preparing an actual-size trajectory diagram of a freely-falling object. The time interval between positions is 0.02 s. How long should you draw the acceleration arrows on your diagram?

A. 9.8 m
B. 0.20 m
C. 0.04 m
D. 3.9 cm
E. 0.39 cm
F. other (specify)

N5T.6 At time $t = 0$, a person is sliding due east on a flat, frictionless plane of ice. The net force on this person is due to a steady wind blowing from the north. The eastward component of the person's velocity is completely unaffected by this force (T or F).

N5T.7 Consider the person in the previous problem. The person's trajectory will look most like which of the following? (The dot shows the person's position at $t = 0$, and east is to the right and north to the top.)

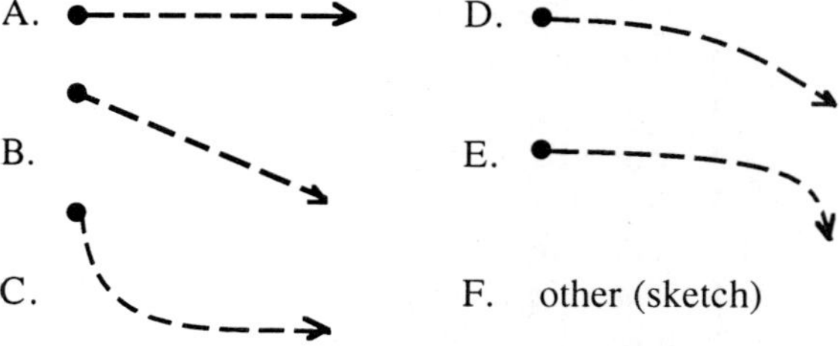

HOMEWORK PROBLEMS

BASIC SKILLS

N5B.1 Construct a graph of the object's x-velocity as a function of time for an object whose x-acceleration is as given below. Assume that $v_x(0) = 0$.

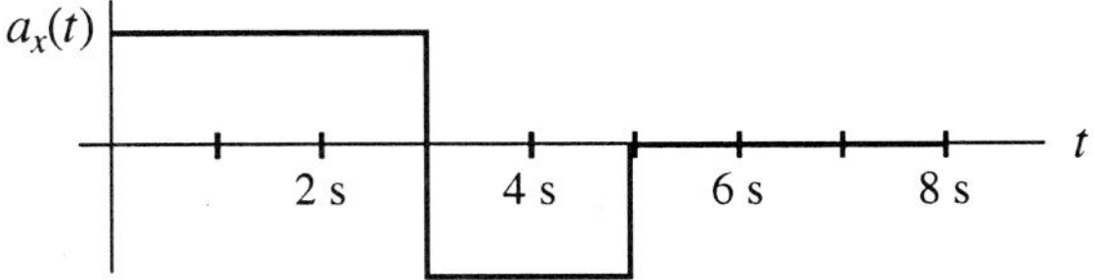

N5B.2 Construct a graph of x-position as a function of time for an object whose x-velocity is as given in problem N5B.5. Assume that $x(0) = 0$.

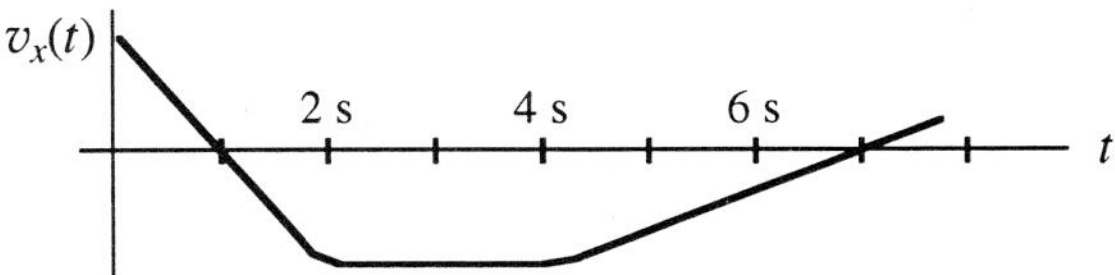

N5B.3 Integrate the following functions from 0 to t.
(a) $f(t) = bt$ **(b)** $f(t) = b(t-T)$ (b, T are constants)

N5B.4 Integrate the following functions from 0 to t.
(a) $f(t) = bt^3$ **(b)** $f(t) = b/(t+T)^{1/2}$ (b, T are constants)

N5B.5 Imagine that you are trying to construct an actual-size motion diagram of a falling object with an initial horizontal velocity of 2.0 m/s. Take the time interval between position points to be 0.05 s. How long should you draw the object's initial velocity arrow on the diagram? Its acceleration arrows?

N5B.6 Imagine that you are trying to construct an actual-size motion diagram of a falling object with an initial horizontal velocity of 4.0 m/s. Take the time interval between position points to be 0.15 s. How long should you draw the object's initial velocity arrow on the diagram? Its acceleration arrows?

N5B.7 A stone dropped from rest from the middle of a bridge hits the water below 2.5 s later. How far is the bridge above the water? (Ignore air resistance.)

N5B.8 An object freely falling from rest for 5.0 s will have what final speed? (Express your result in both meters per second and miles per hour, and ignore air resistance.)

SYNTHETIC

N5S.1 Imagine that the net force on a car moving in one dimension is initially large and forward, but subsequently decreases linearly to zero and then remains zero thereafter. Draw graphs of $a_x(t)$, $v_x(t)$, and $x(t)$ for this car, assuming that it starts from rest at $x = 0$.

N5S.2 A car starts at $t = 0$ from rest at $x = 0$ and accelerates at a constant rate until reaching a cruising speed of 15 m/s. After maintaining that speed for a while, the driver of the car, seeing that a bridge ahead is out, brakes suddenly, and the car comes to rest in a relatively short period of time. After remaining at rest for a few seconds, the driver then backs up at about 3 m/s. Draw qualitatively accurate graphs of the car's x-position, x-velocity, and x-acceleration as functions of time. Assume that the road is straight and the car initially travels in the positive x direction.

N5S.3 Imagine that the x-position of one car is given by the function $x_1(t) = bt^2$ (with b = 2.5 m/s²), whereas the position of another car is given by the function $x_2(t) = ct$ (where c = 18 m/s). Note that both cars have a position equal to zero at $t = 0$. At time t = 5.0 s, which car is moving faster? Which one is further ahead?

N5S.4 Imagine that a car's x-acceleration during a certain time period is given by the function $a_x(t) = -bt$. If the car's x-velocity at $t = 0$ is 32 m/s, and b = 1.0 m/s³, how long will it take the car to come to rest?

N5S.5 The z-velocity of deep-sea probe descending into the ocean is computer-controlled to be $v_z(t) = c + bt^2$ (where c is a constant and b = 0.040 m/min³) until the probe comes to rest. At $t = 0$, $z(0)$ = −22 m (that is 22 m below the ocean surface) and $v_z(0)$ = −120 m/min.
(a) What are the units of c? What value must c have?
(b) Find the probe's z-acceleration as a function of time.
(c) At what time and position will the probe come to rest?

N5S.6 A spaceship is approaching Starbase Beta at an initial velocity of 130 km/s in the $+x$ direction. The ensign sets its computer to initiate a braking program starting at time $t = 0$. After this time, the computer controls the ship until the ship docks at time $t = T$ so that its x-acceleration is $a_x(t) = b(t-T)$, where b = 0.26 m/s³. Note that $a_x < 0$ for $t < T$, but gets smaller as t approaches the docking time T. **(a)** What must T be so that the ship's x-velocity is also zero at the time of docking? **(b)** How far is the ship from the station at $t = 0$? (*Hint*: Use equation N5.30d.)

N5S.7 Imagine that we throw a marble with an initial speed of 2.2 m/s at an angle of 60° above the horizontal. Draw a trajectory diagram of the marble's subsequent trajectory using Δt = 1/30 s = 0.033 s for at least 10 time steps. Estimate from your diagram how long it takes the marble to reach the peak of its trajectory.

N5S.8 Imagine that an outfielder throws a baseball with an initial speed of 12 m/s in a direction 30° up from the horizontal. Use a trajectory diagram to determine the height of the peak of its trajectory. I suggest using a time step of Δt = 0.2 s and a scale of 1.0 m (in reality) = 2.0 cm (on diagram).

N5S.9 You are traveling in the fabled Mines of Moria when you come across an open water well in the stone floor of a tunnel. To find out how deep the well is, you drop a stone into the well and count the seconds until you hear the splash. If you hear the splash 3.0 s after you drop the stone, roughly how deep is the well? What kinds of approximations do you have to make to do this problem?

N5S.10 Imagine an air puck slides frictionlessly on a level air table. The puck is connected to a string going through a hole in the center of the table. A clever student holds the string in exactly the right way so that it exerts a force of constant magnitude on the puck, and as a result, the magnitude of the puck's acceleration is always 1.0 m/s². Assume that at $t = 0$, the puck is exactly 4.0 cm from the center of the table in the $-x$ direction, and is moving at a speed of 13 cm/s in the $+y$ direction. Construct a trajectory diagram that shows the puck's position at subsequent instants of time separated by Δt = 0.1 s until t = 1.15 s. [*Hint*: the main difference between Figure N5.5 and what you will draw is that the acceleration arrow that you should draw through each position dot during step 1 should point from the puck's position at that instant toward the center of the table, since the string exerts its tension force in that direction.]

RICH-CONTEXT

N5R.1 You are a police officer. Your squad car is at rest on the shoulder of an interstate highway when you notice a car fitting the description given in an all-points bulletin passing you at its top speed of 85 mi/h. You jump in your car, start the engine, and find a break in the traffic, a process which takes 25 s. You know from the squad car's manual that when it starts from rest with its accelerator pressed to the floor, the magnitude of its acceleration is $a = a_0 - bt^2$ (where a_0 = 2.5 m/s^2 and b = 0.0028 m/s^4) until $a_0 = bt^2$, and then remains zero thereafter. Can you catch the car before it reaches the next exit 5.3 mi away?

ADVANCED

N5A.1 Prove (without using the chain rule) that

$$\frac{d}{dt}(t+c)^{n+1} = (n+1)(t+c)^n \qquad \text{(N5.31)}$$

for integer values of $n \geq 0$ as follows. (This is part of the general proof of equation N5.30d.) **(a)** Prove that this is true for $n = 0$. **(b)** Prove using the product rule that if it is true for any certain $n = n_0 \geq 0$, then it is true for $n = n_0+1$. (Equation N5.31 then follows by induction.)

ANSWERS TO EXERCISES

N5X.1 Graphs of $z(t)$ and $v_z(t)$ for the basketball look like this (note that I used the slope method to construct the upper graph):

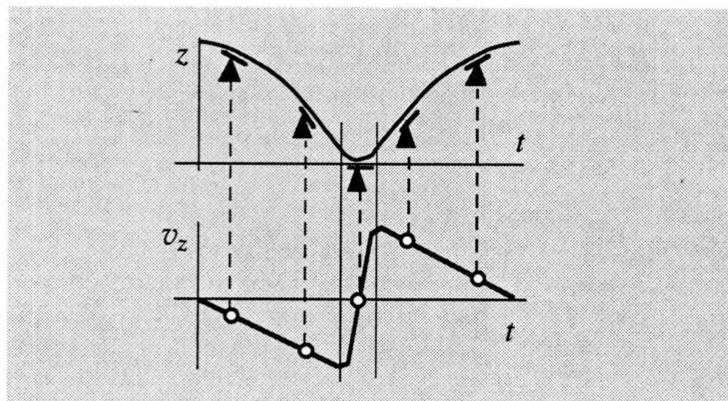

N5X.2 The car's x-acceleration at t = 0 is the same for both cases, but in the second example the acceleration decreases with time and thus is generally less than in the first case. Therefore it makes sense that it should reach a smaller final x-velocity and not go as far in the same time interval.

N5X.3 At time t = 0, we have

$$v_x(0) = \frac{b}{2T^2} - \frac{b}{2(0+T)^2} = \frac{b}{2T^2} - \frac{b}{2T^2} = 0 \qquad \text{(N5.32a)}$$

$$x(0) = \frac{b}{2}\left[\frac{0}{T^2} + \frac{1}{0+T} - \frac{1}{T}\right] = \frac{b}{2}\left[\frac{1}{T} - \frac{1}{T}\right] = 0 \qquad \text{(N5.32b)}$$

N5X.4 $v_x(t) = v_{0x} - at$, $x(t) = v_{0x}t - \frac{1}{2}at^2$. At t = 5.0 s, v_x(5.0 s) = 2.0 m/s, x(5.0 m) = +35 m.

N5X.5 We can find when the ball will hit the ground by setting $z(t) = 0$ and solving for t:

$$0 = z(t) = -\tfrac{1}{2}gt^2 + z_0 \quad \Rightarrow \quad \tfrac{1}{2}gt^2 = z_0 \quad \Rightarrow \quad t^2 = \frac{2z_0}{g}$$

$$t = \sqrt{\frac{2z_0}{g}} = \sqrt{\frac{2(25\ \text{m})}{9.8\ \text{m/s}^2}} = 2.26\ \text{s} \qquad \text{(N5.33)}$$

N5X.6 The downward component of a falling object's velocity is $-v_z(t)$ [since $v_z(t)$ is the upward component]. The rate of change of this downward component is, then

$$\frac{d[-v_z]}{dt} = \frac{d}{dt}[-v_{0z} + gt] = 0 + g = g \qquad \text{(N5.34)}$$

So g expresses the constant rate at which the downward component of a falling object's velocity increases. Since

$$g = 9.8\frac{\text{m}}{\text{s}^2}\left(\frac{2.24\ \text{mi/h}}{1\ \text{m/s}}\right) = 22\frac{\text{mi/h}}{\text{s}} \qquad \text{(N5.35)}$$

a falling object's downward velocity component increases at a rate of 22 mi/h per second.

N5X.8 For a freely falling object, $\vec{a} = \vec{g} = [0, 0, -g]$. So equation N5.18 becomes

$$\begin{bmatrix} v_x(t) \\ v_y(t) \\ v_z(t) \end{bmatrix} = \begin{bmatrix} \int_0^t a_x(t)dt \\ \int_0^t a_y(t)dt \\ \int_0^t a_z(t)dt \end{bmatrix} + \begin{bmatrix} v_{0x} \\ v_{0y} \\ v_{0z} \end{bmatrix} = \begin{bmatrix} \int_0^t 0\,dt \\ \int_0^t 0\,dt \\ \int_0^t -g\,dt \end{bmatrix} + \begin{bmatrix} v_{0x} \\ v_{0y} \\ v_{0z} \end{bmatrix}$$

$$\Rightarrow \begin{bmatrix} v_x(t) \\ v_y(t) \\ v_z(t) \end{bmatrix} = \begin{bmatrix} 0 \\ 0 \\ -gt \end{bmatrix} + \begin{bmatrix} v_{0x} \\ v_{0y} \\ v_{0z} \end{bmatrix} = \begin{bmatrix} v_{0x} \\ v_{0y} \\ -gt + v_{0z} \end{bmatrix} \qquad \text{(N5.36)}$$

Equation N5.19 then becomes:

$$\begin{bmatrix} x(t) \\ y(t) \\ z(t) \end{bmatrix} = \begin{bmatrix} \int_0^t v_x(t)dt \\ \int_0^t v_y(t)dt \\ \int_0^t v_z(t)dt \end{bmatrix} + \begin{bmatrix} x_0 \\ y_0 \\ z_0 \end{bmatrix} = \begin{bmatrix} \int_0^t v_{0x}\,dt \\ \int_0^t v_{0y}\,dt \\ \int_0^t [-gt + v_{0z}]dt \end{bmatrix} + \begin{bmatrix} x_0 \\ y_0 \\ z_0 \end{bmatrix}$$

$$= \begin{bmatrix} v_{0x}t - 0 \\ v_{0y}t - 0 \\ -\frac{1}{2}gt^2 + v_{0z}t - 0 \end{bmatrix} + \begin{bmatrix} x_0 \\ y_0 \\ z_0 \end{bmatrix} = \begin{bmatrix} v_{0x}t + x_0 \\ v_{0y}t + y_0 \\ -\frac{1}{2}gt^2 + v_{0z}t + z_0 \end{bmatrix} \qquad \text{(N5.37)}$$

N5X.9 The acceleration arrow should be 9.8 cm long and the velocity arrow should be 10 cm long.

N5X.10 The trajectory diagram should look like this:

N6

TORQUE AND STATICS

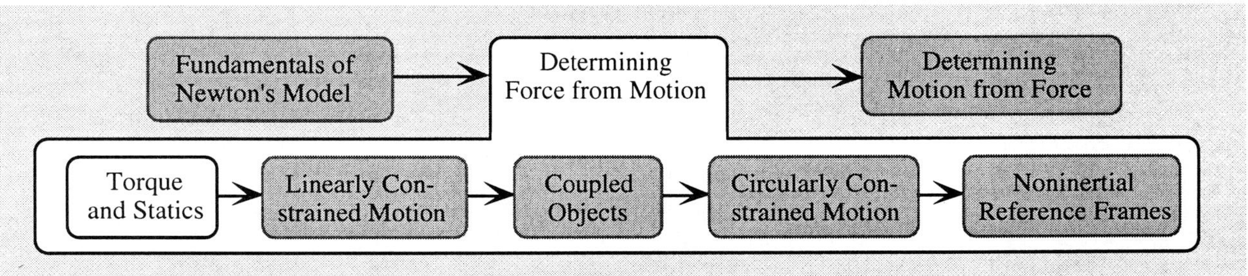

N6.1 OVERVIEW

In the last five chapters, we have discussed the basic principles of the newtonian model, showing how we can use Newton's second law to either (1) infer the forces that act on an object by observing its motion, or (2) predict the motion of an object knowing the forces acting on it. During the remainder of the unit, we will be studying a variety of example applications in each of these two general categories.

This chapter opens the subunit exploring the first of these categories. Our particular focus in this chapter will be *statics problems*, where the object under investigation is known to be completely at rest (this is the simplest possible motion an object can have!). "Completely at rest" implies that not only is the object's center of mass at rest (and thus not accelerating, which in turn means that the net external force on the object is zero) but also that the object is not *rotating* (which as we will see, means that the net *torque* on the object has to be zero also). Part of the purpose of this chapter is to define the concept of torque and illustrate its use.

Here is an overview of this chapter's sections.

N6.2 *TORQUE* explores the mathematical definition of torque and its physical meaning when applied to point particles.

N6.3 *EXTERNAL TORQUES ON EXTENDED OBJECTS* discusses how we can determine the net torque exerted on an extended object due to forces that act at certain localized points on the object.

N6.4 *STATICS PROBLEMS* discusses why both the net force *and* the net torque on an extended object must be zero if the object is to be completely at rest, and looks at several basic examples.

N6.5 *A FRAMEWORK FOR STATICS PROBLEMS* presents a framework (like the frameworks in unit *C*) that helps guide you through the solution of statics problems, and illustrates its use with several examples.

N6.6 *TORQUES ON ROTATING OBJECTS* (optional) explains why applying a torque to a rotating object causes it to *precess*, and looks at a few practical applications of this idea.

N6.2 TORQUE

We have previously defined the force that an interaction exerts on an object to be the rate at which that interaction transfers linear momentum to that object:

$$\vec{F} \equiv \frac{đ\vec{p}}{dt} \tag{N6.1}$$

(where, as you may remember, the bar through the $đ$ indicates that $đ\vec{p}$ is not the total change in the object's momentum but rather the part of the change due to this particular interaction). Analogously, we define the **torque** that an interaction exerts on an object to be the rate at which that interaction transfers *angular momentum* to that object:

Definition of torque

$$\vec{\tau} \equiv \frac{đ\vec{L}}{dt} \tag{N6.2}$$

where $đ\vec{L}$ is the interaction's contribution to the object's total angular momentum during the time dt. (The small, curly "t" that we use as the standard symbol of torque is the Greek letter *tau*.) I like to think of torque as expressing the interaction's "twisting effectiveness" around a certain axis: that is, the rate at which the interaction causes the object's angular momentum around the axis to *change*.

Derivation of the link between torque and force

If an object participates in several interactions 1, 2, 3, ..., the total changes in the linear and angular momenta of such a particle during a very short dt are

$$d\vec{p} = đ\vec{p}_1 + đ\vec{p}_2 + \ldots \quad \text{and} \quad d\vec{L} = đ\vec{L}_1 + đ\vec{L}_2 + \ldots \tag{N6.3}$$

respectively. This means that

$$\frac{d\vec{p}}{dt} = \frac{đ\vec{p}_1}{dt} + \frac{đ\vec{p}_2}{dt} + \ldots = \vec{F}_1 + \vec{F}_2 + \ldots \equiv \vec{F}_{\text{net}} \tag{N6.4a}$$

$$\frac{d\vec{L}}{dt} = \frac{đ\vec{L}_1}{dt} + \frac{đ\vec{L}_2}{dt} + \ldots = \vec{\tau}_1 + \vec{\tau}_2 + \ldots \equiv \vec{\tau}_{\text{net}} \tag{N6.4b}$$

Now, we can use the cross-product definition of $\vec{L}$ to express the torque on a point particle in a form that will help us develop a more intuitive understanding of the idea of torque. As we saw in chapter C13, the angular momentum $\vec{L}$ of a point particle at an instant of time can be written

$$\vec{L} \equiv \vec{r} \times \vec{p} \tag{N6.5}$$

where $\vec{r}$ is the particle's position vector relative to the point around which the angular momentum is being measured. Now, the derivative of a cross product of two vectors obeys a product rule analogous to the product rule for the derivative of a product of two numbers: if $\vec{u}$ and $\vec{w}$ are two arbitrary vectors, then

$$\frac{d}{dt}(\vec{u} \times \vec{w}) = \frac{d\vec{u}}{dt} \times \vec{w} + \vec{u} \times \frac{d\vec{w}}{dt} \tag{N6.6}$$

(Note that since the cross product is not commutative, it is important that the two vectors appear in each term on the right in the same order as on the left.)

Exercise N6X.1: Use the product and rule sum rules for ordinary scalar functions and the fact that $(\vec{u} \times \vec{w})_x = u_y w_z - u_z w_y$, prove that the x-component of equation N6.6 is correct. (The proof for each of the other components of the equation is analogous.)

Equations N6.4 through N6.6 then imply that

$$\vec{\tau}_{\text{net}} \equiv \frac{d\vec{L}}{dt} = \frac{d}{dt}(\vec{r} \times \vec{p}) = \vec{v} \times \vec{p} + \vec{r} \times \vec{F}_{\text{net}} \tag{N6.7}$$

Exercise N6X.2: Prove equation N6.7.

Since $\vec{v}$ and $\vec{p}$ are parallel for any point particle, and since the cross product of parallel vectors is zero, this expression reduces to:

$$\vec{\tau}_{\text{net}} = \vec{r} \times \vec{F}_{\text{net}} \tag{N6.8}$$

By equations N6.4 and the distributive property of the cross product, this can also be written

$$\vec{\tau}_{\text{net}} \equiv \vec{\tau}_1 + \vec{\tau}_2 + \ldots = \vec{r} \times (\vec{F}_1 + \vec{F}_2 + \ldots) = \vec{r} \times \vec{F}_1 + \vec{r} \times \vec{F}_2 + \ldots \tag{N6.9}$$

We can therefore identify the torque exerted by each individual interaction as being equal to $\vec{r} \times \vec{F}$ for that interaction:

$$\vec{\tau}_1 = \vec{r} \times \vec{F}_1, \quad \vec{\tau}_2 = \vec{r} \times \vec{F}_2, \text{etc.} \tag{N6.10}$$

Therefore, we can say quite generally (whether we are talking about individual interactions or the net effect of all interactions) that

Torque of an interaction expressed in terms of that interaction's force

$$\vec{\tau} = \vec{r} \times \vec{F} \tag{N6.11}$$

This equation also applies to the net torque exerted *on* the center of mass of an extended object by external interactions, since the center of mass of an extended object responds to any net external force as if it were a point particle.

Physical implications

If we take the magnitude of both sides of this equation, we get

$$\tau = rF\sin\theta \tag{N6.12}$$

This formula tells us that an interaction's effectiveness in increasing the particle's angular momentum around a certain point O is proportional to the magnitude F of the *force* that it exerts on the particle, the *distance* r that particle is from the point around which the angular momentum is being determined, and the *sine of the angle* θ between the directions of the force and the particle's position vector relative to O. Specifically, the interaction is *most* effective at transferring angular momentum when the angle between the force and the particle's position vector is $\pi/2$ (= 90°), and totally ineffective at doing so when the angle between the force and position vector is either 0 (= 0°) or π (= 180°).

Because of equations N6.11 and N6.12, torque is most often expressed in units of N·m. (While 1 N·m = 1 J, expressing torque in terms of an energy unit would be misleading, since energy and torque are completely different concepts.)

Exercise N6X.3: Show that the basic definition of torque given by equation N6.2 is also consistent with the idea that torque has units of N·m.

Exercise N6X.4: Imagine that you hold a 0.10-kg ball at rest a vertical distance of 50 cm above the floor and 40 cm horizontally away from your left big toe. If you release the object, what is the torque that gravity exerts on the object around your big toe just after you release it? Does this torque change with time?

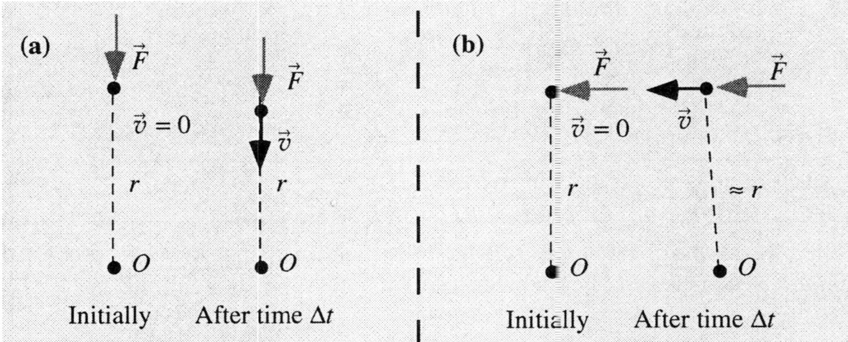

Figure N6.1: (a) A radial force does not change the particle's perpendicular velocity and thus does not change its angular momentum. **(b)** The momentum transferred by a perpendicular force goes entirely into changing the particle's perpendicular velocity, thus providing the maximum change in its angular momentum per time.

Intuitive meaning of torque

You may be able to to understand the meaning of $\vec{\tau} = \vec{r} \times \vec{F}$ more intuitively if you consider the drawings shown in Figure N6.1. Consider a particle initially at rest a distance r from the point O. A force exerted on the particle radially toward or away from point O will accelerate the particle toward or away from O, but does not change the component $v_\perp$ of the object's velocity that is perpendicular to the object's position vector $\vec{r}$: $v_\perp$ remains zero (Figure N6.1a). Therefore, the magnitude $L = mr|v_\perp|$ of the particle's angular momentum (see equation C13.7) remains zero, no matter what the magnitude of the force is. Thus if the force is parallel to the particle's position vector it causes no change in the particle's angular momentum: so $\vec{\tau} = 0$. On the other hand, a force exerted perpendicular to the particle's radius vector $\vec{r}$ (Figure N6.1b) will go exclusively to increasing the particle's perpendicular velocity $v_\perp$ and thus produce the maximum rate of change in the particle's angular momentum $\vec{L}$.

The rate at which such a force increases $v_\perp$ is independent of how far the particle is from O ($|dv_\perp/dt| = F/m$ if $\vec{F}$ is perpendicular to $\vec{r}$), so for a given force, the rate $dL/dt = mr(dv_\perp/dt)$ at which the particle's angular momentum changes is proportional to r. Therefore, the torque exerted by a given force must be directly proportional to r, consistent with the equation $\vec{\tau} = \vec{r} \times \vec{F}$.

The *direction* of torque

Note that torque is a vector quantity. The direction of the torque exerted by an interaction is found by the right hand rule for the cross product: if you point your right fingers in the direction of the vector $\vec{r}$ and curl them toward $\vec{F}$; your thumb indicates the direction of $\vec{\tau}$ (which is perpendicular to both $\vec{r}$ and $\vec{F}$).

In Figure N6.1b, the direction of the torque is perpendicular to the plane of the drawing, pointing directly towards us as we look at the drawing. We can interpret this direction as the axis around which the particle would increase its rotation rate if it were affected by this torque alone.

N6.3 EXTERNAL TORQUES ON EXTENDED OBJECTS

Door example

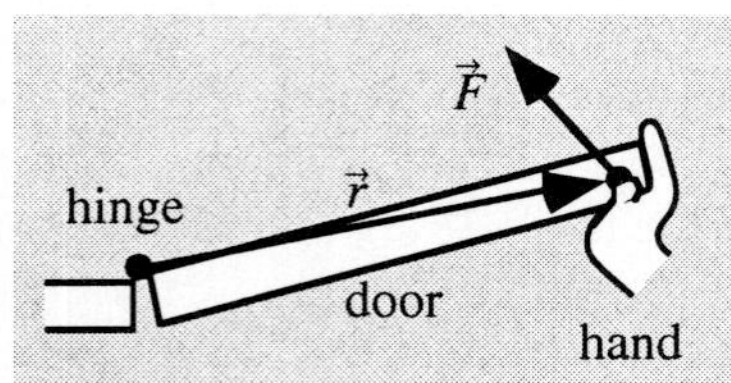

Figure N6.2: Pushing on a door with one's hand.

Imagine a door that is initially at rest, but that is free to rotate around the axis defined by its hinges. Imagine that we push on the door with our hand, exerting a force $\vec{F}$ on the door that is fairly localized at a point whose position relative to the nearest point O on the door's axis is $\vec{r}$ (see Figure N6.2). What will this do to the door?

When you push on a particular part of the door, the force that you exert transfers angular momentum to the particles in the door that your hand touches. The total rate at which your push transfers angular momentum around the point O to these particles is, according to equation N6.11, simply $\vec{\tau} = \vec{r} \times \vec{F}$. If the door is at all rigid, the particles in the door that you touch will transfer this angular momentum to the rest of the door via internal interactions, causing the door to turn on its hinges. Since these *internal* interactions do not change the total angular momentum of the door around O, any change in the door's total angular momentum around O ultimately comes from the angular momentum that you initially supply at a rate $\vec{\tau} = \vec{r} \times \vec{F}$ to a few of the door's particles at a specific location.

Is this model consistent with your many years of experience with doors? Think about it! You *know* that the most effective way to open a door is to exert a force $\vec{F}$ perpendicular to the plane of the door, and that pulling or pushing on the door directly toward or away from its hinges (parallel to the plane of the door) does nothing to help turn the door on its hinges. This is consistent with the formula $\tau = rF\sin\theta$ for the magnitude of $\vec{\tau} = \vec{r} \times \vec{F}$: according to this formula, the "turning effectiveness" around the hinges of the force that you exert depends on the sine of the angle θ between the force $\vec{F}$ and the position vector $\vec{r}$ (relative to the hinges) of the point where you are pushing. This means that a force exerted perpendicular to the plane of the door (and thus $\vec{r}$) should be more effective than one exerted parallel to the plane of the door.

Similarly, you know that if you push a door at a point far from the hinges it is easier to open than if you push at a point very close to the hinges. This is again consistent with $\tau = rF\sin\theta$: to transfer angular momentum to the door at a given desired rate, the strength of the force that you have to exert is inversely proportional to the distance from the hinges of the point on which you push.

Torque exerted by a force on an extended object

This is a specific illustration of a general principle. *A force $\vec{F}$ applied to an extended object at a specific point whose position is $\vec{r}$ relative to some point O transfers angular momentum around O to the extended object at the rate*

$$\vec{\tau} = \frac{d\vec{L}}{dt} = \vec{r} \times \vec{F} \tag{N6.13}$$

This equation has the same form but not quite the same meaning as equation N6.11. In equation N6.11, we were talking about the torque exerted on a *point particle*; here we are talking about the force exerted on an *extended object.* In equation N6.11, $\vec{r}$ referred to the *position of the particle* relative to the reference point O; here it refers to the localized *position* (relative to O) *where the force is applied* to the object.

The following examples illustrate applications of this idea.

EXAMPLE N6.1:

Problem: A 120-kg object hangs from a cable connected to a winch. When the winch operates, it winds the cable on a spool 16 cm in diameter. How much torque does the motor have to apply to the spool in order to lift the object at a constant speed?

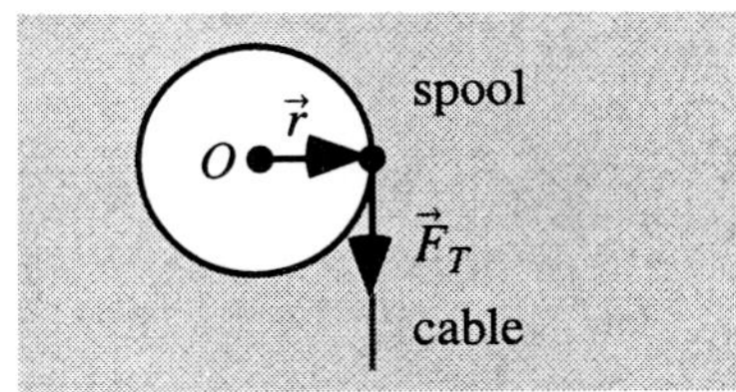

Figure N6.3: Cable applying torque to a winch spool.

Solution Figure N6.3 shows a drawing of the situation. The cable exerts a tension force on the winch spool equal in magnitude to the hanging object's weight: $F_T = mg$. Since the tension force is exerted on the spool at a distance of $r = 0.08$ m from the spool's center, and since this tension force is perpendicular to the radius vector from the spool's center to the point of application (so that $\sin\theta$ between $\vec{F}$ and $\vec{r}$ is 1) the cable applies to the spool a clockwise torque with magnitude

$$\tau = rF_T\sin\theta = rF_T = rmg = (0.08\text{ m})(120\text{ kg})(9.8\text{ m/s})$$

$$= 94\ \text{kg}\cdot\text{m}^2/\text{s}^2\left(\frac{1\text{ N}}{1\ \text{kg}\cdot\text{m/s}^2}\right) = 94\text{ N}\cdot\text{m} \tag{N6.14}$$

Since the spool's angular momentum does not change as the weight is lifted at a constant speed, the motor must supply angular momentum to the spool at exactly the same rate as the tension force on the spool drains it away. Therefore the motor must provide a *counterclockwise* torque of 94 N·m.

EXAMPLE N6.2:

Problem: A disk with a mass of 3.0 kg and a radius of 12 cm is freely spinning at a speed of 15 turns per second. Imagine that you grab the rim of the disk with your hand. What is the magnitude of the friction force you would have to apply to bring the disk to rest within 2.0 s?

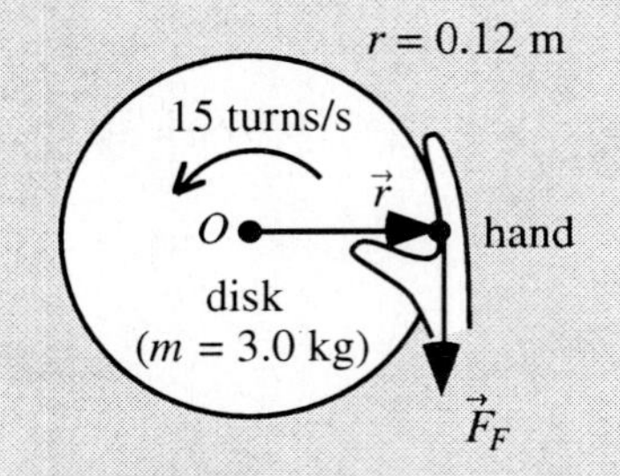

Figure N6.4: Slowing a rotating disk by hand.

Solution Figure N6.4 shows a diagram of the situation. Take the reference point O to be the disk's center of mass. Since the friction force is applied at the disk's rim, the distance between O and the place where the force is applied is $r =$ 12 cm. The direction of the friction force is tangent to the rim and thus perpendicular to the position vector $\vec{r}$ between O and where the force is applied. The torque exerted by the friction force thus has a magnitude of

$$\tau = rF_F \sin\theta = rF_F \tag{N6.15}$$

The direction of this torque, according to the right-hand rule for the cross product, is perpendicular to the plane of the drawing and *away* from the viewer (note that this is opposite to the direction of $\vec{L}$ and thus will act to *decrease* it).

The disk's initial angular momentum has a magnitude of

$$\vec{L}_i = I\vec{\omega}_i = \tfrac{1}{2}mr^2\vec{\omega}_i \tag{N6.16}$$

where $\vec{\omega}_i$ is its angular velocity and I is the moment of inertia for a disk (which I have looked up in chapter C9). The direction of this angular momentum is perpendicular to the plane of the drawing and toward the viewer. If we assume that the applied torque is constant (so that $d\vec{L}/dt = \Delta\vec{L}/\Delta t$), the torque needed to bring the disk to rest in 2.0 s is

$$\vec{\tau} \equiv \frac{d\vec{L}}{dt} = \frac{\Delta\vec{L}}{\Delta t} = \frac{\vec{L}_f - \vec{L}_i}{\Delta t} = \frac{0 - \vec{L}_i}{\Delta t} = \frac{-\frac{1}{2}mr^2\vec{\omega}_i}{\Delta t} \tag{N6.17}$$

The direction of the torque that we *need* to apply is perpendicular to the plane of the drawing and away from the viewer (opposite to $\vec{L}_i$), which is the same as the direction of torque *actually applied* by the friction force (it is good to check this!). Therefore we simply need to make the magnitudes consistent:

$$rF_F = \tau = \frac{\frac{1}{2}mr^2\omega_i}{\Delta t} \quad \Rightarrow \quad F_F = \frac{mr\omega_i}{2\Delta t} \tag{N6.18}$$

We know everything on the right of the last equation, so it is simply a matter of plugging in numbers:

$$F_F = \frac{(3.0\ \text{kg})(0.12\ \text{m})(15\ \text{rev/s})}{2(2.0\ \text{s})}\left(\frac{2\pi\ \text{rad}}{\text{rev}}\right)\left(\frac{1}{1\ \text{rad}}\right)\left(\frac{1\ \text{N}}{1\ \text{kg}\cdot\text{m/s}^2}\right) = 8.5\ \text{N} \tag{N6.19}$$

Exercise N6X.5: Imagine that you wrap a string around a 5.0-kg spool whose radius is 5.0 cm and that is free to rotate around its axis of symmetry. If you then pull on this string with a constant force whose magnitude is 2 N, how long will it take to spin the spool up from rest to a rotation rate of 10 turns/s?

N6.4 STATICS PROBLEMS

Conditions for an object to be completely at rest

Imagine an extended object that has various external forces ($\vec{F}_1, \vec{F}_2, \ldots$) applied to it at various different places. If the object is completely at rest, it means (1) that the object's center of mass is at rest, implying that *the net external force on the object is zero*:

$$\vec{F}_1 + \vec{F}_2 + \ldots = m\vec{a} = 0 \qquad \text{(N6.20a)}$$

and (2) that the object is not rotating, so its angular momentum is not changing, implying that *the net torque on the object is zero*:

$$\vec{r}_1 \times \vec{F}_1 + \vec{r}_2 \times \vec{F}_2 + \ldots = \frac{d\vec{L}}{dt} = 0 \qquad \text{(N6.20b)}$$

around any origin you might choose. Both of these equations must apply simultaneously to any extended object at rest.

A problem involving an extended object at rest is called a **statics problem**. Application of both equations N6.20 in a statics problem often enables us to find the magnitude of contact forces that would seem impossible to determine at first glance. The examples below illustrate such problems.

EXAMPLE N6.3:

Problem: Two objects with masses m_1 and m_2 hang from the ends of a comparatively lightweight rod of length L. At what position (relative to m_1) would you want to suspend the rod so that the rod could hang horizontally at rest?

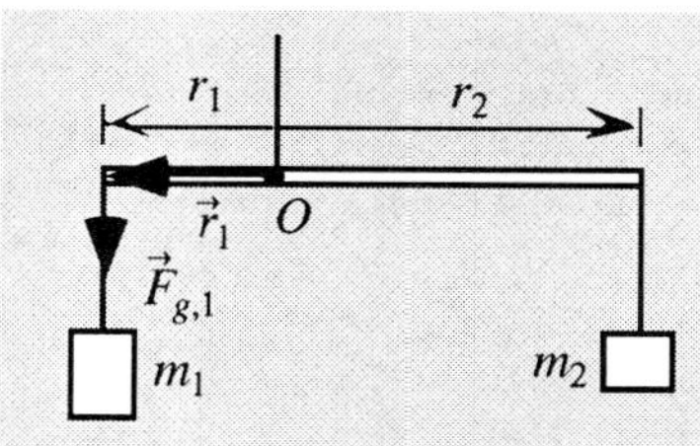

Figure N6.5: Where should we attach the wire to suspend this rod?

Solution Figure N6.5 illustrates the situation. Imagine that we suspend it from a point O a distance r_1 from where the object with mass m_1 is suspended (the distance to the point from which the object with mass m_2 is suspended is then $r_2 = L - r_1$). The rod will only remain at rest if the net torque on the rod around the suspension point O is zero.

Now, you can see that the weight of object 1 will exert a counterclockwise torque on the rod (that is, $\vec{\tau}$ points directly out of the plane of the picture) while the other weight exerts a clockwise torque (that is, directly into the plane of the picture). Since they point in opposite directions, these torques will add to zero if they have the same magnitude. If the rod is horizontal, the force exerted on the rod by each hanging mass is perpendicular to the rod, and thus perpendicular to the position vector from O to the place that the force is applied. This means that the sine of the angle θ between $\vec{r}$ and $\vec{F}$ in each case is 1, and the magnitudes of each torque is simply the product of the magnitude of the force and the distance between O and the point where the force is applied. Requiring that these torques have equal magnitudes therefore means that:

$$r_1 F_1 = r_2 F_2 \qquad \text{(N6.21)}$$

But the magnitudes of the applied forces in this case are simply the magnitudes of the objects' weights ($F_1 = m_1 g$ and $F_2 = m_2 g$), so equation N6.21 becomes

$$r_1 m_1 g = r_2 m_2 g \quad \Rightarrow \quad r_1 m_1 = r_2 m_2 \qquad \text{(N6.22)}$$

So we want the suspension point to be where the distance to each mass is inversely proportional to the mass. Plugging $r_2 = L - r_1$ into N6.22, we get:

$$r_1 m_1 = (L - r_1) m_2 \quad \Rightarrow \quad r_1(m_1 + m_2) = L m_2$$

$$\Rightarrow \quad r_1 = \frac{m_2}{(m_1 + m_2)} L \qquad \text{(N6.23)}$$

If you look back at chapter C4 you will see that this is the same as the position of the system's center of mass! This perhaps should come as no big surprise.

The external forces on the rod are the upward tension force F_T exerted by the supporting wire and the downward forces exerted by the weights. Requiring that the total external force be zero means that the total upward force should balance the total downward force: $F_T = (m_1 + m_2)g$ (again, this is no surprise).

EXAMPLE N6.4:

Problem: A uniform plank of mass m and length L sits on two supports, one a distance $2L/5$ and the other a distance $L/5$ from the plank's center. What are the magnitudes of the normal forces that each exerts on the plank (in terms of mg)?

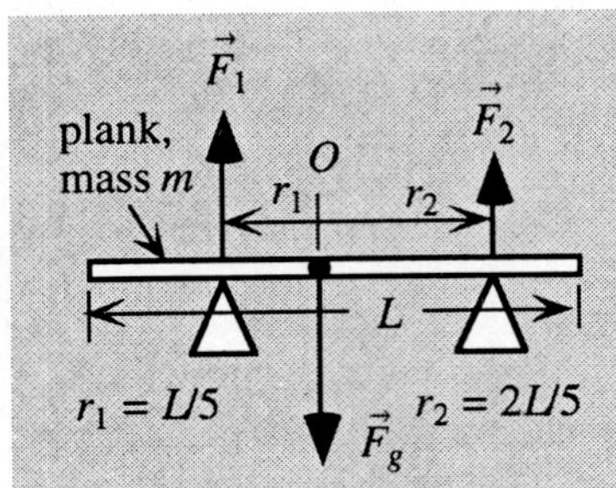

Figure N6.6: How to support a plank.

Solution The situation is shown in Figure N6.6. There are three forces acting on the plank: the two supporting forces and the plank's weight (which can be considered to act on the plank's center of mass). If we take origin O to be the plank's center of mass, then the torque that weight exerts around the center of mass is zero (the weight essentially acts *on* the plank's center of mass, so $\vec{r} = 0$ for that force). The left support exerts a torque into the plane of the paper, while the right support exerts a torque out of the plane of the paper, so these torques will cancel if their magnitudes are equal. Since the normal forces exerted by the supports act perpendicular to the plank, the magnitude of the torque exerted by each normal force will be simply rF, where F is the magnitude of the force and r is the distance between its point of application and the origin. So if we require that the magnitudes of the torques exerted by the supports be equal, we have

$$r_1 F_1 = r_2 F_2 \tag{N6.24}$$

We know that $r_1 = L/5$ and $r_2 = 2L/5$, but this is not enough information to find F_1 and F_2 separately. How can we proceed?

Well, since the center of mass of the plank is not moving, the three forces acting on the plank must add to zero: $\vec{F}_1 + \vec{F}_2 + \vec{F}_g = 0$. Since $\vec{F}_1$ and $\vec{F}_2$ both act upward and $\vec{F}_g$ acts downward, this will happen if

$$F_1 + F_2 = F_g = mg \tag{N6.25}$$

If we solve this for F_2 and plug the result into N6.24, we can solve for F_1:

$$r_1 F_1 = r_2(mg - F_1) \quad \Rightarrow \quad F_1(r_1 + r_2) = r_2 mg$$

$$F_1 = \frac{r_2}{r_1 + r_2} mg = \frac{2L/5}{3L/5} mg = \frac{2}{3} mg \tag{N6.26}$$

Plugging this back into equation N6.25, we find that $F_2 = \frac{1}{3} mg$.

Exercise N6X.6: We do not *have* to choose the origin O to be the plank's center of mass. Redo the plank problem, taking point O to be the position of the left-hand support. (You should get the same answer.)

Exercise N6X.7: A plank 4.0 m long and having a negligible mass is supported by its ends. A person with a weight of 480 N stands 1.0 m from one end. What is the magnitude of the upward force exerted by each support?

N6.5 A FRAMEWORK FOR STATICS PROBLEMS

Comparison of statics worksheet to others

Examples N6.5 and N6.6 on the next two pages illustrate a framework for solving statics problems (see the left-hand column in each). Please study the framework outline carefully. You will see that it is similar in structure to the frameworks we used in unit C, but there are some important differences.

In particular, drawing a free-body diagram is *very* helpful for these kinds of problems, and, in statics problems, it is most efficient to draw this diagram in the context of its surroundings as part of the pictorial representation. Note also that it is essential to define clearly the origin that you will use to calculate torques. Finally, note that for complex problems like these, expressing $\vec{F}_{net} = 0$ in full component form is almost *essential* for keeping things straight. I also think that it helps if you write things out first using abstract components (for example, $F_{T1,x}$) and then rewrite each such component in terms of a vector magnitude and an angle (for example, $F_{T1} \cos\theta_1$).

EXAMPLE N6.5

OUTLINE OF THE FRAMEWORK for STATICS problems

1. Pictorial Representation

a. Draw a picture of the situation that includes:

① (1) a free-body drawing of the object in the context of its surroundings

② (2) a clearly-labeled origin

③ (3) reference frame axes

④ (4) labels defining symbols for relevant quantities

⑤ b. List values for all known labeled quantities and specify which quantities are unknown.

2. Conceptual Representation

The general task is to construct a conceptual model of the situation and link it to an abstract physics model or principle. In statics problems, we do most of this work when constructing the free-body diagram. In addition, we should

⑥ a. Explain (if it is not completely obvious) why you have drawn the force arrows on your free-body diagram with the locations and directions that you have

⑦ b. Describe any approximations or assumptions you have to make to solve the problem

3. Mathematical Representation

⑧ a. Apply the appropriate component of $\vec{\tau} = 0$ (this will almost always be the component in a direction perpendicular to the drawing)

⑨ b. Apply $\vec{F}_{net} = 0$ *in vector component form*

⑩ c. Solve for any unknown quantities symbolically

⑪ d. Plug in numbers and units and calculate the result and its units.

4. Evaluation

Check that the answer makes sense:

⑫ a. Does it have the correct units?

⑬ b. Does it have the right sign?

⑭ c. Does it seem reasonable?

Problem: A 65-kg rock-climber hangs from two ropes on the face of a vertical cliff. One end of each rope is tied to the climber's harness; the other ends are anchored 5.5 m to the left and 2.1 m vertically up, and 7.2 m to the right and 1.5 m vertically up from the climber's harness, respectively. If the climber's full weight were to be supported by the ropes, what tension force must each rope exert? If the ropes can supply 2500 N of tension force without breaking, is the climber safe?

Solution A drawing and free-body diagram of the situation looks like this: ①

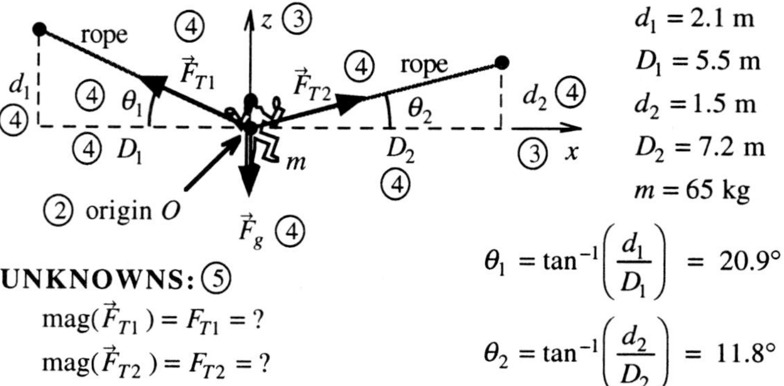

KNOWNS: ⑤

$d_1 = 2.1$ m
$D_1 = 5.5$ m
$d_2 = 1.5$ m
$D_2 = 7.2$ m
$m = 65$ kg

$$\theta_1 = \tan^{-1}\left(\frac{d_1}{D_1}\right) = 20.9°$$

$$\theta_2 = \tan^{-1}\left(\frac{d_2}{D_2}\right) = 11.8°$$

UNKNOWNS: ⑤

$\text{mag}(\vec{F}_{T1}) = F_{T1} = ?$

$\text{mag}(\vec{F}_{T2}) = F_{T2} = ?$

⑥ In this particular case, the forces all act directly on the harness at the origin, so the distances between the origin and the points where these ⑦ forces are applied are essentially zero. This means that no matter what ⑧ the magnitudes of the forces might be, the torque that they exert on the climber is essentially zero, so we don't need to consider torque further. We will assume that the ropes don't stretch very much when the climber puts weight on them, so that the distances given in the drawing re- ⑦ main accurate (this is probably not a very good assumption, but what else can we do?). Even if we can ignore the torque, the net force on the climber has to be zero too; in component form, this tells us that

$$⑨ \quad 0 = \vec{F}_{net} = \begin{bmatrix} F_{T1,x} \\ F_{T1,y} \\ F_{T1,z} \end{bmatrix} + \begin{bmatrix} F_{T2,x} \\ F_{T2,y} \\ F_{T2,z} \end{bmatrix} + \begin{bmatrix} F_{g,x} \\ F_{g,y} \\ F_{g,z} \end{bmatrix}$$

$$\Rightarrow \begin{bmatrix} 0 \\ 0 \\ 0 \end{bmatrix} = \begin{bmatrix} -F_{T1}\cos\theta_1 \\ 0 \\ F_{T1}\sin\theta_1 \end{bmatrix} + \begin{bmatrix} F_{T2}\cos\theta_2 \\ 0 \\ F_{T2}\sin\theta_2 \end{bmatrix} + \begin{bmatrix} 0 \\ 0 \\ -mg \end{bmatrix} \quad (1)$$

The top line of this equation implies that

$$F_{T1}\cos\theta_1 = F_{T2}\cos\theta_2 \Rightarrow F_{T2} = F_{T1}\frac{\cos\theta_1}{\cos\theta_2} \quad ⑩ \quad (2)$$

Plugging this into the bottom line of equation (1), we get

$$mg = F_{T1}\sin\theta_1 + F_{T1}\frac{\cos\theta_1}{\cos\theta_2}\sin\theta_2 = F_{T1}(\sin\theta_1 + \cos\theta_1\tan\theta_2) \quad (3)$$

Solving this for F_{T1} (and noting that $\tan\theta_2 = d_2/D_2$) we get:

$$F_{T1} = \frac{mg}{\sin\theta_1 + \cos\theta_1(d_2/D_2)} = \frac{(65\ \text{kg})(9.8\ \text{m/s}^2)}{\sin(20.9°) + \cos(20.9°)(1.5\ \text{m}/7.2\ \text{m})}\left(\frac{1\ \text{N}}{1\ \text{kg}\cdot\text{m/s}^2}\right) = 1160\ \text{N} \quad (4)$$

⑩ ⑪

Plugging this into equation (2), we get $F_{T2} = (1160\ \text{N})\cos(20.9°)/\cos(11.8°) = 1100$ N. ⑪ These both come out with the right units ⑫ and sign ⑬ (positive for magnitudes), and seem reasonable ⑭. The ropes *should* hold!

EXAMPLE N6.6

OUTLINE OF THE FRAMEWORK for STATICS problems

1. Pictorial Representation

a. Draw a picture of the situation that includes:

(1) (1) a free-body drawing of the object in the context of its surroundings

(2) (2) a clearly-labeled origin

(3) (3) reference frame axes

(4) (4) labels defining symbols for relevant quantities

(5) b. List values for all known labeled quantities and specify which quantities are unknown.

2. Conceptual Representation

The general task is to construct a conceptual model of the situation and link it to an abstract physics model or principle. In statics problems, we do most of this work when constructing the free-body diagram. In addition, we should

(6) a. Explain (if it is not completely obvious) why you have drawn the force arrows on your free-body diagram with the locations and directions that you have

(7) b. Describe any approximations or assumptions you have to make to solve the problem

3. Mathematical Representation

(8) a. Apply the appropriate component of $\vec{\tau} = 0$ (this will almost always be the component in a direction perpendicular to the drawing)

(9) b. Apply $\vec{F}_{net} = 0$ *in vector component form*

(10) c. Solve for any unknown quantities symbolically

(11) d. Plug in numbers and units and calculate the result and its units.

4. Evaluation

Check that the answer makes sense:

(12) a. Does it have the correct units?

(13) b. Does it have the right sign?

c. Does it seem reasonable?

Problem: A 580-kg drawbridge 6.6 m long spans a moat around a castle. One end of the drawbridge is connected by a hinge to the castle wall. The drawbridge is also held up by two chains, which are both connected to a point 4.4 m from the hinge. The bridge is raised by reeling in the chains, which enter the castle wall 4.4 above the hinge. When the bridge is raised just above its support on the far side of the moat but is still essentially horizontal, what is the magnitude of the tension force that each chain must exert on the drawbridge? Also find the components of the force that the hinge must exert on the drawbridge.

Solution: Here is a drawing / free-body diagram of the situation: (1)

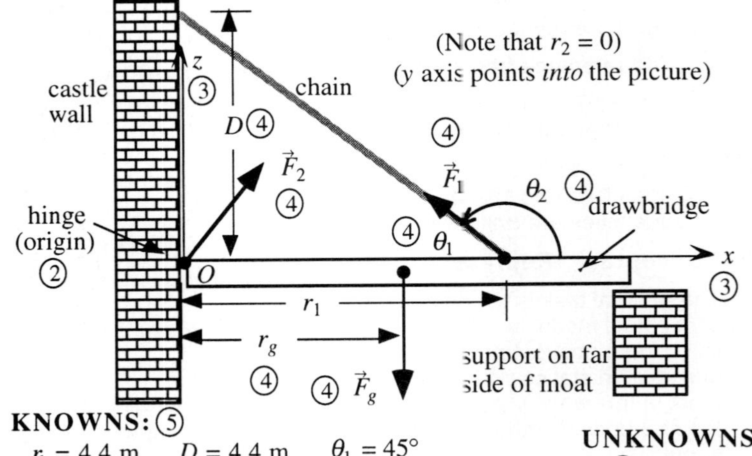

KNOWNS: (5)

$r_1 = 4.4$ m $\quad D = 4.4$ m $\quad \theta_1 = 45°$

$r_2 = 0 \quad m = 580$ kg $\quad \theta_2 = 180°–45° = 135°$

$r_g = \frac{1}{2}(6.6\text{ m}) = 3.3$ m

UNKNOWNS: (5) $F_1, \vec{F}_2$

(6) $\vec{F}_2$ here is the contact force exerted by the hinge on the drawbridge: we know nothing about its magnitude or direction. I am assuming that the (7) drawbridge is uniform, so that its center of mass is located halfway along its length. The net torque on the drawbridge has to be zero, so

$$(8) \quad 0 = \vec{r}_1 \times 2\vec{F}_1 + \vec{r}_2 \times \vec{F}_2 + \vec{r}_g \times \vec{F}_g \tag{1}$$

(The factor of 2 is because we have two chains.) We have conveniently located the origin so that $r_2 = 0$, which means that the second term in equation (1) is zero. The chains exert a torque that points *out* of the picture (the $-y$ direction), while the drawbridge's weight exerts a torque in the opposite ($+y$ direction). Therefore, $\tau_y = 0$ in this case means that $0 = -r_1(2F_1)\sin\theta_2 + r_g mg$, so

$$(10) \quad F_1 = \frac{r_g mg}{2r_1 \sin\theta_2} = \frac{(3.3\text{ m})(580\text{ kg})(9.8\text{ m/s}^2)}{2(4.4\text{ m})\sin(135°)}\left(\frac{1\text{ N}}{1\text{ kg}\cdot\text{m/s}^2}\right) = 3000\text{ N} \quad (11) \tag{2}$$

(13) (12)

This is positive (as a magnitude should be), has the right units, and (14) seems plausible. We must also have $0 = \vec{F}_{net} = 2\vec{F}_1 + \vec{F}_2 + \vec{F}_g$ for the drawbridge to remain at rest, which in component form reads

$$(9) \quad \begin{bmatrix} 0 \\ 0 \\ 0 \end{bmatrix} = 2\begin{bmatrix} F_{1x} \\ F_{1y} \\ F_{1z} \end{bmatrix} + \begin{bmatrix} F_{2x} \\ F_{2y} \\ F_{2z} \end{bmatrix} + \begin{bmatrix} F_{gx} \\ F_{gy} \\ F_{gz} \end{bmatrix} = \begin{bmatrix} -2F_1\cos\theta_1 \\ 0 \\ 2F_1\sin\theta_1 \end{bmatrix} + \begin{bmatrix} F_{2x} \\ F_{2y} \\ F_{2z} \end{bmatrix} + \begin{bmatrix} 0 \\ 0 \\ -mg \end{bmatrix}$$

$$(10) \Rightarrow \begin{bmatrix} F_{2x} \\ F_{2y} \\ F_{2z} \end{bmatrix} = -\begin{bmatrix} -2F_1\cos\theta_1 \\ 0 \\ 2F_1\sin\theta_1 \end{bmatrix} - \begin{bmatrix} 0 \\ 0 \\ -mg \end{bmatrix} = \begin{bmatrix} +2(3000\text{ N})\cos(45°) \\ 0 \\ (580\text{ kg})(9.8\text{ m/s}^2) - 2(3000\text{ N})\sin(45°) \end{bmatrix} = \begin{bmatrix} 4200\text{ N} \\ 0 \\ 1400\text{ N} \end{bmatrix} \quad (11) \tag{3}$$

(12) The units all come out right (since $1\text{ kg}\cdot\text{m/s}^2 = 1\text{ N}$), and $\vec{F}_2$ is in the first quadrant (as we guessed in the drawing: this is the only way that the vectors can add to zero). The magnitudes are comparable to F_1. (14) (13)

N6.6 TORQUES ON ROTATING OBJECTS (optional)

When a net torque around point O acts on a non-rotating object at rest, the effect of the torque is to give the object an angular momentum around the point O (which may show up as angular momentum of the object's center of mass, angular momentum of rotation around the object's center of mass, or both). The angular momentum the object gains will point in the same direction as the applied torque. So, for example, a counterclockwise torque applied to a disk at rest will cause the disk to rotate counterclockwise at an ever-increasing rate.

A torque can change the orientation of an object's axis of rotation

If the object is *already* rotating, things are more complicated. The definition $\vec{\tau} = d\vec{L}/dt$ implies that the direction of a torque specifies the direction in which the object's angular momentum *changes*. Therefore if the torque is not aligned with the object's initial angular momentum, the object's angular momentum will change *in the direction of the applied torque*. This can have some strange consequences, as the examples below illustrate.

Spinning top example

For example, you know that a spinning top will not fall over, whereas the same top will simply topple over if it is *not* spinning. Why is this? Imagine that you set a top on a table so that its bottom touches the table and its axis of symmetry makes an angle θ with the vertical. Then you release the top. What happens if the top is not spinning? What happens if the top is spinning?

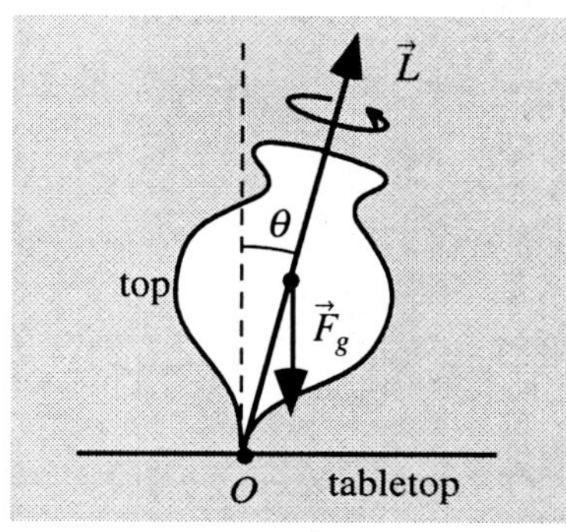

Figure N6.7: A leaning top. The torque around O exerted by the graviational force points directly *into* the plane of the picture.

Figure N6.7 shows a picture of the situation. Let us take the point where the top touches the tabletop to be the point O we use as a reference for computing torques. After you release the top, there are two forces acting on it: a gravitational force which can be considered to act on the top's center of mass and a compression force exerted by the table on the top at point O. The latter force exerts zero torque around point O (since the distance between O and the point where this force is applied is zero). The gravitational force, on the other hand, exerts a torque on the top around point O whose direction points directly *into* the plane of the drawing in Figure N6.7.

If the top is *not* spinning, this torque will simply give the top an angular momentum in the same direction, which (by the right hand rule) will cause it to rotate clockwise around point O with increasing angular speed. We describe what happens to the top as it responds to this torque as "falling over."

On the other hand, if the top is spinning, it already has an angular momentum $\vec{L}$ parallel to its axis of symmetry. In an infinitesimal time dt, the torque applied by the gravitational force will thus *change* the top's angular momentum by $d\vec{L} = \vec{\tau}dt$, causing the top's axis of rotation to shift slightly *into the plane of the drawing* in Figure N6.7.

But by equation N6.11, the torque exerted by the gravitational force is always strictly perpendicular to the vertical force of gravity and the top's axis of symmetry, which is also the direction of its angular velocity $\vec{L}$. So the torque always acts perpendicular to the vertical plane that contains $\vec{L}$, and as $\vec{L}$ shifts direction in response to the torque, the direction of $\vec{\tau}$ shifts as well, always remaining both horizontal and perpendicular to $\vec{L}$.

You should be able to convince yourself that the net effect of this changing torque is to move the tip of the angular momentum vector $\vec{L}$ around in a circle as time passes (see Figure N6.8). This means that instead of falling over, the top's axis of rotation will "wobble" around the vertical direction (maintaining a constant angle θ with the vertical direction). The technical term for this kind of orderly wobbling is called **precession** around the vertical direction. The rate that the top precesses depends on the magnitude of the applied torque.

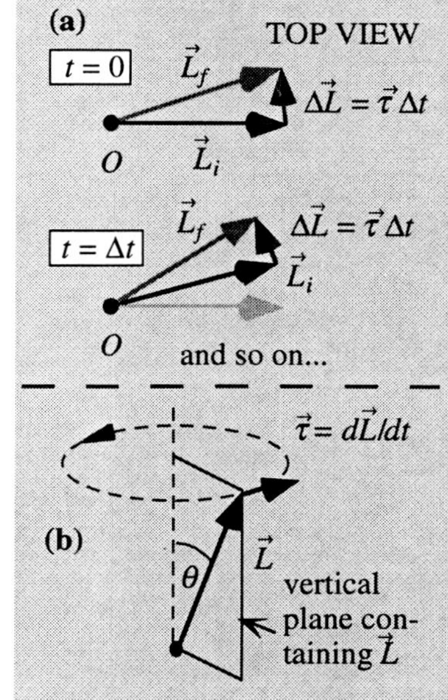

Figure N6.8: (a) At each instant of time, the torque on the top causes the top's $\vec{L}$ to change in a direction perpendicular to $\vec{L}$. **(b)** This causes the tip of the $\vec{L}$ vector to move in a circle (*precess*) about the vertical direction.

Exercise N6X.8: Do all tops necessarily precess counterclockwise, as shown in Figure N6.8? Explain your response.

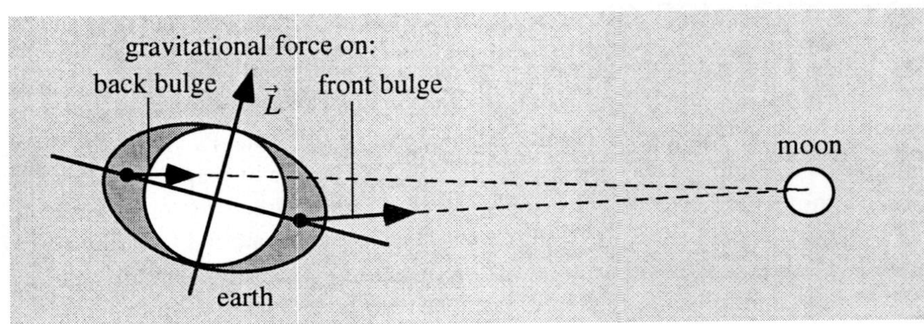

Figure N6.9: The gravitational interaction between the moon and the earth's equatorial bulges exerts a torque on the earth (that points directly out of the plane of this drawing). This torque causes the earth's axis of rotation to precess relative to the distant stars.

This idea has a number of interesting applications in realistic physical situations. For example, it has been known for many centuries that the earth's axis of rotation moves very slowly relative to the distant stars. Isaac Newton was able to offer the first explanation of this phenomenon in the *Principia*, as part of his general effort to illustrate how his laws of physics explained celestial as well as terrestrial phenomena. (If this explanation had been his *only* contribution to physics, his name would be recalled and honored by students of physics and astronomy today, but this important theoretical discovery has been all but overshadowed by his even greater contributions to physics.)

Newton's explanation of the earth's precession

Newton's explanation goes like this. The earth is not *quite* spherical: it bulges slightly at the equator. Since the moon orbits in a plane that is inclined at an angle relative to the earth's equator, the gravitational pull of the moon actually exerts a small torque on the earth around the earth's center of mass (because it pulls the closer front bulge somewhat harder and at a more perpendicular angle than the back bulge, the gravitational torques around the center of mass do not quite cancel out). This is illustrated (and greatly exaggerated) in Figure N6.9. The gravitational pull of the sun exerts another torque on the earth for the same reason.

Because the earth is rotating, this torque (which you can see acts perpendicular to the earth's axis) causes the earth to precess like a top. Because the torque is weak and the earth is very massive, this precession is rather slow: the combined effects of the moon and sun's gravitational torque cause the earth to precess once every 26,000 years.

This precession makes the axis around which the stars appear to rotate when viewed from the earth shift with time. In the northern hemisphere, the stars *currently* seem to rotate around the star we call Polaris (the North Star): this star currently marks the direction of the north celestial pole. But the North Star has only been a good approximation to the location of this pole for a handful of centuries: the precession of the earth causes the location of the pole to shift around the sky as time passes. About 3000 years ago a star named Thuban was very close to the north celestial pole. In about 14,000 years the bright star Vega in the constellation Lyra will be very close to the north celestial pole. Of course, about 26,000 years from now, Polaris will again be the North Star.

This effect is also called "the precession of the equinoxes" because as the axis of the earth shifts, the two places in the sky where the earth's equator intersects the plane of its orbit (which are called the *equinoxes*) also shift.

Skipping a frisbee

The ideas presented in this section also explain why it is possible to skip a frisbee off a hard surface. If a right-handed thrower throws a frisbee at a slight angle so that its *left* edge (as seen by the thrower) hits the ground first, then the frisbee will angle upward back into the air after it hits. Why does this work? What would happen if the right edge were to hit first?

Figure N6.10a shows the situation just as the frisbee hits the ground. The normal force between the ground and the frisbee exerts a brief torque around the frisbee's center of mass whose direction (by the right-hand rule) is directly *into* the plane of the picture. When a frisbee is thrown in the normal (backhanded)

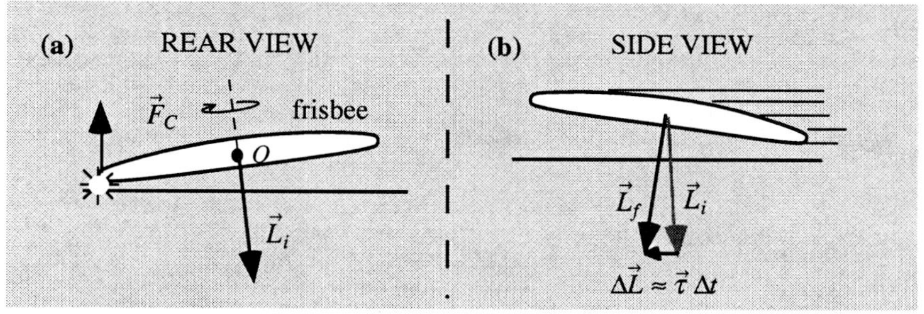

Figure N6.10: (a) When the frisbee hits the ground on its left edge, the force of impact exerts a torque around O that points forward into the plane of the diagram. **(b)** This causes the frisbee's angular momentum to shift forward, which causes it to climb back into the air.

manner by a right-handed thrower, it spins clockwise, so its angular momentum is downward, as shown in the drawing. The torque due to the impact causes the frisbee's angular momentum to shift forward, as shown in Figure N6.10b. This lifts the leading edge of the frisbee, allowing it to climb back into the air. The torque applied to the frisbee must point in the *forward* direction to do this, so the frisbee *must* strike the ground on its left edge. (For example, if the frisbee hits with its right edge, the torque will point directly *out* of the plane of Figure N6.10a, which causes the frisbee to dig into the ground.)

SUMMARY

I. TORQUE
- A. *Torque* expresses the rate at which an interaction transfers angular momentum to an object: $\vec{\tau} \equiv d\vec{L}/dt$ (analogous to $\vec{F} \equiv d\vec{p}/dt$)
- B. For a point particle, $\vec{\tau} \equiv \vec{r} \times \vec{F}$, where $\vec{r}$ is the particle's position relative to a given origin, $\vec{F}$ is the force applied by the interaction, and $\vec{\tau}$ is torque around the origin due to that interaction
  1. This follows from $\vec{L} \equiv \vec{r} \times \vec{p}$ and product rule for cross product
  2. Because of this formula, the SI units of torque are N·m
  3. The magnitude of $\vec{\tau}$: $\tau = rF\sin\theta$
     - a) θ here is the angle between the directions $\vec{r}$ and $\vec{F}$
     - b) thus forces acting parallel to $\vec{r}$ do nothing, while forces acting perpendicular to $\vec{r}$ have the maximum effect
  4. The *direction* of the torque vector is defined by a right-hand rule
     - a) if you point your right fingers along $\vec{r}$ and curl them toward $\vec{F}$, then your right thumb indicates the direction of $\vec{\tau}$
     - b) this indicates the axis around which the torque would cause an object to rotate if it acted alone on the object at rest
- C. $\vec{\tau} \equiv \vec{r} \times \vec{F}$ also applies to the CM of an extended object if we interpret $\vec{F} = \vec{F}_{\text{net,ext}}$ (since the CM responds to $\vec{F}_{\text{net,ext}}$ like a particle would)

II. EXTERNAL TORQUES ON EXTENDED OBJECTS
- A. When a force is applied to a specific point on an extended object:
  1. The torque applied around a given origin O is $\vec{\tau} \equiv \vec{r} \times \vec{F}$, where $\vec{r}$ is the position (relative to O) where force is applied
  2. We can choose any convenient origin O
- B. Looks like the equation for a particle, but doesn't mean quite the same thing ($\vec{r}$ here is position where force is applied, not particle position)

III. STATICS PROBLEMS
- A. When an extended object is completely at rest
  1. All external forces must add to zero (since $\vec{p}_{\text{tot}}$ does not change)
  2. All external torques must add to zero (since $\vec{L}$ does not change)
- B. We can use this information to find unknown forces, locate balance points, and so on
- C. In such problems, we can treat gravity as if it acts at center of mass
- D. See examples N6.5 and N6.6 for a framework for solving statics problems (similar to the frameworks we used in unit *C*)

GLOSSARY

torque $\vec{\tau}$**:** the torque exerted by an interaction on an object is defined to be the rate at which that interaction transfers to that object angular momentum around some specified origin O: $\vec{\tau} \equiv d\vec{L}/dt$. This turns out to be equal to $\vec{\tau} = \vec{r} \times \vec{F}$, where $\vec{F}$ is the force exerted by the interaction on the object and $\vec{r}$ is the position where the force is applied relative to the origin. The direction of the torque vector specifies the axis around which the torque would cause the object to rotate if the object were at rest and the torque in question were the only one acting.

statics problem: a problem involving an extended object that is completely at rest, where we determine the magnitudes and/or directions of unknown forces by applying the conditions that the net force and net torque on the object must be zero.

precession: In some cases, a spinning object's angular momentum vector responds to an external torque by rotating around a certain axis while keeping its projection on that axis constant. We call this kind of motion *precession*.

TWO-MINUTE PROBLEMS

N6T.1 Imagine that you are the pitcher in a baseball game. The batter hits a foul ball vertically in the air. If the ball has a weight of 2 N and an initial upward velocity of about 30 m/s, and you are 40 m from where the ball is hit, what is the gravitational torque (magnitude and direction) on the ball around you just after it is hit?
A. 2400 N·m upward
B. 2400 N·m to your left
C. 2400 N·m to your right
D. 80 N·m upward
E. 80 N·m to your left
F. 80 N·m to your right
T. other (specify)

N6T.2 As the ball described in the previous problem continues to rise, the magnitude of the torque on the ball around you due to the ball's weight
A. increases B. remains the same C. decreases

N6T.3 A cylinder rolls without slipping down an incline directly toward you. The contact interaction between the cylinder and the incline exerts a static friction torque on the cylinder around the cylinder's center of mass. What is the direction of this torque?
A. toward you
B. away from you
C. to your right
D. to your left
E. upward
F. downward

N6T.4 Imagine that you are looking down on a turntable that is spinning counterclockwise. If an upward torque is applied to the turntable, its angular speed
A. increases B. decreases C. remains the same

N6T.5 A wheel of radius 50 cm rotates freely on an axle of radius 0.5 cm. If you wanted to slow the wheel to rest with your hand, you can either exert a friction force with your hand on the wheel's rim (call the magnitude of this force F_{rim}) or exert a friction force on the wheel's axle (call the magnitude of this force F_{axle}). If you had to bring the wheel to rest in 2.0 s either way, how would the force that you'd have to exert on the rim compare to the force that you'd have to exert on the axle?
A. $F_{\text{rim}} = 100F_{\text{axle}}$
B. $F_{\text{rim}} = 10F_{\text{axle}}$
C. $F_{\text{rim}} = F_{\text{axle}}$
D. $10F_{\text{rim}} = F_{\text{axle}}$
E. $100F_{\text{rim}} = F_{\text{axle}}$
F. other (specify)
T. depends (specify on what)

N6T.6 A board of mass m lays on the ground. What is the magnitude of the force that you would have to exert to lift *one end* of the board barely off the ground (assuming that the other end still touches the ground)?
A. $2mg$
B. mg
C. Depends on the length of the board
D. $\frac{1}{2}mg$
E. other (explain)

N6T.7 Imagine that you continue to lift the board described in the previous problem. Assume that the force you exert is always perpendicular to the board, and that one end of the board always remains on the ground. What happens to the magnitude of the force you exert on the end as the angle between the board and the ground increases? It
A. increases B. decreases C. remains the same

HOMEWORK PROBLEMS

BASIC SKILLS

N6B.1 A sky diver whose mass is 65 kg drops vertically past a motionless balloon. If the diver is 120 m away from the balloon at the point of closest approach, what is the torque on the diver around the balloon due to gravity at that time? Does the value of the torque increase, decrease or remain the same as the diver continues to fall? Explain.

N6B.2 A diver whose mass is 42 kg jumps vertically off the high diving board into the water while you watch from the side of the pool. When the diver enters the water, you are 12 m away. What is the torque around you that gravity exerts on the diver just before he or she enters the water? The diving board is 3.0 m above the water. What is the torque around you that gravity exerts on the diver just after he or she leaves the board?

N6N.3 Using the concept of torque, explain why using a long wrench makes it easier to loosen a stubborn nut or bolt than using a short wrench would.

N6B.4 A certain electric motor can exert a maximum torque of 110 N·m. Imagine that you lift a weight with this motor by wrapping a cable connected to the weight around the motor's axle. If the axle has a diameter of 1.0 cm, what is the maximum mass that the motor can lift this way? Please explain your response.

N6B.5 A wheel 60 cm in radius is connected to an axle that is 1.5 cm in radius. A 25-kg object hangs from a thin cord that is wrapped around the axle. How much force do you have to exert on the wheel's rim to hold it still? Please explain your response.

N6B.6 A 5.0-kg disk with a radius of 0.40 m increases its rotational speed from 2.0 turns/s to 3.5 turns/s during a time period of 3.0 s. What average torque was supplied to the disk during this time period?

N6B.7 A 6-kg bowling ball with a radius of 14 cm is spinning around a vertical axis (with its center of mass motionless) at an initial rate of 3 turns/s. Friction with the floor, however, causes the bowling ball to rotate slower and slower until it stops after about 22 s. What was the average torque that the friction interaction applied to the bowling ball?

SYNTHETIC

N6S.1 Consider a 0.2-kg hockey puck at rest 5.0 m due north of point O. Imagine that you give it a whack that exerts 10 N of force due west that lasts 0.1 s.
(a) How fast does the puck's perpendicular velocity $v_\perp$ *change* during this time (that is, what is the component of its acceleration perpendicular to the puck's position)?
(b) What is the total change in its perpendicular velocity?
(c) Use $L = mr|v_\perp|$ to find the magnitude of the change in the puck's angular momentum around O during this time.
(d) Use equation N6.12 to find the torque on the object.
(e) Use your answer to part *d* to find the total change in the puck's angular momentum during this time. Does your answer agree with part *c*? *Should* it agree?

N6S.2 A basketball with a mass of 0.15 kg and a radius of 0.15 m is thrown to you rotating 5 times a second. What is the magnitude of the frictional force that you have to exert on it with your hands to stop it rotating within 0.1 s when you catch it? (*Hint:* The moment of inertia of a spherical shell is $\frac{2}{3}MR^2$.)

N6S.3 A 3.0-kg sphere with a radius of 5.0 cm rolls from rest without slipping 3.0 m down an incline, reaching a speed of 3.0 m/s after rolling for 2.0 s. What is the average torque that the static friction interaction between the sphere and the incline exerts on the sphere?

N6S.4 A plank 3.0 m long and having a mass of 20 kg is supported at its ends. Imagine that a person with a mass of 50 kg stands 1.0 m from one end. What is the magnitude of the force exerted by the supports?*

N6S.5 Imagine that you have a plank 2.0 m long. You put a small piece of wood under the plank 60 cm from its far end, making the plank into a lever. A friend with a mass of 65 kg stands on that end of the plank. How much downward force do you have to exert on your end to lift your friend?*

N6S.6 Imagine that a pole with a mass of 8 kg and a length 1.8 m is connected to a wall so that the pole sticks out horizontally from the wall. One end of the pole is connected directly to the wall, while the other end is connected to a higher point on the wall by a chain that makes a 45° angle with respect to the pole. If a 65-kg person hangs from the center of the pole, what is the tension on the chain?*

N6S.7 A certain board is 4.0 m long and rests horizontally and somewhat above the ground on two cylinders of wood, each supporting the board 0.5 m from the corresponding end of the board. (The axis of each cylinder is perpendicular to the length of the board.) The board has a mass of 21 kg. If a person with a mass of 68 kg steps on one end of the board, will it support that person or not?*

N6S.8 A board has one end wedged under a rock with a mass of 380 kg, and is supported by another rock that contacts the bottom side of the board at a point 85 cm from the end under the rock. The board is 4.5 m long, has a mass of about 22 kg and projects essentially horizontally out over a river. Is it safe for an adult with a mass of 62 kg to stand at the unsupported end of the board? If not, how far out on the board can one safely go?*

N6S.9 A 5-m ladder of negligible mass leans against the side of a house, with the bottom of the ladder 2.5 m from the wall. The contact interaction between the ladder and the ground is able to exert a horizontal static friction force of no more than $0.4mg$, where mg is the weight of the person on the ladder. Assume that the contact interaction between the wall and the top of the ladder is essentially frictionless. How high can a person safely climb on this ladder without running the risk of the ladder slipping out from under him or her? (*Hint*: the answer is independent of the person's mass.)*

N6S.10 A ladder of length L and mass m leans against the side of a house, making an angle of θ with the vertical. Assume that the ladder is free to slide at the point where it touches the side of the house (there is no significant friction). Find an expression for the normal force that the side of the house exerts on that end of the ladder in terms of m, g, L, and θ.*

*For these starred problems, be sure to follow the steps in the problem-solving framework described in section N6.5 as you write up your solution.

RICH-CONTEXT

N6R.1 A person's forearm consists of a bone (which we can model by a rigid rod) that is free to rotate around the elbow. When the forearm is held level with the elbow bent at a 90° angle, the bone is supported by a the biceps muscles, which are attached by a tendon to the bone a few centimeters away from the axis represented by the elbow. By making measurements on your own forearm and appropriate approximations and estimations (be sure that you describe them), estimate the tension force on the biceps tendon when you hold an object with a mass of 10 kg in your hand (with your forearm level and elbow bent at 90°).

N6R.2 You are part of a research team in the deepest jungles of South America. At one point, the team has to cross a 20-m wide gorge by stringing a rope across it and then crossing hand over hand while hanging from the rope. After someone has managed to hook the rope on a rock on the other side of the gorge, a relatively new member of the team ties the near end of the rope to a tree in such a way that the rope is very tight. A more experienced member makes the neophyte untie the rope and retie it so that there is some slack. Why? If the rope can exert a tension force of 3500 N without breaking and the heaviest member of the group has a mass of 110 kg, how much slack do you need? Is the tension on the rope worse when a person hanging near one side of the gorge or in the middle? Assume that the rope doesn't stretch much when under tension. (*Hint*: This *is* a statics problem but it does *not* involve torque.)

ADVANCED

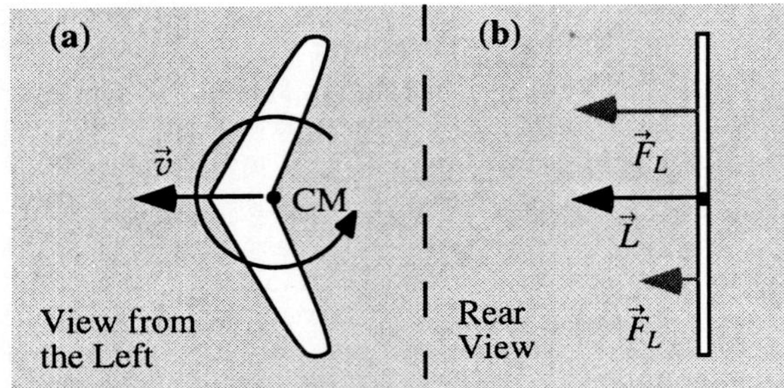

Figure N6.11: (a) A boomerang just as it is released, as viewed by a person on the thrower's left. **(b)** The same boomerang as seen by the thrower (showing the different lift forces on the two wings).

N6A.1 (Study section N6.6 before doing this problem.) Imagine that you throw a boomerang vertically so that its axis of rotation is horizontal and the boomerang rotates counterclockwise when viewed by someone to your left (see Figure N6.11). Because each blade moves faster with respect to the air when it is on top than it does when it is on the bottom, whichever blade is on top will exert more lift than the blade on the bottom. This exerts a torque on the boomerang around its center of mass. What will this do to the boomerang's axis of rotation? (*Hint*: Due to this effect, the boomerang travels in a horizontal circle back to the thrower. See "The Aerodynamics of Boomerangs" in the November 1968 issue of *Scientific American* for more details.)

N6A.2 (Study section N6.6 before doing this problem.) Consider a top consisting of a massive disk of mass M and radius R pierced through the center by lightweight rod perpendicular to the disk. The rod has a length of $2L$ and its center coincides with the center of mass of the disk (so that the rod sticks symmetrically out of each face of the disk). Imagine that you give this top an initial angular velocity $\vec{\omega}$ around the axis defined by the rod, and then place one end of the rod on the ground so that the rod makes an angle θ with respect to the vertical. Determine the period of this top's precession (that is, the number of seconds required for each precession) as a function of the quantities M, R, L, ω, and θ. (*Hint*: Study Figure N6.8 carefully. How does the fraction of the circle that the top precesses in a given short time Δt depend on the magnitude of the top's angular momentum and the applied torque?)

ANSWERS TO EXERCISES

N6X.1 Using $(\vec{u}\times\vec{w})_x = u_y w_z - u_z w_y$, we have

$$\frac{d}{dt}(\vec{u}\times\vec{w})_x = \frac{d}{dt}(u_y w_z - u_z w_y)$$

$$= \frac{du_y}{dt}w_z + u_y\frac{dw_z}{dt} - \frac{du_z}{dt}w_y - u_z\frac{dw_y}{dt} \qquad \text{(N6.27)}$$

according to the product rule of ordinary calculus. Reordering terms, we get

$$= \frac{du_y}{dt}w_z - \frac{du_z}{dt}w_y + u_y\frac{dw_z}{dt} - u_z\frac{dw_y}{dt}$$

$$= \left(\frac{d\vec{u}}{dt}\times\vec{w}\right)_x + \left(\vec{u}\times\frac{d\vec{w}}{dt}\right)_x \qquad \text{N6.28)}$$

which is the x component of equation N6.6.

N6X.2 By the product rule for the cross product, we get

$$\frac{d}{dt}(\vec{r}\times\vec{p}) = \frac{d\vec{r}}{dt}\times\vec{p} + \vec{r}\times\frac{d\vec{p}}{dt} \qquad \text{(N6.29)}$$

But $d\vec{r}/dt = \vec{v}$ by definition and $d\vec{p}/dt = \vec{F}_{net}$ by Newton's second law. Plugging these into the equation above, we get equation N6.7.

N6X.3 $\vec{L}$ has units of kg·m²/s, so $d\vec{L}/dt$ must have units of kg·m²/s² = (kg·m/s²)·m = N·m.

N6X.4 The gravitational force $\vec{F}_g$ is vertically downward. The object's initial $\vec{r}$ points upward at an angle θ such that $\theta = \tan^{-1}(50\text{ cm}/40\text{ cm}) = 51°$, and this $\vec{r}$ and its angle change with time as the object drops. But the component of $\vec{r}$ that is perpendicular to $\vec{F}_g$ (that is, the horizontal component) is a constant 40 cm, so $\tau = rF_g\sin\theta = r_\perp F_g = (0.40\text{ m})(0.10\text{ kg})(9.8\text{ m/s}^2) \approx 0.4$ N·m, *independent* of time.

N6X.5 about 3.9 s.

N6X.6 Answer is given. (*Hints:* Note that the torque of left-hand support is now zero, but also that gravity now exerts a torque around the new C. Take the gravitational force to act at the plank's center of mass.)

N6X.7 near support exerts 360 N; far support, 120 N.

N6X.8 If top is initially rotating clockwise instead of counterclockwise, then it will also precess clockwise. To see this, consider Figure N6.8a: in this top view, $\vec{L}$ now points to the *left* of O, and an upward $\Delta\vec{L}$ causes $\vec{L}$ and thus axis of rotation change toward the clockwise direction.

N7

LINEARLY CONSTRAINED MOTION

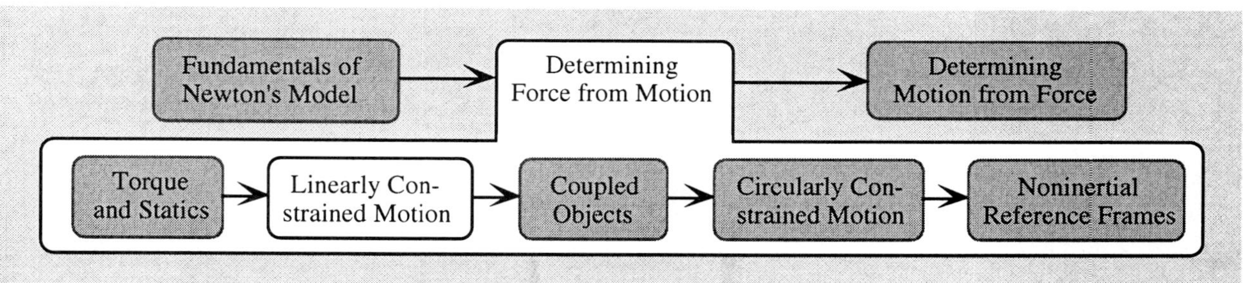

N7.1 OVERVIEW

In the last chapter, we examined problems where an object of interest was completely at rest, and used the implication that both the net force and net torque on the object are zero to determine the forces acting on the object. In this chapter, we continue our study of how we can determine forces from motion by examining cases where the object of interest is either at rest or *constrained to move in a straight line*. We will see that the constraint of linear motion *also* allows us to draw conclusions about the forces acting on the object.

Here is an overview of the sections in this chapter.

N7.2 *FREE-PARTICLE DIAGRAMS* presents a variant of the *free-body diagram* introduced in chapter N3 that in complicated situations can help us to apply Newton's second law more accurately.

N7.3 *MOTION AT A CONSTANT VELOCITY* argues that since an object observed to be moving at a constant velocity is not accelerating, the forces acting on the object must *still* add up to zero, and explores an example of this kind of problem.

N7.4 *STATIC AND KINETIC FRICTION FORCES* discusses the quantitative behavior of these common forces.

N7.5 *DRAG FORCES* explores the quantitative behavior of the contact force that opposes an object's motion through a fluid (liquid or gas).

N7.6 *LINEARLY ACCELERATED MOTION* looks at cases where the object of interest is accelerating in one dimension, and shows how a judicious choice of coordinate system can greatly simplify such problems.

N7.7 *A CONSTRAINED-MOTION FRAMEWORK* presents a variant of our problem-solving framework that you can use as a guide to solving constrained motion problems in this chapter (and through chapter N9).

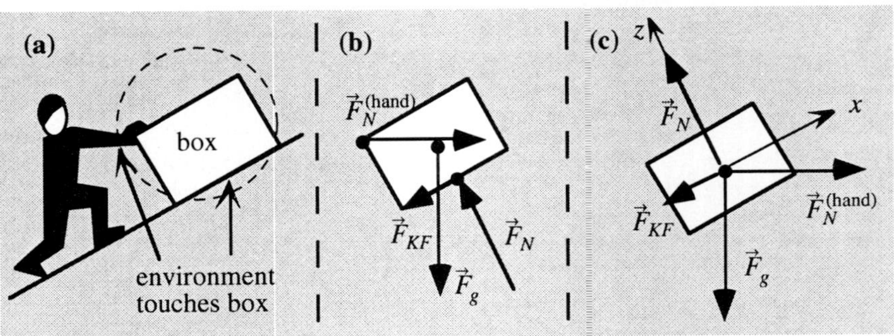

Figure N7.1: (a) A person pushes a large box up an incline. Seeing where the environment touches the box helps us determine and classify the forces acting on it. **(b)** A free-body diagram of the box. **(c)** A free-particle diagram of the box. Note that the tails of all vectors are placed at the box's center of mass, even though that is not actually where the force is applied.

N7.2 FREE-PARTICLE DIAGRAMS

Newton's second law

In this chapter and in much of the rest of the text, we will be exploring the consequences of applying Newton's second law to the center-of-mass motion of objects large and small. Newton's second law $\vec{F}_{\text{net,ext}} = m\vec{a}_{\text{CM}}$ tells us that the *net external force* acting on an object causes its center of mass to accelerate in inverse proportion to its total mass m. In the interests of simpler notation, in the future I will simply write Newton's second law as

$$\vec{F}_{\text{net}} = m\vec{a} \tag{N7.1}$$

and we will automatically assume that when we are talking about an extended object, $\vec{F}_{\text{net}}$ refers to the net *external* force on that object and $\vec{a}$ is the acceleration of its *center of mass*.

We will work with the component form of the law

It is crucial to note that this is a *vector* equation, and thus is equivalent to three completely independent component equations:

$$\begin{bmatrix} F_{\text{net},x} \\ F_{\text{net},y} \\ F_{\text{net},z} \end{bmatrix} = m \begin{bmatrix} a_x \\ a_y \\ a_z \end{bmatrix} \quad \text{or} \quad \begin{aligned} F_{\text{net},x} &= ma_x &\text{(N7.2a)} \\ F_{\text{net},y} &= ma_y &\text{(N7.2b)} \\ F_{\text{net},z} &= ma_z &\text{(N7.2c)} \end{aligned}$$

where $F_{\text{net},x}$ is the sum of the x components of all the (external) forces acting on the object, $F_{\text{net},y}$ is the sum of the y components of the same and so on.

A free-particle diagram helps us determine force components

To determine the components of the net external force on an object, we need to know the components of *each* of the forces acting on that object. Starting in this chapter, we will use a variant of the free-body diagram to help us do this. A **free-particle diagram** combines aspects of the free-body and net-force diagrams introduced in chapters N3 and N4 in a way that is especially well-suited to this task. A free-particle diagram consists of the following items:

1. A sketch of the object, with a dot representing its center of mass.
2. A labeled arrow representing *each* (external) force acting on the object *with its tail attached to the center of mass.* (If multiple forces act in the same direction, draw them in sequence to display the net force acting in that direction.)
3. A set of reference-frame axes

The differences between free-body and free-particle diagrams

A free-particle diagram differs from a free-body diagram in several crucial ways. First, a free-particle diagram makes *explicit* the idea that we are modeling the object as if it were a particle located at its center of mass. Second, drawing the force arrows so that they are attached to the center of mass (rather than the point where the forces are applied) makes it easier to visually read the components of these forces. It is also usually easier to see what the net force on an object will be from a free-particle diagram than a free-body diagram, making drawing a separate net-force diagram less crucial. Finally, we did not worry about the reference frame axes in chapter N3 and N4 because they were not needed for a qualitative discussion of forces, but we will need them from now on. Figure N7.1 illustrates these differences.

In the last chapter, we were concerned with the necessary conditions for an object to be at rest and also *not rotating*. Free-body diagrams, which show exactly where the external forces act on an object, are helpful in such cases because they make it easier to compute torques. From now on, however, we will be more interested in the motion of an object's *center of mass*. Information about where forces act on the object is thus not relevant (and can be distracting), and so is removed from a free-particle diagram. Because of this, a free-particle diagram is more abstract than a free-body diagram, and so you might find it easiest (particularly at first) to draw a free-*body* diagram before you make the abstractions involved in drawing a free-*particle* diagram: I encourage you to do so.

We will not be concerned with torque in what follows

We generally label the forces in a free-particle using the force-classification scheme in chapter N3. Sometimes forces arising from distinct interactions end up with the same symbol, which could be confusing. In such cases, let us follow the convention of attaching to the force symbol a superscript (in parentheses) that specifies the other object involved in the interaction exerting the force. For example, in Figure N7.1 there is a normal force acting on the box due to its contact interaction with the incline and another normal force arising from its contact interaction with the hand of the person pushing it. We distinguish the two by attaching a superscript "(hand)" to the symbol for the latter force.

N7.3 MOTION AT A CONSTANT VELOCITY

When an object moves at a constant velocity, its acceleration is zero. Newton's second law then implies that the net force acting on that object must be *zero*, just as if the object were at rest. Indeed, we can consider *rest* to be simply a special case ($\vec{v} = 0$ = constant) of constant velocity.

Constant-$\vec{v}$ problems are like statics problems

EXAMPLE N7.1

Problem: Imagine that you push a 50-kg cart with frictionless wheels at a constant speed up a ramp that makes an angle of 30° with the horizontal. If you exert your push parallel to the ramp, how hard do you have to push?

Solution Figure N7.2 shows a picture of the situation and a free-particle diagram of the cart. The two normal forces are perpendicular to each other, so if we tilt our reference frame axes so that the x axis is parallel to the incline, then each normal force has only one nonzero component. Since the cart's velocity is constant, the net force on it must be zero, which means that

$$\begin{bmatrix}0\\0\\0\end{bmatrix} = \begin{bmatrix}F_{\text{net},x}\\F_{\text{net},y}\\F_{\text{net},z}\end{bmatrix} = \begin{bmatrix}F_{g,x}\\F_{g,y}\\F_{g,z}\end{bmatrix} + \begin{bmatrix}F_{N,x}\\F_{N,y}\\F_{N,z}\end{bmatrix} + \begin{bmatrix}F_{N,x}^{(\text{hand})}\\F_{N,y}^{(\text{hand})}\\F_{N,z}^{(\text{hand})}\end{bmatrix}$$

$$= \begin{bmatrix}-mg\sin\theta\\0\\-mg\cos\theta\end{bmatrix} + \begin{bmatrix}0\\0\\+F_N\end{bmatrix} + \begin{bmatrix}+F_N^{(\text{hand})}\\0\\0\end{bmatrix} \Rightarrow \begin{matrix}0 = -mg\sin\theta + F_N^{(\text{hand})}\\0 = 0 + 0\\0 = -mg\cos\theta + F_N\end{matrix} \qquad \text{(N7.3)}$$

where $F_N^{(\text{hand})} \equiv \text{mag}(\vec{F}_N^{(\text{hand})})$, $mg = \text{mag}(\vec{F}_g)$, and $F_N \equiv \text{mag}(\vec{F}_N)$. Note that since $\phi + 90° + \alpha = 180° = 90° + \phi + \theta$, θ must be the same as the angle of the ramp: $\theta = \phi = 30°$. The top line (x component) of equation N7.3 thus tells us that

$$F_N^{(\text{hand})} = mg\sin\theta = (50\ \cancel{\text{kg}})(9.8\ \text{N}/\cancel{\text{kg}})\sin(30°) = 245\ \text{N} \qquad \text{(N7.4)}$$

This is what we were looking for. (Note that the bottom line [z component] of equation N7.3 tells us that $F_N = mg\cos\theta < mg$: the magnitude of the normal force the contact interaction with the incline exerts on the cart is thus *smaller* than the cart's weight here!)

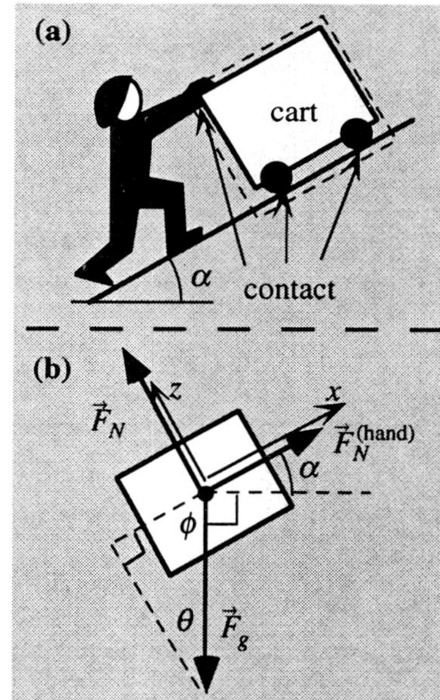

Figure N7.2: **(a)** A person pushing a cart up a ramp. **b)** A free-particle diagram for the cart in this situation.

The key to solving force-from-motion problems

The most important thing to note in example N7.1 is how I expressed Newton's second law first in column-vector form, expressed the components of each force vector in terms of the *magnitude* of that vector (keeping careful track of signs) and finally reduced the vector equation to three separate *component* equations (equation N7.3). This process is really the *key* to solving quantitative force-from-motion problems.

How to orient frame axes when $\vec{a} = 0$

Note also that when we solve a problem where the object's acceleration is zero, it is usually most convenient to orient our reference frame so that as many force components as possible are zero. I did this in example N7.1 by tilting the reference frame so that the x axis is parallel to the incline.

N7.4 STATIC AND KINETIC FRICTION FORCES

Many realistic situations involving objects at rest or objects moving at constant speeds involve *friction* forces. The purpose of this section is to explore the quantitative nature of static and kinetic friction between two solid objects.

Static friction

Imagine the following situation. A large and heavy box sits on the floor of your room. Imagine that you push on the box in a direction parallel to the floor. What happens? You know from experience that if you don't push hard enough, you can exert a steady horizontal force on the box and yet it doesn't move.

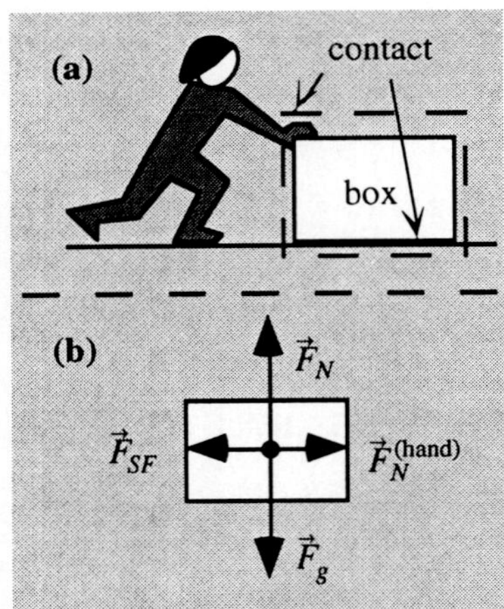

Figure N7.3: (a) Pushing on a box at rest. **(b)** A free-particle diagram of the same.

How can we reconcile this behavior with Newton's second law? Since the box remains at rest, it is not accelerating, and therefore there must be *zero* net force on the box. Since you are pushing horizontally forward on the box, there must be another force acting backwards on the box that exactly balances the force that you exert. But what exerts this force, and why does it seem to automatically adjust its value to match however hard we push on the box?

The following model helps us understand this force. When you place a box on the floor, the box's atoms and the floor's atoms come into contact, microscopic hills in the box's surface become interlocked with microscopic valleys in the floor and vice versa, and some of the box's atoms actually become "cold welded" to floor atoms. When you then push the box parallel to the floor, the interactions between the atoms of the box and the atoms of the floor under the box push the floor atoms forward, and thus (by Newton's third law) also push the box atoms backward. We call the total force exerted on the box due to this interaction between cold-welded atoms a *static friction* force.

This force adjusts its magnitude to keep the box at rest. If you push the box harder forward, the interaction intensifies, pushing the floor atoms farther forward and the box atoms more strongly backward. If you ease up, the interaction becomes less intense, allowing atoms on the surface of both the floor and box to relax back closer to their original positions. Thus, however hard you push, the static friction force adjusts to exactly cancel the sum of horizontal components of the other forces acting on the box, keeping the box at rest.

Kinetic friction

However, this only works up to a point. If you push hard enough, the box will suddenly start to slide. Once it starts sliding, you can keep the box moving at a constant velocity by exerting a steady (and usually smaller) force.

Again, how do we reconcile this behavior with Newton's second law? Here we are exerting a steady horizontal force on the box, and yet it is only moving with a constant velocity, not accelerating (there may be only a brief moment of acceleration as the box began to move). This means that there *still* must be some horizontal force acting to cancel the effect of your push. This force is the *kinetic friction force between* the rubbing surfaces.

Here is a model for this force. When you finally get the box moving, the microscopic hills in the box's surface have to push past microscopic hills in the floor's surface. When two hills come in contact, they momentarily weld together and deform as the surfaces continue to move before ultimately breaking apart. As the hills on the box and floor surfaces meet and deform, their interaction exerts a force on the box's hills that opposes the deformation. The net result of the tiny

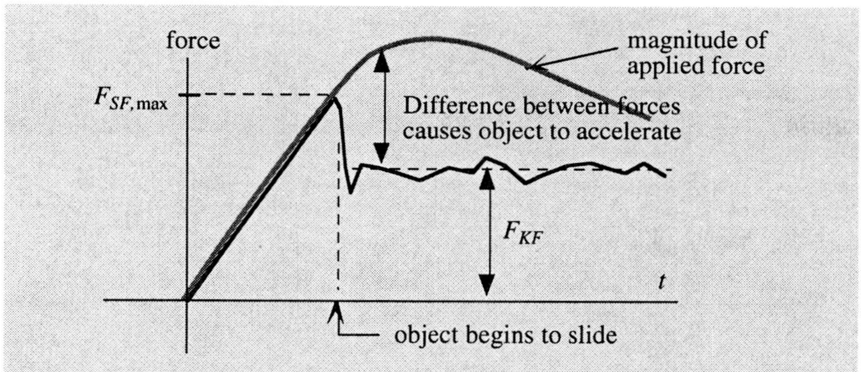

Figure N7.4: A graph illustrating the behavior of the static and kinetic friction forces. If we manually apply a horizontal force to an object sitting on a surface, the magnitude of the static friction will at first increase in step with that of the applied force until a certain maximum is reached, after which the object suddenly begins to slide and the friction force drops to the smaller kinetic friction value for sliding surfaces. (Usually a person pushing will ease up after the object starts to move, which is why the applied force shown turns over.)

forces exerted by untold numbers of interacting hills is a (roughly) steady force opposing box's motion.

F_{KF} is usually less than the maximum value of F_{SF}

It turns out that while the magnitude of this friction force can depend weakly on the speed of the box, it is generally *less* than the magnitude of force required to start the box moving at first (the welds that form on the fly are not as strong as ones that can form when surfaces are at rest relative to each other). This means that you have to exert a lot of force to get the box going, but once it is, not as much force is required to *keep* it going.

This phenomenon is illustrated by the graph shown in Figure N7.4. This graph illustrates what happens when you slowly and steadily increase a manually-applied horizontal force $\vec{F}$ to a motionless box. At first, the magnitude of the static friction force due the box's interaction with the floor increases exactly in step with that of applied horizontal force, keeping the box at rest. But as the magnitude of your applied force continues to increase, there comes a point where the bonds between floor atoms and box atoms cannot stretch any further without breaking: at this point the atoms are exerting the maximum possible static friction force that they can (we call the magnitude of this force $F_{SF,\max}$). If you push still harder, these bonds break, and the horizontal force exerted by the contact interaction suddenly drops to the lower kinetic friction value. There is now a difference in magnitude between the horizontal applied force and the opposed friction force that allows the box to accelerate from rest. The box will continue to accelerate until either the kinetic friction force has increased (with increasing speed) to match your applied force or (more likely) you ease up automatically on the box so as to match your push to its new lower friction force.

This experimentally-observed behavior has some interesting implications. *Antilock brakes* on cars prevent the brakes from locking the cars' wheels, so that the tires stay locked to the road instead of skidding. Since the maximum static friction force that the road can exert on the tires is greater than the kinetic friction force that it can exert if the tires are skidding, antilock brakes enable the driver to brake the car more rapidly (as well as keep the car under better control).

$F_{SF,\max}$ is proportional to F_N (for same interaction)

Empirically, the maximum magnitude of the static friction force that a contact interaction can exert seems to be roughly *proportional* to the magnitude of the normal force exerted by the same contact interaction:

$$\text{mag}(\vec{F}_{SF}) \leq \text{mag}(\vec{F}_{SF,\max}) \approx \mu_s \text{mag}(\vec{F}_N) \tag{N7.5}$$

where μ_s is a constant of proportionality we call the **coefficient of static friction**. This makes some sense, because the normal force reflects how strongly the surfaces are pressed together and thus how extensively the surface atoms are likely to interpenetrate and bond together.

Note that μ_s is unitless (since $\vec{F}_{SF}$ and $\vec{F}_N$ have the same units). Its value depends on the nature of the interacting surfaces and ranges from values like 1.1 (for two nickel surfaces in contact) to 0.04 (for two Teflon surfaces in contact). Values of μ_s for common interacting surfaces are typically between 0.4 and 0.6.

Since the normal force on an object sitting on a surface typically (but not always!) cancels the object's weight, the applied force needed to get an object moving is thus often proportional to the object's weight. This is the newtonian explanation for the intuitive idea that one has to exert a force strong enough to "overcome an object's inertia" before it starts to move. This aristotelian notion thus works well enough in many cases, but it fails to explain why even a tiny force suffices to accelerate an object in space or in an otherwise frictionless environment. If a freight train car could be mounted on sufficiently frictionless wheels, even a toddler could cause it to accelerate!

F_{KF} is proportional to F_N (for same interaction)

A *similar* relation seems to adequately model the relatively steady magnitude of the kinetic friction forces exerted by surfaces moving relative to each other

$$\text{mag}(\vec{F}_{KF}) \approx \mu_k \,\text{mag}(\vec{F}_N) \tag{N7.6}$$

where we call μ_k the **coefficient of kinetic friction**. Again the value of μ_k depends on the nature of the interacting surfaces (and sometimes weakly on speed) and is generally somewhat less than the coefficient of static friction.

Though they look very similar, there is an important difference between equations N7.5 and N7.6. Equation N7.6 expresses the *actual value* of the kinetic friction force (which has a fairly fixed value), while equation N7.5 expresses an *upper limit* on the adjustable static friction force.

EXAMPLE N7.2

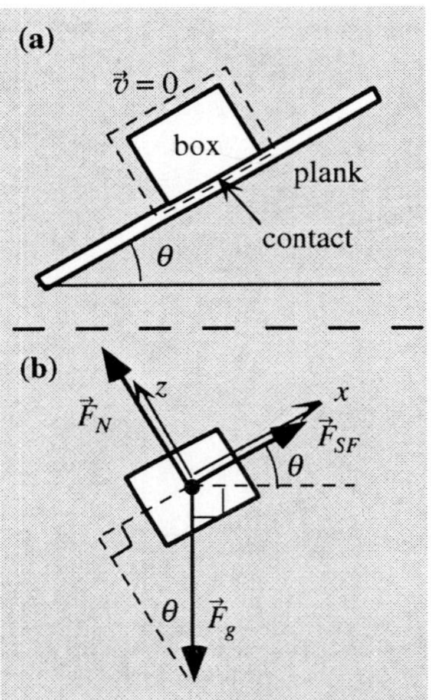

Figure N7.5: (a) A box sitting on an inclined plank. **(b)** A free-body diagram for the situation.

Problem: Imagine that a box sits on a plank of wood. If we gradually lift one end of the plank, we find that the box suddenly starts to slide down the plank when the plank's angle with the horizontal reaches 28°. What is the coefficient of static friction between the box and the plank?

Solution Figure N7.5 illustrates the situation where the plank is inclined at an angle of $\theta < 28°$ and the box is still at rest. Since the box is at rest, it is not accelerating, so the net force acting on it must be zero: $\vec{F}_{\text{net}} = 0$. If we orient our reference frame as shown in Figure N7.5b, then $0 = \vec{F}_{\text{net}}$ becomes:

$$\begin{bmatrix} 0 \\ 0 \\ 0 \end{bmatrix} = \begin{bmatrix} F_{g,x} \\ F_{g,y} \\ F_{g,z} \end{bmatrix} + \begin{bmatrix} F_{N,x} \\ F_{N,y} \\ F_{N,z} \end{bmatrix} + \begin{bmatrix} F_{SF,x} \\ F_{SF,y} \\ F_{SF,z} \end{bmatrix} = \begin{bmatrix} -mg\sin\theta \\ 0 \\ -mg\cos\theta \end{bmatrix} + \begin{bmatrix} 0 \\ 0 \\ +F_N \end{bmatrix} + \begin{bmatrix} +F_{SF} \\ 0 \\ 0 \end{bmatrix} \tag{N7.7}$$

where $F_{SF} \equiv \text{mag}(\vec{F}_{SF})$, $F_N \equiv \text{mag}(\vec{F}_N)$, and $mg = \text{mag}(\vec{F}_g)$. Solving the top and bottom lines (the x and z components) of this vector equation for the unknown force magnitudes F_{SF} and F_N, we find that

$$F_{SF} = mg\sin\theta \quad \text{and} \quad F_N = mg\cos\theta \tag{N7.8}$$

The magnitude of the static force will reach its maximum value at $\theta_{\max} = 28°$. According to equation N7.5,

$$F_{SF,\max} = \mu_s F_N \quad \Rightarrow \quad \mu_s = \frac{F_{SF\max}}{F_N} \tag{N7.9}$$

So setting $\theta = \theta_{\max}$ in equations N7.8 and plugging them into N7.9, we get

$$\mu_s = \frac{\cancel{mg}\sin\theta_{\max}}{\cancel{mg}\cos\theta_{\max}} = \tan\theta_{\max} = \tan(28°) = 0.53 \tag{N7.10}$$

Exercise N7X.1: When the plank in example N7.2 is inclined at 20°, what is the ratio of the magnitude of the actual static friction force on the box to the maximum possible value $\mu_s F_N$ for a box sitting on a plank at this angle?

Exercise N7X.2: Imagine that you have to exert a horizontal force of 49 N on a 10-kg box to get it to begin to slide along a level floor. What is the coefficient of static friction between the box and the floor?

N7.5 DRAG FORCES

The drag on large object moving quickly through air

While the kinetic friction force between two solid objects is fairly independent of relative speed, the drag force experienced by an object as it moves through a fluid depends sharply on speed. Empirically, the magnitude of the drag force on most objects moving through air is given by the formula

$$F_D = \text{mag}(\vec{F}_D) = \tfrac{1}{2}C\rho A v^2 \qquad \text{(N7.11)}$$

where ρ is the density of air (1.20 kg/m^3 at normal pressure and room temperature), A is the object's cross-sectional area, v is its speed relative to the air, and C is an empirical quantity called the **drag coefficient** that depends on the shape of the object. ($C \approx 0.5$ for a sphere, is smaller for more streamlined shapes, and can be as large as 2 for irregular shapes.)

Exercise N7X.3: What are the SI units for C?

Exercise N7X.4: A certain streamlined car has a frontal area of about 3.0 m^2 and a drag coefficient of 0.3. What is the approximate magnitude of the drag force exerted on the car when it travels at a speed of 25 m/s (55 mi/h)?

The drag on objects that are very small, move very slowly, and/or are in a viscous medium (like water)

Equation N7.11 accurately models the drag force on most sports projectiles and other large objects moving through air. The drag force on objects that are very small, move very slowly, and/or move through a more viscous fluid (like water) turns out to be proportional to v and thus is better described by

$$F_D = \text{mag}(\vec{F}_D) = bv \qquad \text{(N7.12)}$$

where b is some constant with units of N·s/m = kg/s.

Exercise N7X.5: If the forward force on a motorboat has to be 550 N to drive it through the water at a constant speed of 5 m/s, what is the the constant b for that particular motorboat?

N7.6 LINEARLY ACCELERATED MOTION

Frame alignment is key to solving problems with $\vec{a} \neq 0$

Consider now an object moving along a straight line with nonzero acceleration. Newton's second law $\vec{F}_{\text{net}} = m\vec{a}$ implies that the net force on the object must point in the same direction as its observed acceleration. This means that analyzing the motion of an object becomes *much* simpler if we *orient our reference frame axes so that one axis coincides with the direction of the observed acceleration*. Then the two components of the net force perpendicular to this axis are zero (just as in a constant velocity problem) and analysis of the object's motion becomes a one-dimensional problem.

Some constrained-motion problems involving accelerating objects are *hybrid* problems in that we use information about the motion to determine things about the forces acting on the object, which in turn we use to determine things about the linear motion of the object (such as its acceleration). The following example illustrates such a problem.

EXAMPLE N7.3

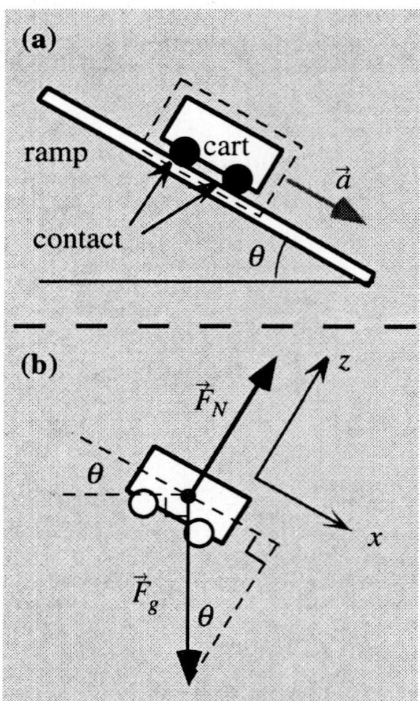

Figure N7.6: **(a)** A cart rolling down an incline. **(b)** A free-body diagram of this situation.

Problem: Find the acceleration of a cart rolling on small frictionless wheels down a ramp inclined at an angle θ with respect to the horizontal. How does the magnitude of the normal force exerted on the cart by the contact interaction with the ramp compare to the magnitude of the object's weight?

Solution Figure N7.6 illustrates this situation. If there are no significant friction forces acting on the object, then the only forces acting on the cart are the normal force (which points perpendicular to the incline) and the cart's weight (which points vertically downward). Since the object is confined to moving on the surface of the incline, its acceleration has to point parallel to the incline. Let us orient our reference frame so that the x axis points down the incline and the z axis is perpendicular to the incline. Newton's second law $\vec{F}_{\text{net}} = m\vec{a}$ then reads

$$\begin{bmatrix} ma_x \\ 0 \\ 0 \end{bmatrix} = \begin{bmatrix} F_{g,x} \\ F_{g,y} \\ F_{g,z} \end{bmatrix} + \begin{bmatrix} F_{N,x} \\ F_{N,y} \\ F_{N,z} \end{bmatrix} = \begin{bmatrix} +mg\sin\theta \\ 0 \\ -mg\cos\theta \end{bmatrix} + \begin{bmatrix} 0 \\ 0 \\ +F_N \end{bmatrix} \tag{N7.13}$$

where $mg = \text{mag}(\vec{F}_g)$ and $F_N \equiv \text{mag}(\vec{F}_N)$. Solving the first and third lines of this equation for a_x and F_N, respectively, we get

$$a_x = +g\sin\theta \quad \text{and} \quad F_N = mg\cos\theta = F_g\cos\theta \tag{N7.14}$$

We see that an object moving frictionlessly down an incline will have a constant acceleration whose magnitude is $a = g\sin\theta$ and the magnitude of the normal force is $\cos\theta$ times that of the object's weight. Since $\cos\theta < 1$, this means that $F_N < F_g$. (This last statement is true whether the object is accelerating or not.)

Exercise N7X.6: How long would it take such a cart to travel from rest down a 3-m ramp inclined at an angle of 15°?

Exercise N7X.7: Imagine that we analyzed the situation here using a reference frame where the x axis was horizontal. Why would this be more difficult?

N7.7 A CONSTRAINED-MOTION FRAMEWORK

We can easily adapt our problem-solving framework to "constrained-motion" problems. (**Constrained motion** problems are problems where the object in question is constrained by its environment to move in a certain way (for example, with a constant velocity along a straight line, constant acceleration along a straight line, or constant speed in a circle) and we use this information to determine the forces acting on the object.) Let us look at the adaptations to each major section of the framework in turn.

Changes to the sections of our generalized problem-solving framework

In the *pictorial representation* section, in addition to the the usual drawing of the situation with symbolic labels and our list of knowns and unknowns, it is crucial to define and draw our reference-frame axes, and also draw an arrow indicating the direction of the object's acceleration (if any).

In the *conceptual representation* section, in addition to describing any approximations that we have to make, it is essential to draw at least a free-particle diagram of the object (draw a free-body and/or net force diagram too, if that helps). You should also write something describing the constraint on the object's motion and what this means for the direction of the acceleration (if any).

In the *mathematical representation* section, always start by writing Newton's second law in column vector form and then express the components of each force in terms of the (positive!) magnitude of that force, as I have done in all of the examples so far. This makes solving complicated problems much easier.

Examples N7.4 and N7.5 show the complete framework and how it is used.

EXAMPLE N7.4

OUTLINE OF THE FRAMEWORK for CONSTRAINED-MOTION problems

1. Pictorial Representation

a. Draw a picture of the situation that includes:

① (1) a drawing of the object in the context of its surroundings

② (2) a labeled arrow indicating the direction of the object's acceleration (or say "$\vec{a} = 0$")

③ (3) reference frame axes, with one axis aligned with the object's acceleration (if any)

④ (4) labels defining symbols for relevant quantities

⑤ b. List values for all known labeled quantities and specify which quantities are unknown.

2. Conceptual Representation

The general task is to construct a conceptual model of the situation and link it to an abstract physics model or principle. In constrained motion problems, we do the following:

⑥ a. Draw a free-particle (and/or a free-body) diagram of the object

⑦ b. Describe the constraint on the object's motion and check that the net force on the object is consistent with the acceleration implied by that constraint (use a net-force diagram if necessary)

⑧ c. Describe any approximations or assumptions you have to make to solve the problem

3. Mathematical Representation

⑨ a. Apply Newton's second law *in vector component form*

⑩ b. Solve for any unknown quantities symbolically

⑪ c. Plug in numbers and units and calculate the result and its units.

4. Evaluation

Check that the answer makes sense:

⑫ a. Does it have the correct units?

⑬ b. Does it have the right sign?

⑭ c. Does it seem reasonable?

Problem: A 14-kg box sits in the back of a 2800-kg pickup truck waiting at a stoplight. When the stoplight turns green, the driver of the truck drives forward with an acceleration whose magnitude is 5.0 m/s^2. If the coefficient of static friction between the box and the truck bed is 0.40, will the box be able to accelerate with the truck or will it slide backward relative to the truck bed?

Solution: A picture of the situation appears below:

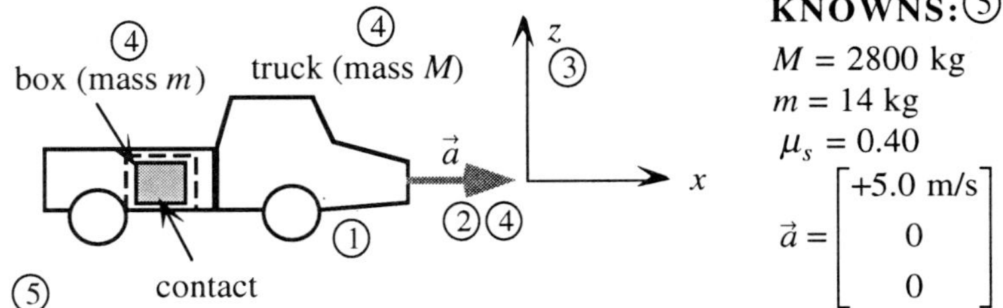

KNOWNS:⑤

$M = 2800$ kg

$m = 14$ kg

$\mu_s = 0.40$

$$\vec{a} = \begin{bmatrix} +5.0\text{ m/s} \\ 0 \\ 0 \end{bmatrix}$$

UNKNOWN: Is the magnitude of $\vec{F}_{SF}$ required to keep the box accelerating with the truck larger than $\vec{F}_{SF,\max}$?

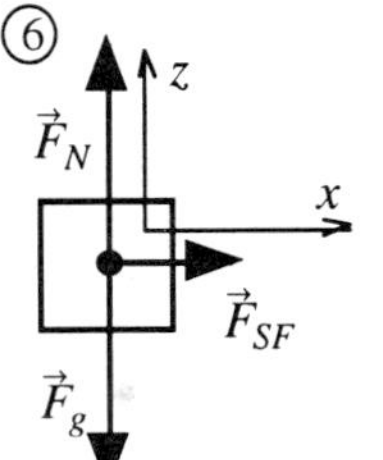

The only thing the box touches is the truck bed ⑧ (we will ignore its contact interaction with the air): this contact interaction exerts a normal force and a static friction force (assuming the box isn't ⑧ slipping). A free-particle diagram of the box therefore looks as shown to the left. Note that since the box is *supposed* to be constrained to ⑦ move forward with the truck, it must be accelerating in the horizontal direction along with the truck. This means that the static friction force on the box (which is the only horizontal force acting on it) must act *forward* on the box so that the net force on the box is forward and thus is consistent with the box's acceleration (again assuming that it is keeping up with the truck).

⑨ Newton's second law in this case then implies that:

$$\begin{bmatrix} ma_x \\ 0 \\ 0 \end{bmatrix} = \begin{bmatrix} F_{g,x} \\ F_{g,y} \\ F_{g,z} \end{bmatrix} + \begin{bmatrix} F_{N,x} \\ F_{N,y} \\ F_{N,z} \end{bmatrix} + \begin{bmatrix} F_{SF,x} \\ F_{SF,y} \\ F_{SF,z} \end{bmatrix} = \begin{bmatrix} 0 \\ 0 \\ -mg \end{bmatrix} + \begin{bmatrix} 0 \\ 0 \\ +F_N \end{bmatrix} + \begin{bmatrix} +F_{SF} \\ 0 \\ 0 \end{bmatrix} \quad (1)$$

The top and bottom rows (x and z components) of this equation therefore imply (respectively) that

$$ma_x = F_{SF} \quad (2a)$$

$$F_N = mg \quad (2b)$$

Since $F_{SF,\max} = \mu_s F_N$, equation (2b) means that

$$F_{SF,\max} = \mu_s F_N = \mu_s mg \quad (3)$$

Therefore, according to equations 2a and 3, the condition that the static friction force be enough to keep the box accelerating with the truck is:

$$F_{SF} = F_{SF,\max} \quad \Rightarrow \quad ma_x < \mu_s mg \quad \Rightarrow \quad a_x < \mu_s g \quad ⑩ \quad (4)$$

Since $\mu_s g = (0.40)(9.8\text{ m/s}) = 3.9\text{ m/s}^2$ ⑪ and the truck's actual acceleration $a_x = 5.0\text{ m/s}^2$ exceeds this, the box will indeed slip backwards as the truck accelerates forward. Note that this result is independent of the ⑬ mass of the box! The result for a_x also does have the correct sign ⑫ (positive) and the correct units. The result is also believable: an acceler- ⑭ ation of 5.0 m/s^2 is pretty large ($\approx \frac{1}{2}g$) and so might cause slipping.

EXAMPLE N7.5

OUTLINE OF THE FRAMEWORK for CONSTRAINED-MOTION problems

1. Pictorial Representation

a. Draw a picture of the situation that includes:

① (1) a drawing of the object in the context of its surroundings

② (2) a labeled arrow indicating the direction of the object's acceleration (or say "$\vec{a} = 0$")

③ (3) reference frame axes, with one axis aligned with the object's acceleration (if any)

④ (4) labels defining symbols for relevant quantities

⑤ b. List values for all known labeled quantities and specify which quantities are unknown.

2. Conceptual Representation

The general task is to construct a conceptual model of the situation and link it to an abstract physics model or principle. In constrained motion problems, we do the following:

⑥ a. Draw a free-particle (and/or a free-body) diagram of the object

⑦ b. Describe the constraint on the object's motion and check that the net force on the object is consistent with the acceleration implied by that constraint (use a net-force diagram if necessary)

⑧ c. Describe any approximations or assumptions you have to make to solve the problem

3. Mathematical Representation

⑨ a. Apply Newton's second law *in vector component form*

⑩ b. Solve for any unknown quantities symbolically

⑪ c. Plug in numbers and units and calculate the result and its units.

4. Evaluation

Check that the answer makes sense:

⑫ a. Does it have the correct units?

⑬ b. Does it have the right sign?

⑭ c. Does it seem reasonable?

Problem: A 55-kg person in an elevator traveling upward is standing on a spring scale that reads 420 N. What is the magnitude and direction of the elevator's acceleration? [*Note:* An ordinary scale does not directly register your weight, since the weight is a force that acts directly on the you and cannot be intercepted by the scale. Rather the scale registers the magnitude of the upward normal force that the scale's spring has to exert to support you. This same contact interaction puts pressure on your feet that gives you a sense of what your "perceived weight" is. When you are at rest, this normal force is equal to your actual weight, but if you are accelerating vertically, the net vertical force on you will *not* be zero and thus the normal force (and thus your perceived weight) will not be equal to your actual weight (which does not change significantly no matter how the elevator moves). This is why your perceived weight changes as you ride an elevator.]

Solution: A diagram of the situation is shown below (and left):

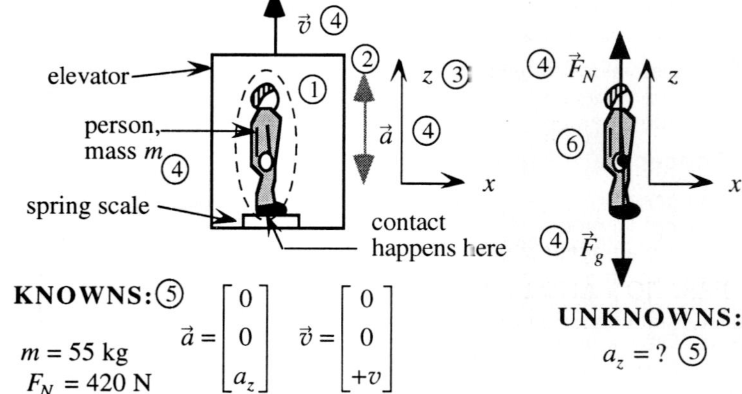

KNOWNS: ⑤

$m = 55$ kg
$F_N = 420$ N

$$\vec{a} = \begin{bmatrix} 0 \\ 0 \\ a_z \end{bmatrix} \quad \vec{v} = \begin{bmatrix} 0 \\ 0 \\ +v \end{bmatrix}$$

UNKNOWNS:

$a_z = ?$ ⑤

⑦ A free-particle diagram for the person appears above and to the right, next to the picture of the situation. Since the person is constrained to move vertically with the elevator, the person's acceleration (if there is any must be vertical. This is consistent with the free-particle diagram: there are no horizontal forces and the relative magnitudes of the vertical normal and gravitational forces will determine the person's (and thus the elevator's) acceleration. In this case, then, the component form of Newton's second law $m\vec{a} = \vec{F}_{\text{net}}$ looks like this:

$$⑨ \quad \begin{bmatrix} 0 \\ 0 \\ ma_z \end{bmatrix} = \begin{bmatrix} F_{N,x} \\ F_{N,y} \\ F_{N,z} \end{bmatrix} + \begin{bmatrix} F_{g,x} \\ F_{g,y} \\ F_{g,z} \end{bmatrix} = \begin{bmatrix} 0 \\ 0 \\ +F_N \end{bmatrix} + \begin{bmatrix} 0 \\ 0 \\ -mg \end{bmatrix} \qquad (1)$$

Only the bottom line (z component) of this equation tells us anything useful. If we solve it for a_z, we find that

$$a_z = \frac{F_N}{m} - g \overset{⑩}{=} \frac{(420\,\cancel{\text{N}})}{(55\,\cancel{\text{kg}})}\left(\frac{1\,\cancel{\text{kg}}\cdot\text{m/s}^2}{1\,\cancel{\text{N}}}\right) - 9.8\ \text{m/s}^2 \overset{⑪}{=} -2.2\ \text{m/s}^2 \qquad (2)$$

Since this is negative, the elevator must be accelerating *downward* at a rate of 2.2 m/s^2 (perhaps as it slows when it approaches an upper-floor destination). This makes the person apparently weigh less, since the normal force of 420 N is lower than the person's actual unchanging ⑭ ⑬ weight ($mg \approx 540$ N). This coincides with our experience: when an elevator slows down while going up, we feel momentarily lighter. ⑫ The acceleration also has the right units and has ⑭ a reasonable magnitude ($< g$). [Note that the full three-dimensional machinery of equation (1) is probably overkill here, but it is *essential* in more complicated problems.]

SUMMARY

I. FREE-PARTICLE DIAGRAMS
   A. Our basic task is to apply Newton's second law $\vec{F}_{\text{net}} = m\vec{a}$ (N7.1)
      1. we will express in component form (see equation N7.2)
      2. so we need to find the components of $\vec{F}_{\text{net}}$
   B. This is the point of a free-*particle* diagram:
      1. This diagram shows the object as a particle responding to external forces, whose directions are displayed relative to the frame axes
      2. This facilitates finding force components and applying $\vec{F}_{\text{net}} = m\vec{a}$
   C. Necessary parts of a free-body diagram:
      1. A sketch of object with central dot representing its center of mass
      2. A labeled arrow (with its tail attached to the object's CM) for each external force exerted on that object
      3. A sketch of reference frame axis directions
   D. We will distinguish forces that would have the same symbols by superscript indicating external thing with which the object is interacting
   E. In the remainder of unit *N*, we will no longer be concerned with torque
   F. The right way to apply Newton's second law from now on
      1. Write $m\vec{a} = \vec{F}_1 + \vec{F}_2 + \ldots$ in column vector form
      2. Rewrite all the force components appearing in this equation in terms of their magnitudes (paying careful attention to signs!)
      3. Separate into three distinct component equations

II. MOTION AT A CONSTANT VELOCITY
   A. $\vec{a} = 0$ implies that $\vec{F}_{\text{net}} = 0$, just like for objects at rest
   B. Choose frame axes so as many force components as possible are zero

III. FRICTION AND DRAG FORCES
   A. Static Friction (example: box sitting on floor)
      1. This force arises because atom in box lock to atoms on the floor
      2. It *adjusts itself* in response to other forces to keep box at rest, up to a maximum magnitude of $F_{SF,\max} = \mu_s \text{mag}(\vec{F}_N)$ (N7.5)
         a) $\vec{F}_N$ is normal force exerted by same contact interaction
         b) mag($\vec{F}_N$) expresses how hard the surfaces are pressed together
         c) we call unitless μ_s the *coefficient of static friction*
         d) μ_s depends on nature of surfaces: a typical value ≈ 0.5
   B. Kinetic Friction
      1. If object is pushed beyond static force's ability to resist, it will start to slide and the magnitude of the friction force drops
      2. Again, the magnitude of friction force involved here is proportional to mag($\vec{F}_N$): $F_{KF} = \mu_k \text{ mag}(\vec{F}_N)$ (N7.6)
         a) μ_k is the *coefficient of kinetic friction*, and is typically $< \mu_s$
         b) Equation N7.6 gives the actual steady value of F_{KF}, *not* a maximum possible value (as equation N7.5 does)
   C. The drag on object moving through a fluid with speed v
      1. For large object moving through air $F_D = \frac{1}{2} C\rho A v^2$ (N7.11)
         a) ρ is the density of air, *A* is the object's cross-sectional area
         b) C is a *drag coefficient* that depends on shape (0.5 for sphere)
      2. Formula only works for large *A*, high v, thin fluid
         a) most sports projectiles obey this law
         b) $F_D \propto v$ for small objects moving slowly in thick fluids

IV. LINEARLY ACCELERATED MOTION
   A. The key to simple solution: *align one reference frame axis with* $\vec{a}$
   B. Some such problems are hybrids: use motion (or lack thereof) to find *forces* which allows one to find *motion* in another component direction

V. A CONSTRAINED MOTION FRAMEWORK
   A. This is an adaptation of our framework for constrained-motion problems
   B. See the extended outline of the framework in examples N7.4 and N7.5

GLOSSARY

linearly-constrained motion: the motion of an object that is constrained to move along a line.

constrained-motion problem: a physics problem where we use an observed constraint on an object's motion to determine the forces acting on that object.

coefficient of static friction μ_s: expresses the constant of proportionality between the magnitude of the maximum possible static friction force that a contact interaction can provide and the magnitude of the normal force exerted by the same contact interaction (F_N essentially expresses how tightly the interacting surfaces are pressed together): $\text{mag}(\vec{F}_{SF,\max}) = \mu_s \text{mag}(\vec{F}_N)$. The unitless values of μ_s range from 0.04 for very slippery surfaces to 1.1 or more for very sticky surfaces, and are typically about 0.5.

coefficient of kinetic friction μ_k: expresses the constant of proportionality between the magnitude of the steady kinetic friction force exerted by a contact interaction between sliding surfaces and the magnitude of the normal force exerted by the same interaction (which expresses how tightly the interacting surfaces are pressed together): $\text{mag}(\vec{F}_{KF}) = \mu_k \text{mag}(\vec{F}_N)$. The unitless values of μ_k are generally somewhat smaller than values of μ_s for the same type of surfaces.

drag coefficient C: the quantity appearing in the drag force equation $\text{mag}(\vec{F}_D) = \frac{1}{2}C\rho Av^2$ for an object moving through air. This empirical coefficient depends in the shape of the object (it has a value ≈ 0.5 for a sphere).

TWO-MINUTE PROBLEMS

N7T.1 The magnitude of the normal force on a box sitting on an incline is equal to that of its weight (T or F).

N7T.2 A certain crate sits on a rough floor. You find that you have to apply a horizontal force of 200 N to get the crate moving. If you put some massive objects in the crate so that its mass is doubled, how much force does it take to get the crate moving now?
A. still 200 N C. 800 N
B. 400 N D. it depends (specify)

N7T.3 Two boxes of the same mass sit on a rough floor. These boxes are made out of the same kind of cardboard and are identical except that one is twice as large as the other. If it takes 200 N to start moving the smaller box, how much force does it take to start moving the larger one?
A. still 200 N C. 800 N
B. 400 N D. it depends (specify)

N7T.4 The coefficient of static friction between Teflon and scrambled eggs is about 0.1. What is the smallest tilt angle from the horizontal that will cause the eggs to slide across the surface of a tilted Teflon-coated pan?
A. 0.002° C. 15° E. other (specify)
B. 5.7° D. 33°

N7T.5 If you want to stop a car as quickly as possible on an icy road, you should
A. jam on the brakes as hard as you can
B. push on the brakes as hard as you can without locking the car's wheels and thus making the car skid.
C. pump the brakes D. something else (specify)

N7T.6 Putting wider tires on your car will clearly give you more traction (T or F).

N7T.7 Assume that the coefficient of static friction between your car's tires and the road is about 0.6. Your car can climb a 45° slope (T or F).

N7T.8 Imagine that an external force of 100 N must be applied to keep a bicycle and rider moving at a constant speed of 12 mi/h against opposing air drag. To double the bike's speed to 24 mi/h, we must increase the magnitude of the force exerted on the bike to:
A. 141 N C. 400 N E. other (specify)
B. 200 N D. it depends on the bike's shape, area

N7T.9 Imagine that a certain engine can cause the road to exert a certain maximum forward force F_{SF} on a certain car. If we change the car's design to reduce its coefficient of friction by a factor of 2, by what factor will the car's maximum speed increase (other things being equal)?
A. no increase C. 2 E. depends on (specify)
B. 1.41 D. 4 F. other (specify)

N7T.10 A truck is traveling down a steady slope such that for each meter the truck goes forward it goes down 0.04 m (we call this a 4% grade). Imagine that the truck's brakes fail. What is the approximate increase in the truck's speed after 30 s, assuming the engine is not used and there is little drag or other friction? (g = 22 [mi/h]/s.)

HOMEWORK PROBLEMS

BASIC SKILLS

N7B.1 Imagine that you have to exert 200 N of horizontal force on a 30-kg crate to get it moving on a level floor. What is the value of μ_s for the surfaces involved here?

N7B.2 Imagine that the coefficient of static friction μ_s between a 25-kg box and the floor is 0.55. How hard would you have to push on the box to get it moving?

N7B.3 Imagine that you have to exert a horizontal force of magnitude 80 N to push a 20-kg box at a constant speed of 3 m/s. What is the coefficient of kinetic friction μ_k between the box and the floor in this case?

N7B.4 Imagine that the coefficient of kinetic friction between a certain 15-kg box and the floor is 0.35. How hard would you have to push on it to move it at a constant speed of 2 m/s across the floor?

N7B.5 Assuming that air drag is the main opposing force here, that the area the bike and rider present to the wind is about 0.75 m^2, and that C for such an irregular shape is about 1, what is the approximate forward force required to give the bicycle a constant speed of 10 m/s?

N7B.6 The drag force on a car moving at 65 mi/h is how many times bigger than the drag force on a car moving at 45 mi/h?

N7B.7 Imagine that a cart rolls without friction down a slope that makes an angle of 6° with respect to the horizontal. If the cart rolls for 12 s, how far does it go? (*Hint*: you can use the result of example N7.3)

N7B.8 Imagine that a glider slides without friction down a tilted air track. If it takes 3.0 s to slide the 1.5 m length of the track, at what angle was the track inclined? (*Hint*: you can use the result of example N7.3.)

SYNTHETIC

N7S.1 The coefficient of static friction between a certain car's tires and the road is about 0.60. Only the rear tires are powered. The magnitude of the car's acceleration can be at most what value? Explain your response carefully.

N7S.2 Why do the tires of a car grip the road better on level ground than when the car is going up or down an incline? Explain carefully. (You may find that a couple of force diagrams will help you make your case.)

N7S.3 A 2.0-kg box slides down a 25° incline at a constant velocity of 3.0 m/s. What is the magnitude and direction of the kinetic friction force acting on this box?*

N7S.4 Imagine that the coefficient of static friction between the tires of a certain 2250-kg all-terrain vehicle and a typical gravel roadbed is 0.45. What is the maximum possible incline that the vehicle can climb?*

N7S.5 A 12-kg box sits at rest on a 15° incline. If the coefficient of static friction is 0.3 between the box and the incline, what additional force would you have to exert directly down the incline to get the box to start sliding?*

N7S.6 A certain car has a drag coefficient of 0.32. Looking at it from the front, the car's cross-section looks roughly like a rectangle that is 1.5 m high by 2.1 m across. What is the *minimum* horsepower of its engine if its top cruising speed is 45 m/s (≈ 100 mi/h)?*

N7S.7 A 200-kg motorboat is cruising in the $+x$ direction at a speed of v_0. The motor suddenly dies at a time we will call $t = 0$ and the boat's x-velocity thereafter is observed to be described by the following equation $v_x(t) = v_0 e^{-qt}$, where $q = 0.5\ \text{s}^{-1}$. Does the observed motion of this boat seem consistent with the idea that the drag force on the boat is given by $F_D = bv$? If so, determine the value of b for this boat. [*Hint:* The time derivative of e^{-qt} is $-qe^{-qt}$.]

N7S.8 A crane hauls a crate (mass 250 kg) upward at a constant acceleration of 2.2 m/s^2. What is the magnitude of the tension force exerted on the crate by the crane's cable?*

N7S.9 A 65-kg person is standing on a bathroom scale in an elevator moving downward. If the scale reads 720 N, what is the magnitude and direction of the elevator's acceleration?*

N7S.10 A railroad flatcar is loaded with crates having a coefficient of static friction of 0.25 with respect to the car's floor. If a train is moving at 22 m/s (≈ 48 mi/hr), within how short a distance can the train be stopped without letting the crates slide?*

N7S.11 A pickup truck carries cans of paint in the back bed. The coefficient of static friction between the cans and the bed of the truck is 0.28. There is no back gate to the truck. How long should the driver take to accelerate to a speed of 55 mi/h to avoid losing the paint cans out of the rear of the truck?*

N7S.12 An 1100-kg car with a frontal cross-sectional area of 3.5 m^2 and a drag coefficient of 0.42 rolls down a slope that makes a constant angle of 8° with respect to the horizontal. After accelerating a while, the car will eventually reach a maximum constant speed. What is this speed? (assume that the wheels rotate frictionlessly)?*

N7S.13 A 1800-kg car with four-wheel drive travels up a 12° incline at a speed of 15 m/s. What can you infer about the coefficient of static friction between the tires of this car and the road? Which pieces of information provided are relevant, and which are not?*

N7S.14 Antilock brakes keep a car's tires from skidding on a road surface. A certain 1500-kg car equipped with such brakes and initially traveling at 27 m/s (≈59 mi/hr) is able to come to rest within a time interval of 6.0 seconds.
(a) What is the coefficient of gripping friction between the tires and the road in this case? **(b)** How far does the car travel before it stops?*

*For these starred problems, please make sure that your solution follows the constrained-motion framework whose outline appears in examples N7.4 and N7.5.

RICH-CONTEXT

N7R.1 A bicyclist (whose mass is 54 kg and whose bike has a mass of 11 kg) coasting down an essentially endless 5° slope is observed to reach a maximum speed of 42 mi/h. At these kinds of speeds, air drag dominates over all other kinds of friction. *Estimate* the drag coefficient C for the bike and rider.

N7R.2 You are driving a 12,000-kg truck at a constant speed of 65 mi/h down a 6% slope (that is, an incline that goes down 0.06 m for every meter that one goes along the incline). You suddenly see that a bridge is out 425 ft ahead and jam on the brakes. The coefficient of static friction between your tires and the road is 0.45, the coefficient of kinetic friction is 0.30, and the cross-sectional area of your truck is 6.6 m^2. Can you stop in time? (Make plausible estimates for quantities you do not know.)

ADVANCED

N7A.1 We can understand equation N7.11 using the following model. Imagine that we assume that an object with cross-sectional area A moving through air of density ρ has to accelerate all of the air molecules that it touches from rest up to its speed. Show that if this were so, the drag force on the object would be

$$\text{mag}(\vec{F}_D) = \rho A v^2 \qquad \text{(N7.15)}$$

(The actual drag force will be somewhat smaller, since some of the air will slip around the object without being fully accelerated to its speed.)

N7A.2 (Requires using some calculus involving exponentials and logarithms.) Imagine that an object of mass m is initially moving in the $+x$ direction through a viscous fluid at a speed v_0. If the only force acting on this object is a viscous drag force whose magnitude is given by $F_D = bv$, prove that the object's x-velocity is given by

$$v_x(t) = v_0 e^{-qt} \qquad \text{(N7.16)}$$

and determine how the constant q depends on b and m. *Hint*: This is actually a motion-from-force problem. Divide both sides of the x component of Newton's second law by v_x, and take the indefinite integral of both sides.

ANSWERS TO EXERCISES

N7X.1 According to equation N7.8, $F_{SF} = mg\sin\theta$. Equations N7.8 and N7.9, on the other hand, also indicate that $F_{SF,\max} = \mu_s F_N = \mu_s mg\cos\theta$. Therefore

$$\frac{F_{SF}}{F_{SF,\max}} = \frac{\cancel{mg}\sin\theta}{\mu_s \cancel{mg}\cos\theta} = \frac{\tan\theta}{\mu_s} = 0.69 \qquad \text{(N7.17)}$$

N7X.2 Since $F_{SF,\max} = 49$ N and $F_N = F_g = mg = 98$ N in this situation, $\mu_s = F_{SF,\max} / F_N = 0.50$.
N7X.3 C is unitless.
N7X.4 340 N.
N7X.5 The motorboat is not accelerating when it is moving at a constant speed, so the forward force acting on it must cancel the drag force. Solving $F_{\text{forward}} = F_D = bv$ for b, we get $b = F_{\text{forward}} / v = 110$ kg/s.

N7X.6 From chapter N4 we know that

$$x(t) - x_0 = \tfrac{1}{2}a_x t^2 + v_{0x}t \qquad \text{(N7.18)}$$

Since $v_{0x} = 0$ here, $x(t) - x_0 = D = 3.0$ m and $a_x = g\sin\theta$,

$$D = \tfrac{1}{2}g\sin\theta\, t^2 \quad \Rightarrow \quad t = \sqrt{2D/g\sin\theta} = 1.54 \text{ s} \qquad \text{(N7.19)}$$

N7X.7 In this case, Newton's second law would read

$$\begin{bmatrix} ma_x \\ 0 \\ ma_z \end{bmatrix} = \begin{bmatrix} F_{N,x} \\ F_{N,y} \\ F_{N,z} \end{bmatrix} + \begin{bmatrix} F_{g,x} \\ F_{g,y} \\ F_{g,z} \end{bmatrix} = \begin{bmatrix} F_N\sin\theta \\ 0 \\ F_N\cos\theta \end{bmatrix} + \begin{bmatrix} 0 \\ 0 \\ -mg \end{bmatrix} \qquad \text{(N7.20)}$$

But neither a_x or a_z are zero in this case, so it becomes much more difficult to solve this problem. It can be done: we can use what we know about the acceleration's direction to show that $a_x = a\cos\theta$ and $a_z = -a\sin\theta$, plug these into the equations above and solve the two coupled equations for the two unknowns a and F_N. However, this is much more complicated than the process illustrated in the example.

N8

COUPLED OBJECTS

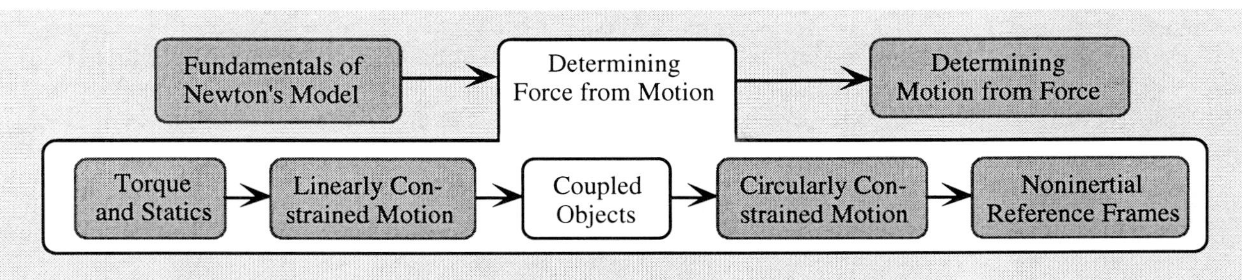

N8.1 OVERVIEW

In the last chapter, we saw how to apply Newton's second law to problems involving objects whose motion (if any) is confined to a line. In this chapter, we will combine Newton's second and *third* laws to analyze the forces that act on pairs of coupled objects that are constrained by some physical connection to move together each along a line. Doing this will help us appreciate more deeply the meaning and utility of Newton's third law and extend our ability to use the second law in a variety of realistic situations.

Here is an overview of the sections in this chapter.

N8.2 *FORCE NOTATION FOR COUPLED OBJECTS* presents a notation that can help us keep track of the forces acting on and between coupled objects. This notation will also help us recognize and describe forces whose magnitudes are linked by Newton's third law.

N8.3 *PUSHING BLOCKS* explores the simple situation of two blocks that interact as they are pushed together across a surface in order to illustrate clearly how we can apply Newton's second and third laws to analyze the behavior of a system of objects.

N8.4 *STRINGS, REAL AND IDEAL* discusses problems involving two objects connected by a string. In the process, we will develop the model of an *ideal string* and discuss how well real strings fit this model.

N8.5 *PULLEYS* looks at what happens when we pass a string over a pulley, illustrating how we can use the ideal string approximation in more complicated situations.

N8.6 *USING THE FRAMEWORK* shows how we can adapt the framework for constrained-motion problems to handle coupled-object problems.

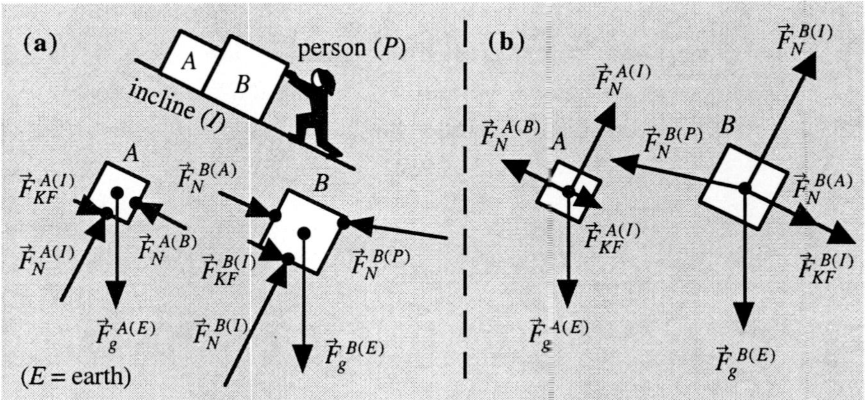

Figure N8.1: **(a)** A sketch of a person pushing two boxes up an incline and free-body diagrams for each of the boxes. **(b)** Free-particle diagrams for each of the boxes. The forces in both the free-body and free-particle diagrams are labeled according to the notation convention established in this section. Note that we typically use the symbol E for the earth in the labels for gravitational forces.

N8.2 FORCE NOTATION FOR COUPLED OBJECTS

What are coupled objects?

Our general task in chapters N6 through N10 is to expand our ability to apply Newton's laws to determine the forces acting on objects whose motion is constrained in various ways. In this chapter, we will focus on situations involving *coupled objects*. A pair of **coupled objects** consists of two objects that are constrained by some kind of connection so that their motions (which we will assume here to be linear) are linked in some well-defined manner. We can use this constraint to analyze the forces acting on the individual objects in the pair and describe their motion.

When we are called on to describe and analyze the forces acting on more than one object, we begin to run into serious problems with notation. For example, imagine that a person pushes a box B up an incline, which in turn pushes box A in front of it up that incline (Figure N8.1a). How can we distinguish the symbol for the normal force exerted on A (by A's contact interaction with the incline) from the symbol for the normal force exerted on B (by B's interaction with the incline) from the symbol for the normal force exerted on B (by B's contact interaction with the pusher's hands) from the symbol for the normal force exerted on A (by A's contact interaction with the box B behind it). We would get very confused if we give all these different forces the same symbol $\vec{F}_N$!

A notation convention for force symbols

Let us adopt the following notation convention for cases like this. We start with the basic symbol for the type of force involved, using the standard symbols described in section N3.6. We add to this a pair of superscripts, the first indicating the object on which this force is exerted, and the second (in parentheses) indicating the other object involved in the interaction that gives rise to this force. For example, we would give the normal force exerted on box A by its contact interaction with box B (which is pushing A up the incline) the symbol $\vec{F}_N^{A(B)}$. The normal force exerted on box B by the person (P) we might call $\vec{F}_N^{B(P)}$, and so on. While it is not necessarily the prettiest notation imaginable, this notation (which is summarized in Figure N8.2) is simple, relatively easy both to typeset and to write by hand, and it is easy to interpret (as long as we have well-defined single letters for every object involved. Figure N8.1b shows free-particle diagrams for the two boxes with all forces labeled using this notation.

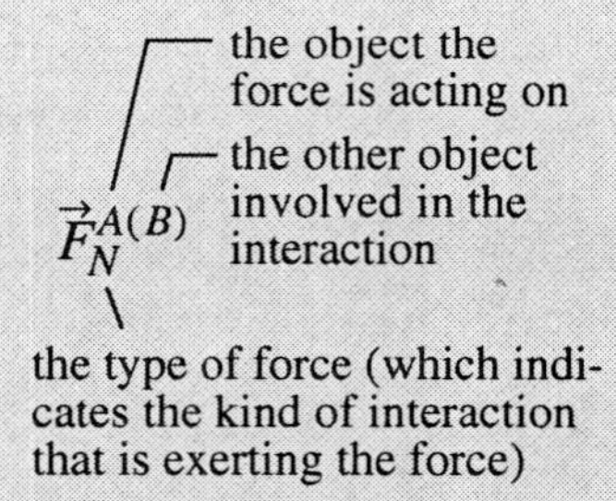

Figure N8.2: A convention for force symbols used in coupled-object problems.

Obviously, this kind of notation is too cumbersome to use all the time. When we are analyzing the forces acting on a single object, this notation is only rarely helpful. In this chapter, however, it is more than helpful; this notation (or something equivalent) is essential for keeping things straight.

Recognizing third-law partners

One advantage of this notation is that it makes it pretty easy to recognize third-law partners in a given situation. Newton's third law states that

> When two objects interact, the force the interaction exerts on each is equal in magnitude and opposite in direction to the force it exerts on the other.

This means (as discussed in chapter N4) that the pair of forces linked by this law (third law partners) have the following characteristics:

1. They always act on *different* objects (when A and B interact, the interaction exerts one force on A and one on B).
2. They always reflect the *same* interaction.

This in turn means that the symbols for third-law partners will have *subscripts* that are the *same* (since both forces must reflect the same interaction) and *superscripts* that are reversed (since the partner to the force exerted on A due to its interaction with B is the force on B due to its interaction with A). For example, if I exert a normal force $\vec{F}_N^{W(H)}$ on the wall (W) by pushing on it with my hand (H), the contact interaction exerts a force $\vec{F}_N^{H(W)}$ back on my hand.

The symbols for third-law partners

$$\vec{F}_N^{W(H)} = -\vec{F}_N^{H(W)} \qquad \text{(N8.1)}$$

wall (note the interchanged order)
hand
(reflect the *same* interaction)

by Newton's third law. The symbols for third-law partners will always have the characteristics noted in equation N8.1.

In the situation shown in Figure N8.1, the only third-law partners among the forces shown are the normal forces $\vec{F}_N^{A(B)}$ and $\vec{F}_N^{B(A)}$ that the boxes exert on each other due to their contact interaction. Note that these symbols display the characteristics shown in equation N8.1. Figure N8.1a also illustrates that it is often particularly easy to spot third-law partners on the free-*body* (as opposed to free-*particle*) diagrams of a pair of objects participating in a *contact* interaction: one only has to look the pair of forces that act on each object on the surface that is in contact with the other object. Check to see that the contact forces $\vec{F}_N^{A(B)}$ and $\vec{F}_N^{B(A)}$ in Figure N8.1a fit this description.

Spotting third-law partners on free-body diagrams

EXAMPLE N8.1

Problem: Consider a book sitting at rest on a table. Draw free-body diagrams (not free-particle diagrams) for both the book and the table and determine which pairs of forces on these diagrams (if any) are third-law partners.

Solution Figure N8.3 shows the free-body diagrams for these objects. The only forces acting on the book are a downward gravitational force $\vec{F}_g^{B(E)}$ and an upward normal force $\vec{F}_N^{B(T)}$ due to its contact interaction with the table. The forces acting on the table are: a downward gravitational force $\vec{F}_g^{T(E)}$, a set of upward normal forces $\vec{F}_N^{T(F)}$ exerted by the contact interaction between the floor (F) on the table legs, and the downward normal force $\vec{F}_N^{T(B)}$ exerted by the contact interaction between the book and the table.

The only third-law partners among these forces are the pair that arise from the contact interaction between the book and the table. Note that this pair have all of the characteristics described in this section: their symbols are consistent with the pattern shown in equation N8.1 and these forces appear on the the free-body diagram acting on the surfaces where the book and table touch.

In contrast, note that the gravitational force on the book $\vec{F}_g^{B(E)}$ and the normal force $\vec{F}_N^{B(T)}$ are *not* third-law partners, even though these forces are opposite and they have the same magnitude. These forces are equal and opposite because of Newton's *second* law: since the book is at rest, its acceleration is zero; so the net force on it is zero and so $\vec{F}_g^{B(E)}$ and $\vec{F}_N^{B(T)}$ must be equal in magnitude and opposite in direction so that they cancel.

Note that even though the book's weight $\vec{F}_g^{B(E)}$ does not act directly on the table, the sum of the normal forces $\vec{F}_{N,\text{tot}}^{T(F)}$ acting on the table legs is larger when the book sits on the table than it would be if the book were not there, because Newton's second law requires that this force cancel both the table's weight $\vec{F}_g^{T(E)}$ and the downward contact force $\vec{F}_N^{T(B)}$, which is equal in magnitude to the book's weight (as discussed in the previous paragraph).

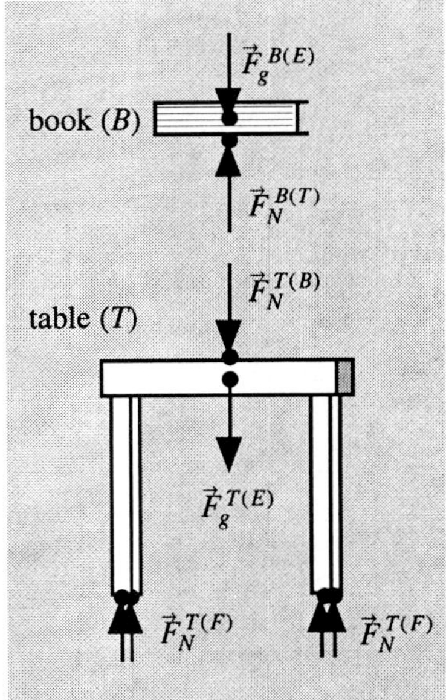

Figure N8.3: Free-body diagrams for the book and the table. Note that E = earth and F = floor.

Exercise N8X.1: Consider someone in an elevator moving at a constant speed. Draw free-body diagrams (not free-particle diagrams) for both the person and the elevator (ignoring friction), label the forces using the convention described in this section, and determine which pairs of forces on these diagrams (if any) are third-law partners. Explain why there is more tension on the elevator's cable when the person is in the elevator than when it is empty, even though the gravitational force (weight) of the person does not act directly on the elevator.

N8.3 PUSHING BLOCKS

When two objects are connected in such a way that they become constrained to move together in a well-defined way, the constraint on their motion itself provides information that we can use to learn something about the forces acting on those objects. The following example illustrates how we can use Newton's second and third laws together to extract information from this constraint.

EXAMPLE N8.2

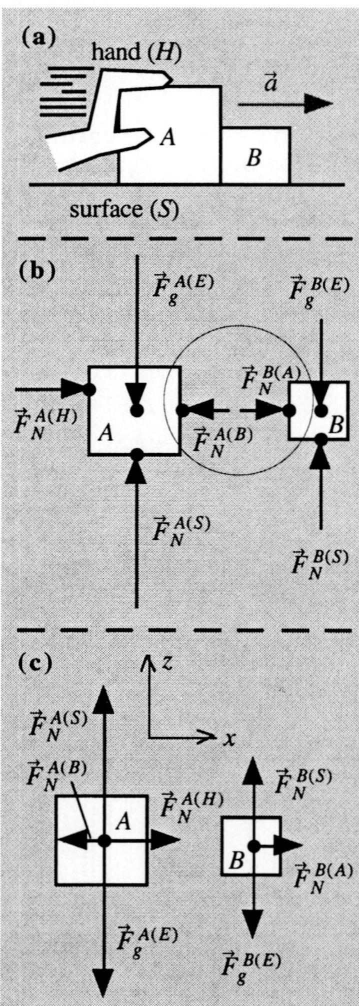

Figure N8.4: (a) Two blocks being pushed along a frictionless surface. **(b)** Free-body diagrams of the blocks, with the single third law pair circled. **(c)** Free-particle diagrams of the blocks.

Problem: Consider two blocks pushed along a frictionless surface by someone who exerts a constant force with his or her hand (see Figure N8.4a). Imagine that block A has a mass of 4.0 kg, block B has a mass of 2.0 kg, and the hand exerts a force of 3.0 N (about 0.65 lb). What is the magnitude of the acceleration of the blocks? What is the magnitude of the force exerted by block B on block A?

Solution The first step in solving virtually any problem is to draw a sketch. Figure N8.4a provides such a sketch. The next crucial step is to specify a coordinate system and draw free-particle diagrams of the two objects in question: these items are shown in Figure N8.4c. To more clearly see the relationships involved, it is often helpful to draw free-body diagrams of the objects (see Figure N8.4b) before drawing the free-particle diagrams.

The contact interaction between blocks A and B exerts opposing normal forces on each block, which are labeled above as being $\vec{F}_N^{A(B)}$ (read " the normal force on A due to its interaction with B") and $\vec{F}_N^{B(A)}$ (read "the normal force on B due to its interaction with A"). Newton's third law asserts that these forces have equal magnitudes, whether the blocks are accelerating or not.

$\vec{F}_N^{A(H)}$ represents the force exerted on A by the interaction with the person's hand, $\vec{F}_N^{A(S)}$ and $\vec{F}_N^{B(S)}$ represents the normal forces exerted on A and B by their contact interaction with the surface on which they slide, and $\vec{F}_g^{A(E)}$ and $\vec{F}_g^{B(E)}$ represent the forces exerted on each by their gravitational interaction with the earth (E = earth). Note that Figure N8.4a explicitly defines the symbols S and H to make these single-letter abbreviations clearer.

The third step in solving this problem is to list "knowns and unknowns". We know that mag($\vec{F}_N^{A(H)}$) = 3.0 N, that m_A = 4.0 kg, and that m_B = 2.0 kg. Knowing these masses would allow us to compute the magnitudes of the weight forces $\vec{F}_g^{A(E)}$ and $\vec{F}_g^{B(E)}$ should we so desire. The implicit constraints in the problem are (1) that the blocks move only in the x direction, which means that $\vec{a} = [a_x, 0, 0]$, and (2) that the two boxes have the *same* horizontal acceleration a_x (since they are moving as a unit). On the other hand, we do not know either the value of a_x or the magnitudes of the forces $\vec{F}_N^{A(B)}$ and $\vec{F}_N^{B(A)}$ (the first of which we are asked to find) or the magnitudes of $\vec{F}_N^{A(S)}$ and $\vec{F}_N^{B(S)}$ (which are probably of no concern).

The next step in most constrained motion problems is to apply the component form of Newton's second law. Newton's second law for block A reads

$$\begin{bmatrix} m_A a_x \\ 0 \\ 0 \end{bmatrix} = \begin{bmatrix} F_{g,x}^{A(E)} \\ F_{g,y}^{A(E)} \\ F_{g,z}^{A(E)} \end{bmatrix} + \begin{bmatrix} F_{N,x}^{A(S)} \\ F_{N,y}^{A(S)} \\ F_{N,z}^{A(S)} \end{bmatrix} + \begin{bmatrix} F_{N,x}^{A(B)} \\ F_{N,y}^{A(B)} \\ F_{N,z}^{A(B)} \end{bmatrix} + \begin{bmatrix} F_{N,x}^{A(H)} \\ F_{N,y}^{A(H)} \\ F_{N,z}^{A(H)} \end{bmatrix}$$

$$\Rightarrow \begin{bmatrix} m_A a_x \\ 0 \\ 0 \end{bmatrix} = \begin{bmatrix} 0 \\ 0 \\ -m_A g \end{bmatrix} + \begin{bmatrix} 0 \\ 0 \\ +F_N^{A(S)} \end{bmatrix} + \begin{bmatrix} -F_N^{A(B)} \\ 0 \\ 0 \end{bmatrix} + \begin{bmatrix} +F_N^{A(H)} \\ 0 \\ 0 \end{bmatrix} \quad \text{(N8.2a)}$$

(Force symbols without arrows or component subscripts refer to *magnitudes*, as usual.) The x and z components of this vector equation imply, respectively, that

$$m_A a_x = F_N^{A(H)} - F_N^{A(B)} \quad \text{(N8.2b)}$$

$$F_N^{A(S)} = m_A g \quad \text{(N8.2c)}$$

Similarly, Newton's second law for block B implies that

$$\begin{bmatrix} m_B a_x \\ 0 \\ 0 \end{bmatrix} = \begin{bmatrix} 0 \\ 0 \\ -m_B g \end{bmatrix} + \begin{bmatrix} 0 \\ 0 \\ +F_N^{B(S)} \end{bmatrix} + \begin{bmatrix} +F_N^{B(A)} \\ 0 \\ 0 \end{bmatrix} \Rightarrow \begin{matrix} m_B a_x = F_N^{B(A)} & \text{(N8.3a)} \\ 0 = 0 & \text{(N8.3b)} \\ F_N^{B(S)} = m_B g & \text{(N8.3c)} \end{matrix}$$

The z-component equations for both blocks tell us that the vertical normal force on each block cancels its weight: no surprises here. We can simplify equations N8.2b and N8.3a by recognizing that $F_N^{A(B)} = F_N^{B(A)}$ by Newton's third law. This bit of information means that we can rewrite equation N8.3a as follows:

$$m_B a_x = +F_N^{A(B)} \quad \text{(N8.4)}$$

We can then eliminate the unknown force $F_N^{A(B)}$ by adding this equation to equation N8.2b: this unknown force cancels out and we are left with

$$(m_A + m_B) a_x = F_N^{A(H)} \quad \text{(N8.5a)}$$

Note that this tells us that the *system* consisting of the two objects accelerates as if it were a particle of mass $M = m_A + m_B$ subject to the net external force acting on the system (which is equal to $\vec{F}_N^{A(H)}$). We already knew that this *should* be the case from chapter C4: the important thing here is to notice how Newton's third law ensures that it *does* happen. Solving for a_x, we get:

The blocks' common acceleration

$$a_x = \frac{F_N^{A(H)}}{m_A + m_B} = \frac{3.0\,\not{\text{N}}}{4.0\,\not{\text{kg}} + 2.0\,\text{kg}} \left(\frac{1\,\not{\text{kg}} \cdot \text{m/s}^2}{1\,\not{\text{N}}} \right) = 0.50\ \text{m/s}^2 \quad \text{(N8.5b)}$$

Plugging this result into equation N8.4 and using $F_N^{A(B)} = F_N^{B(A)}$, we find that

The magnitude of the forces due to the contact interaction between the blocks

$$F_N^{A(B)} = F_N^{B(A)} = m_B a_x = (2.0\,\not{\text{kg}})(0.50\,\not{\text{m/s}^2}) \left(\frac{1\ \text{N}}{1\,\not{\text{kg}} \cdot \not{\text{m/s}^2}} \right) = 1.0\ \text{N} \quad \text{(N8.6)}$$

Note also that the net x-force on object A is $F_N^{A(H)} - F_N^{A(B)} = 3.0\ \text{N} - 1.0\ \text{N} = 2.0\ \text{N}$, which is just the force required to give this 4.0-kg object an x-acceleration of $a_x = 0.5\ \text{m/s}^2$. Note also that the net force on B is 1.0 N, which is just what is required to give it an x-acceleration of $0.5\ \text{m/s}^2$. Everything is self-consistent!

The point of this example is that when objects are coupled so that they are constrained to have the same acceleration, the forces associated with the interaction between the objects will adjust themselves to whatever common magnitude gives each block the *same* horizontal acceleration. (Such forces are invariably contact forces, which *can* adjust themselves in this way.)

Exercise N8X.2: What if the hand pushes with a force of 9.0 N on block A? What is the magnitude of the contact force between the blocks in this case?

N8.4 STRINGS, REAL AND IDEAL

Overview

Objects do not have to be in direct contact to exert forces on each other. Two objects connected by string or cord can exert forces on each other through the string that are similar to the forces that they exert when in direct contact.

Each end of a string should be thought of as exerting a tension force on the object to which it is connected. While these two forces are generally *not* equal in direction, they are often at least approximately equal in magnitude (for reasons that we will discuss shortly). Because of this, we sometimes refer to the common magnitude of these forces as being ***the* tension on** (or **of**) **the string**. It is not *always* the case that the forces exerted by the ends of a string have the same magnitude, and in such cases "the string's tension" is not well defined, but you should understand that when this phrase is used, it refers to the common magnitude of the forces exerted by each end of the string.

A specific example will help us understand these issues more clearly.

EXAMPLE N8.3

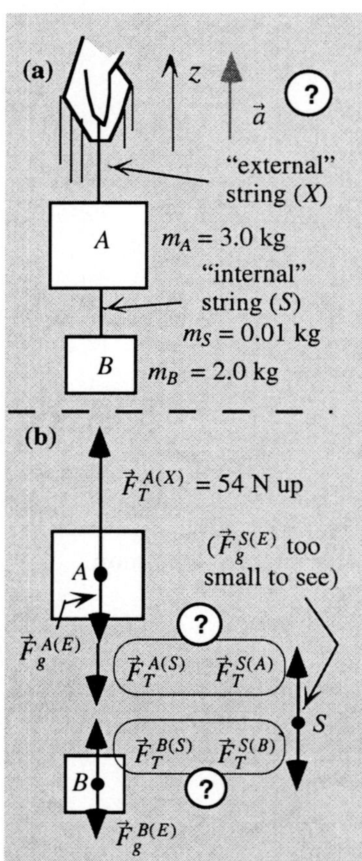

Figure N8.5: (a) A person pulls vertically on two blocks connected by a string. **(b)** Free-particle diagrams for the two blocks and the string. (Note that this diagram also specifies the knowns and unknowns in this problem.)

Problem: Consider two blocks connected by a string (call this the *internal* string) and being pulled vertically upward by another string (the *external* string) attached to block A. Imagine that block A has a mass of 3.0 kg and block B has a mass of 2.0 kg, and imagine that the external string exerts a tension force of magnitude 54 N. How do the tension forces exerted by the ends of the internal string compare if the mass of the internal string is 0.010 kg?

Solution Again, the usual first step is to draw a sketch (see Figure N8.5a) and a set of free-body diagrams (Figure N8.5b). I have defined symbols for the internal string (S) and the external string (X), set up a reference frame (we only need a z-axis for this problem), and indicated knowns and unknowns (the latter with question marks). The implicit constraints on the motion are that the objects only move vertically, and that all three objects have the same z-acceleration a_z.

To solve the problem, we apply Newton's second law in component form to each of the three objects involved (block A, block B, and the internal string S). In this case, only the z component of Newton's second law is interesting (the others all read 0 = 0). So, rather than write a bunch of zeros, we will cut to the chase: the z component of Newton's second law for the three objects reads

$$m_A a_z = F^A_{\text{net},z} = F^{A(E)}_{g,z} + F^{A(X)}_{T,z} + F^{A(S)}_{T,z} = -m_A g + F_T^{A(X)} - F_T^{A(S)} \quad \text{(N8.7a)}$$

$$m_S a_z = F^S_{\text{net},z} = F^{S(E)}_{g,z} + F^{S(A)}_{T,z} + F^{S(B)}_{T,z} = -m_S g + F_T^{S(A)} - F_T^{S(B)} \quad \text{(N8.7b)}$$

$$m_B a_z = F^B_{\text{net},z} = F^{B(E)}_{g,z} + F^{B(S)}_{T,z} = -m_B g + F_T^{B(S)} \quad \text{(N8.7c)}$$

respectively, where (as usual) the force symbols appearing on the right without arrows or component subscripts refer to magnitudes In addition, Newton's third law tells us that forces circled in gray in Figure N8.5b have equal magnitudes:

$$F_T^{A(S)} = F_T^{S(A)} \quad \text{(N8.7d)}$$

$$F_T^{S(B)} = F_T^{B(S)} \quad \text{(N8.7e)}$$

Equations N8.7 represent five equations in the four unknown force magnitudes $F_T^{S(A)}$, $F_T^{A(S)}$, $F_T^{S(B)}$, and $F_T^{B(S)}$ and the one unknown acceleration a_z. Since we have as many equations as unknowns, we can solve this problem. Here is how. First, solve equation N8.7c for $F_T^{B(S)}$. According to equation N8.7e, that force magnitude is equal to $F_T^{S(B)}$ so we can plug this all into equation N8.7b to eliminate $F_T^{S(B)}$ and rearrange things a bit to get:

$$F_T^{S(A)} = m_S a_z + m_S g + m_B a_z + m_B g \quad \text{(N8.8)}$$

Exercise N8X.3: Verify equation N8.8.

Equation N8.7d tells us that $F_T^{S(A)} = F_T^{A(S)}$, so we can plug equation N8.8 into equation N8.7a, eliminating $F_T^{A(S)}$. After a bit more rearrangement, we get

$$(m_A + m_S + m_B)a_z = + F_T^{A(X)} - (m_A + m_S + m_B)g \qquad \text{(N8.9)}$$

Exercise N8X.4: Verify equation N8.9.

This equation says that the system consisting of the two blocks and the internal string accelerates as if it were a single object of mass $M = m_A + m_S + m_B$ acted on by two external forces: (1) the tension force exerted by the external string and (2) the system's total weight. All references to the internal forces acting in the system have disappeared in this equation! This is yet another illustration of the general principle that a system of interacting objects responds to external forces as if it were a single object. Notice again how the third law is required to ensure that this is the case.

We can easily solve equation N8.9 for the unknown acceleration a_z:

$$a_z = \frac{F_T^{A(X)}}{M} - g = \frac{54\,\not{\text{N}}}{5.01\,\not{\text{kg}}}\left(\frac{1\,\not{\text{kg}}\cdot\text{m/s}^2}{1\,\not{\text{N}}}\right) - 9.8\frac{\text{m}}{\text{s}^2} = 0.98\frac{\text{m}}{\text{s}^2} \qquad \text{(N8.10)}$$

Once a_z is known, we can plug its value back into equations N8.7b and N8.7c to find the magnitudes of the tension forces exerted by the string. A simple rearrangement of equation N8.7c tells us that

$$F_T^{B(S)} = m_B(g + a_z) \qquad \text{(N8.11)}$$

Note that this implies that the string exerts a force on block B that is *larger* than the weight of block B. (It is necessary to have a nonzero *net* force on B to accelerate it upward.) You can verify that $F_T^{B(S)} = 21.6$ N.

Exercise N8X.5: Check this result.

Equation N8.7b in combination with Newton's third law (equations N8.7d and N8.7e) implies that the difference in the magnitudes of the tension forces exerted by each end of the string is:

$$F_T^{A(S)} - F_T^{B(S)} = m_S(g + a_z) \qquad \text{(N8.12)}$$

Exercise N8X.6: Verify equation N8.12.

This difference in forces is necessary to provide a nonzero net force to accelerate the string. The difference is 0.11 N when $m_S = 0.010$ kg.

The difference between the tension magnitudes at the string's ends decreases as its mass decreases

Note in example N8.3 that the difference in the tension forces between the two ends of the string is 0.11 N, which is about 0.5% of the magnitude of the tension forces themselves. Therefore, with a string this light, the magnitudes of the tension forces are almost imperceptibly different.

Even 0.010 kg = 10 g is a pretty large mass for a short string. A realistic string having a length of 20 cm or so might actually have a mass of 2 g or less.

Exercise N8X.7: If the string's mass were 1 g, show that the system's acceleration is 0.998 m/s^2, the tension force exerted on block B is still essentially 21.6 N, and the difference in tension across the string is less than 0.05% of this.

The last exercise makes it clear that as the string's mass approaches zero, we can begin to think of 21.6 N as being *the* tension of the string: the difference between the tensions at its ends is very very small.

If the string were completely massless, then equation N8.12 tells us that the tension forces exerted by the string's ends are exactly the same, and *the* tension on the string is a precisely defined number. While no string is truly massless, strings connecting objects usually have masses much smaller than the objects that they connect, so the idea that the string is massless represents an excellent approximation to the real situation.

Treating strings as massless makes problem easy!

If we adopt the fiction that the string is *massless*, then analyzing a situation like the one shown in Figure N8.5 becomes much simpler. We can ignore the free-body diagram of the string altogether and simply assume that the forces exerted by the string's ends have the common magnitude F_T^S. Instead of the *five* equations N8.7a through N8.7e, the problem is completely described by *two* equations in the unknowns a_z and F_T^S that express the z component of Newton's second law as it applies the two blocks:

$$m_A a_z = F^A_{\text{net},z} = F^{A(E)}_{g,z} + F^{A(X)}_{T,z} + F^{A(S)}_{T,z} = -m_A g + F_T^{A(X)} - F_T^S \quad \text{(N8.13a)}$$

$$m_B a_z = F^B_{\text{net},z} = F^{B(E)}_{g,z} + F^{B(S)}_{T,z} = -m_A g + F_T^S \quad \text{(N8.13b)}$$

Simply *adding* these equations eliminates the unknown tension F_T^S, leaving

$$(m_A + m_B)a_z = F^{A(X)}_{T,z} - (m_A + m_B)g \quad \text{(N8.14)}$$

which can be quickly solved for a_z:

$$a_z = \frac{F_T^{A(X)}}{(m_A + m_B)} - g = \frac{54\,\cancel{\text{N}}}{5.0\,\cancel{\text{kg}}}\left(\frac{1\,\cancel{\text{kg}}\cdot\text{m/s}^2}{1\,\cancel{\text{N}}}\right) - 9.8\frac{\text{m}}{\text{s}^2} = 1.0\frac{\text{m}}{\text{s}^2} \quad \text{(N8.15)}$$

Plugging this result into equation N8.13b, we find that

$$F_T^S = m_B(g + a_z) = (2.0\text{ kg})(10.8\text{ m/s}^2) = 21.6\text{ kg}\cdot\text{m/s}^2 = 21.6\text{ N} \quad \text{(N8.16)}$$

Note how using this massless string approximation allows us to analyze the situation shown in Figure N8.5 in a few lines, where it took more than a page of work before. Moreover, we get essentially the same results!

Ideal string model

In physics, the technical term **ideal string** refers to a hypothetical massless, inextensible, and flexible cord binding two objects together. An ideal string is really a *model* that we can apply to any real string, chain, wire, or other cord connecting two objects as long as (1) the cord's mass is very much less than the masses of the objects that it connects, (2) the cord is reasonably inextensible and flexible, and (3) we state that we are using the approximation. **The tension on a** (presumably ideal) **string** is defined to be the tension force exerted on the object at *either* end of the string as a result of its interaction with the string.

N8.5 PULLEYS

The ideal pulley model

Consider the situation shown in Figure N8.6. The pulley in this situation redirects the string so that the tension forces exerted by its ends need no longer be opposite in direction: the string pulls *rightward* on object A and *upward* on object B. An **ideal pulley** simply redirects the string without creating any kind of difference in tension between the string on one side of the pulley and the string on the other. Like an ideal string, an ideal pulley is a mental construct that we use as an approximate model for a real pulley.

A real pulley will be approximately ideal if (1) it has a very small mass (particularly out near the rim) and (2) if it has nearly frictionless bearings. If the pulley has a significant mass, then making significant changes in the pulley's rotational velocity requires applying a significant torque to the pulley, which in turn requires that there be a significant difference in the tension of the string on one side and the string on the other. To see this, imagine tugging on a rope that goes over a very massive pulley. Perhaps you can imagine that you will have to pull very hard on the rope to accelerate the massive pulley wheel, even if the other side of the rope is almost slack. Similarly a difference in tension would be needed to rotate the pulley if its bearings have friction: this difference in tension supplies a net force and thus a net torque to the rim of the pulley that keeps it turning against the opposing torque due to friction.

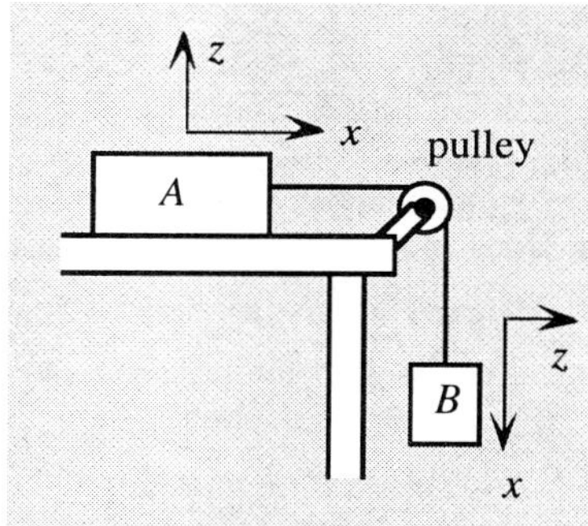

Figure N8.6: A situation involving two objects connected by a string going over a pulley.

Pulleys these days are sufficiently light and frictionless that the ideal pulley model is actually a fairly reasonable approximation in many situations. Like the ideal string model, the ideal pulley model makes problems involving pulleys *much* easier to do.

Separate reference frames are often advantageous

In working problems involving pulleys, you will often find it advantageous to use a separate reference frame for each object, as shown in Figure N8.6. This is perfectly acceptable as long as you (1) keep straight which reference frame applies to which object, and (2) are careful to correctly describe in *each* reference frame any quantities that link the objects. For example, in the situation shown in Figure N8.6, the reference frames have been chosen so that the acceleration of each object points entirely along the x axis and a_x has the same positive value for both objects (since the string has a fixed length and thus any motion of object A along its x axis is exactly duplicated by object B along *its* x axis.)

N8.6 USING THE FRAMEWORK

We can adapt the constrained-motion framework to solve coupled-object problems: these problems are almost exactly like the linearly constrained motion problems discussed in the last chapter.

Adapting the framework for coupled-object problems

The main differences are as follows. First of all, we need to:

1. Draw a *separate* free-particle diagram for each object in the system (and provide single-letter labels for all objects involved)
2. Circle and link any *third-law partners* in these diagrams
3. Describe in the *conceptual representation* section any *joint* constraints on the objects' motion (for example, we might say "Because the objects are connected, a_x is the same for both")
4. Apply Newton's second law in component form to *each* object

In addition (depending on the details of the problem), it *may* be necessary to:

5. Draw individual reference frames for each object
6. Draw separate acceleration arrows for each object
7. Use Newton's third law to connect the magnitudes of third-law partners

You may also find it *helpful* to draw free-body diagrams of the objects involved in the problem before drawing free-particle diagrams. Free-body diagrams are particularly useful in locating third-law partners.

The example on the following page illustrates how to use the adapted constrained motion framework to analyze a pulley problem based on the situation shown in Figure N8.6.

EXAMPLE N8.4

OUTLINE OF THE FRAMEWORK for CONSTRAINED-MOTION problems (adapted for coupled objects)

1. Pictorial Representation

a. Draw a picture of the situation that includes:

(1) (1) a drawing of the objects in the context of their surroundings

(2) (2) a labeled arrow indicating the direction of each object's acceleration (or say "$\vec{a} = 0$")

(3) (3) reference frame axes (perhaps one for each object), with one axis aligned with the object's acceleration (if any)

(4) (4) labels defining symbols for relevant quantities (and single-letter labels for objects)

(5) b. List values for all known labeled quantities and specify which quantities are unknown.

2. Conceptual Representation

The general task is to construct a conceptual model of the situation and link it to an abstract physics model or principle. In constrained motion problems, we do the following:

(6) a. Draw a free-particle (and, if helpful, a free-body) diagram of each object, and indicate any third-law partners

(7) b. Describe the constraint(s) on the motions of the objects and check that the net force on each object is consistent with its observed acceleration (use a net-force diagram if necessary)

(8) c. Describe any approximations or assumptions you have to make to solve the problem

3. Mathematical Representation

(9) a. Apply Newton's second law *in vector component form*

(10) b. Solve for any unknown quantities symbolically (using Newton's third law if necessary)

(11) c. Plug in numbers and units and calculate the result and its units.

4. Evaluation

Check that the answer makes sense:

(12) a. Does it have the correct units?
(13) b. Does it have the right sign?
(14) c. Does it seem reasonable?

Problem: In the situation shown in Figure N8.6, assume that block A has a mass of 0.75 kg, block B has a mass of 0.25 kg, and the table is frictionless. What is the tension on the string connecting the blocks?

Solution: A diagram of the situation looks like this:

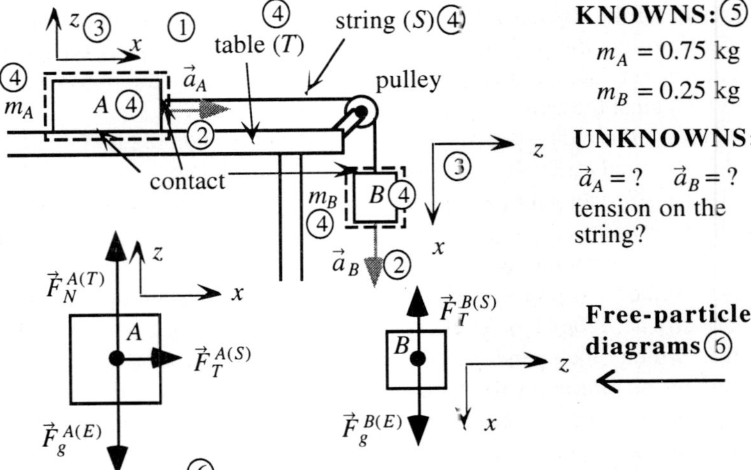

There are no third-law partners in these diagrams, but we will use the (8) ideal string and ideal pulley approximations, which together mean that $F_T^{A(S)} = F_T^{B(S)}$. Each object is constrained to move in its the +x direction (according to its own set of reference frame axes, as defined above). The string also constrains the object's to have the same magnitude of (7) acceleration. So *in each object's respective reference frames*, both $\vec{a}_A$ and $\vec{a}_B$ have the components $[a_x, 0, 0]$. The forces shown in the free-particle diagrams look consistent with an acceleration in the +x direction (as long as $F_g^{B(E)} > F_T^{B(S)}$). Applying Newton's second law in component form to object A, we get:

$$(9)\quad \begin{bmatrix} m_A a_x \\ 0 \\ 0 \end{bmatrix} = \begin{bmatrix} F_{g,x}^{A(E)} \\ F_{g,y}^{A(E)} \\ F_{g,z}^{A(E)} \end{bmatrix} + \begin{bmatrix} F_{N,x}^{A(T)} \\ F_{N,y}^{A(T)} \\ F_{N,z}^{A(T)} \end{bmatrix} + \begin{bmatrix} F_{T,x}^{A(S)} \\ F_{T,y}^{A(S)} \\ F_{T,z}^{A(S)} \end{bmatrix} = \begin{bmatrix} 0 \\ 0 \\ -m_A g \end{bmatrix} + \begin{bmatrix} 0 \\ 0 \\ F_N^{A(T)} \end{bmatrix} + \begin{bmatrix} F_T^{A(S)} \\ 0 \\ 0 \end{bmatrix}$$

The bottom and top lines (z and x components of this equation tell us (respectively) that $m_A g = F_N^{A(T)}$ (which is not really relevant) and

$$m_A a_x = F_T^{A(S)} \qquad (1)$$

Since the forces on object B act entirely in the $\pm x$ direction, when we apply Newton's second law to that object, we will get 0 = 0 for all components except the x component, which will read

$$(9)\quad m_B a_x = F_{g,x}^{B(E)} + F_{T,x}^{B(S)} = m_B g - F_T^{B(S)} \qquad (2)$$

The ideal pulley and string approximations mean that $F_T^{A(S)} = F_T^{B(S)}$, so adding equations (1) and (2) causes the unknown tension force magnitudes to cancel, leaving us with $(m_A + m_B)a_x = m_B g$. Solving this for a_x and plugging the result into equation (1), we get (10)

$$F_T^{A(S)} = m_A\left(\frac{m_B g}{m_A + m_B}\right) = \frac{(0.25\ \cancel{\text{kg}})(0.75\ \text{kg})(9.8\ \text{m/s}^2)}{0.25\ \cancel{\text{kg}} + 0.75\ \cancel{\text{kg}}} = 1.84\ \text{N} \qquad (11)$$

since $1\ \text{kg}\cdot\text{m/s}^2 = 1$ N. This is the string's tension by definition. This has the right sign for a magnitude, the right units, and seems plausible, (13)(12) particularly as $F_g^{B(E)} = m_B g = 2.45\ \text{N} > 1.84\ \text{N} = F_T^{A(S)} = F_T^{B(S)}$, as is (14) required to produce a positive x-acceleration in block B.

SUMMARY

I. FORCE NOTATION FOR COUPLED OBJECTS
   A. A pair of *coupled objects* are objects that are constrained by some connection so that their motions are linked in a well-defined manner
   B. Notation for forces in coupled-object problems (example: $\vec{F}_N^{A(B)}$)
      1. the subscript describes the type of force (and thus the interaction)
      2. the superscripts specify the objects involved in the interaction
         a) the first superscript describes object that the force acts on
         b) the second (in parentheses) describes *other* object in interaction
   C. Recognizing *third-law partners*
      1. Basic rules about third-law partners
         a) each force must act on a *different* object
         b) the forces must represent the two ends of the *same* interaction
      2. The implications for the partners' force symbols: $\vec{F}_N^{A(B)} = -\vec{F}_N^{B(A)}$
         a) the superscripts must be reversed
         b) the subscripts must the same (to reflect the same interaction)
      3. Third-law partners are easy to spot on the free-body diagrams of two objects participating in a contact interaction
         a) look for the surfaces in contact
         b) the partners are the forces applied to these surfaces

II. IMPORTANT FEATURES OF THE PUSHED BLOCKS EXAMPLE
   A. Newton's third law links the forces that the blocks exert on each other
   B. the set behaves as if it were a single object pushed by the hand

III. THE EXAMPLE INVOLVING BLOCKS CONNECTED BY A STRING
   A. Solving the problem the hard way
      1. Applying Newton's second and third laws to the three objects yield five equations in five unknowns (four tensions and an acceleration)
      2. Again, the set's acceleration is as if it were a particle responding to the *external* forces exerted on the set.
      3. the difference between the magnitudes of the forces exerted by each end of the internal string goes to zero as string's mass goes to zero
   B. The *ideal string* model assumes that the string is completely massless, inextensible, and flexible
      1. the tension forces exerted by string ends then will have *equal* magnitudes (we call this common magnitude *the* tension on the string)
      2. two objects tied together by an ideal string move exactly together
   C. The ideal string model makes problems *much* easier to solve

IV. PULLEYS
   A. An *ideal* (massless, frictionless) pulley changes the *direction* of a string without affecting the tension forces it exerts
   B. Real pulleys have both nonzero mass and nonzero friction
      1. rotating a real pulley requires a nonzero torque and thus a difference in the string tension before and after the pulley
      2. this difference is often fairly small for a good pulley
   C. Using a *separate reference frame* for each object can be advantageous

V. ADAPTING THE CONSTRAINED-MOTION FRAMEWORK
   A. When solving coupled-objects problems, we need to:
      1. draw separate acceleration arrows for each object (if necessary)
      2. draw a *separate* free-particle diagram for each object (and maybe set up a separate reference frame for each if this is helpful)
      3. note any third-law partners in the free-particle diagrams (if there are any, you can use Newton's third law to link their magnitudes)
      4. describe the joint constraints on objects' motion
      5. Apply Newton's second law in component form to *each* object
   B. Example N8.3 illustrates the use of this adapted framework

GLOSSARY

coupled objects: a set of interacting objects that are constrained to move together in such a way that their acceleration magnitudes are linked (in almost all cases, the acceleration magnitudes are actually *equal*).

ideal string: a massless, inextensible and perfectly flexible string. The tension forces exerted by each end of such a string are always equal. This model not only is a reasonably accurate description of most real strings (or other connecting cords) but its use greatly simplifies problems involving strings.

***the* tension on a string:** the magnitude of the tension force exerted by *either* end of an ideal string.

ideal pulley: a massless, frictionless pulley that changes the direction of a string without causing a difference in tension between the string on one side of the pulley and the string on the other.

TWO-MINUTE PROBLEMS

N8T.1 A jet airplane flies at a constant velocity through the air. Its jet engines exert a constant force forward on the plane that exactly balances the force of air friction exerted backward on the plane. These forces are equal in magnitude and opposite in direction. Do we know this because of Newton's second law or Newton's third law?
A. Newton's second law
B. Newton's third law
C. both laws
D. neither (explain)

N8T.2 Which of the following pairs of forces are third-law partners? Answer T if the pair of forces described are third-law partners, F if they are not.
(a) A thrust force from its propeller pulls a plane forward; a drag force pushes it backward.
(b) A car exerts a forward force on a trailer; the trailer tugs backward on the car.
(c) a motorboat propeller pushes backward on the water; the water pushes forward on the propeller
(d) gravity pulls down on a person sitting in a chair; the chair pushes back up on the person.

N8T.3 A box (B) sits in the back of a truck (T) as the truck slows down for a stop (the box remains motionless relative to the truck). What is the appropriate symbol for the horizontal force that the contact interaction between the box and the truck exerts on the truck?
A. $\vec{F}_{N}^{B(T)}$
B. $\vec{F}_{N}^{T(B)}$
C. $\vec{F}_{SF}^{B(T)}$
D. $\vec{F}_{SF}^{T(B)}$
E. $\vec{F}_{KF}^{B(T)}$
F. other (specify)

N8T.4 A child (C) pulls on a wagon (W) using a string (S); the wagon moves forward at a constant speed as a result. The third-law partner to the forward force exerted on the wagon is which of the following forces? (R = road.)
A. $\vec{F}_{T}^{S(W)}$
B. $\vec{F}_{T}^{W(S)}$
C. $\vec{F}_{T}^{W(C)}$
D. $\vec{F}_{T}^{C(W)}$
E. $\vec{F}_{KF}^{W(R)}$
F. other (specify)

N8T.5 A small car pushes on a disabled truck, accelerating it slowly forward. Each exerts a force on the other as a result of their contact interaction. Which vehicle exerts the *greater* force on the other?
A. the car
B. the truck
C. both forces have the same magnitude
D. the truck doesn't exert any force on the car
E. not enough information for a meaningful answer

N8T.6 A physicist and a chemist are playing tug-of-war. For a certain length of time during the game, the participants are essentially at rest. During this time, each person pulls on the rope (which can be treated like an ideal string) with a force of 350 N. What is the tension on the rope?
A. 700 N
B. 350 N
C. 175 N
D. other (specify).

N8T.7 Object A (m_A = 1.0 kg) hangs at rest from an ideal string A connected to the ceiling. Object B (m_B = 2.0 kg) hangs at rest from an ideal string B connected to object A. The tension on string A is
A. Twice the tension on string B
B. 3/2 times the tension on string B
C. Equal to the tension on string B
D. 2/3 the tension on string B
E. other (specify)

N8T.8 Two people are attempting to break a rope, which will break if the tension on the rope exceeds 360 N. If each person can exert a pull of 200 N, they can break the rope if
A. they each take an end and pull
B. they tie one end to the wall and both pull on the other
C. they use either of the strategies above
D. (they cannot break the rope)

N8T.9 A spring scale typically indicates the magnitude of the tension force exerted on its bottom hook. What will the scale read in each of the cases shown in the diagram below? (*Hint:* Construct a free-body or free-particle diagram for the scale in each case.)
A. 49 N
B. 98 N
C. 186 N
D. other (specify)

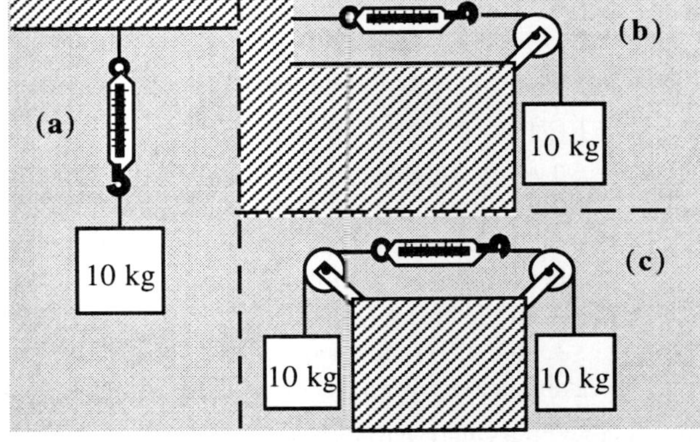

HOMEWORK PROBLEMS

BASIC SKILLS

N8B.1 Which of the following force pairs are third-law partners? Explain your reasoning.
(a) The earth attracts a stone; the stone attracts the earth.
(b) A jet's engine thrusts it forward; drag pushes it back.
(c) You push on a box without moving it; the floor pushes back on the box.

N8B.2 Which of the following force pairs are third-law partners? Explain your reasoning.
(a) A mule pulls on a plow, moving it forward; the ground pulls backward on the plow.
(b) A team of dogs pulls a sled, moving it; the sled pulls backward on the dogs.
(c) Gravity tugs downward on a box sitting on the ground; the ground pushes up on the box.

N8B.3 In example N8.2, what would be the magnitude of the force that the blocks exert on each other if blocks A and B were to have masses of 10 kg and 6.0 kg and we push on the blocks with a force of 4.0 N?

N8B.4 Find the acceleration of the system, the magnitude of the tension force on block *B*, and the difference in the magnitudes of the tension forces exerted by each end of the string in example N8.3 if the string has a mass of 100 g. Compare to the answers that we got before. Is this string even approximately ideal, in your opinion?

Free-particle diagrams and third-law partners. For each of the situations described in problems N8B.5 through N8B.12 draw a separate, isolated free-particle diagram for each object involved (you may also draw free-body diagrams if this is helpful). Assign an appropriate symbol to each force vector, according to the conventions established in this chapter and in chapter N3. Indicate the approximate relative magnitudes of the forces by giving each arrow an appropriate length. Circle and link any third-law partners in the diagrams that you draw.

N8B.5 The moon orbits the earth. (Draw diagrams for both the moon and the earth.)

N8B.6 You jump up off the floor. (Draw a diagram for both you and the earth during the interval of time while you are still in contact with the floor and you are still accelerating upward.)

N8B.7 A little box sits on a bigger box. Both boxes are at rest. (Draw diagrams for both boxes.)

N8B.8 A little box sits on top of a bigger box sitting on an incline. Both boxes are at rest. (Draw diagrams for both of the boxes.)

N8B.9 A little box sits on top of a bigger box. The big box is sliding on a rough but level floor, and as a result is slowing down. (Draw diagrams for both boxes.)

N8B.10 A tractor pulls a plow at a constant velocity in a field. (Draw diagrams for both the tractor and the plow.)

N8B.11 A person hangs from a helicopter by a rope as the helicopter begins to accelerate upward. (Draw diagrams for the person, the helicopter, and the rope.)

N8B.12 A small car pushes a large disabled truck so that both accelerate gently forward. (Draw diagrams for both the car and the truck. Ignore air resistance.)

SYNTHETIC

N8S.1 Two teams of people are involved in a tug-of-war. Since the forces exerted by each team on the other are equal and opposite by the third law, how is it possible for either team to win? Explain carefully, using appropriate free-body diagrams, how one team *can* win.

N8S.2 A block with a substantial mass *M* is suspended from the ceiling by a light string *A*. An identical string *B* hangs from the bottom of the block. If you jerk suddenly on string *B*, it will break, but if you pull steadily on string *B*, string *A* will break. Using a force diagram of the block, carefully explain why.

N8S.3 A 12,000-kg tugboat pushes on a 420,000-kg barge in still water. If the tug and barge accelerate at a rate of 0.20 m/s^2, what is the magnitude of the force that the water has to exert on the tugboat's propellers? What is the magnitude of the force that the tug exerts on the barge? Ignore friction.*

N8S.4 A 32-kg child puts a 15-kg box into a 12-kg wagon. The child then pulls horizontally on the wagon with a force of 65 N. If the box does not move relative to the wagon, what is the static friction force on the box?*

N8S.5 A 32,000-kg spaceship in deep space is hauling a 250,000-kg canister filled with ore from an asteroid mining operation. The canister is connected to the spaceship by a cable that can withstand a tension of 1.0 MN without breaking. What is the maximum acceleration that the spaceship can give to the canister? What thrust would the spaceship's engines have to exert to provide this?*

N8S.6 A 32-kg crate slides on a rough plane inclined upward at an angle of 28°. The crate is hauled up the plane by lightweight rope that goes parallel to the incline and then over a pulley at the top of the incline. A worker standing below the pulley pulls vertically downward on the rope. If the 55-kg worker hangs his or her entire weight from the rope, it is barely sufficient to move the crate up the incline at a constant velocity of 1.2 m/s. What is the magnitude of the sliding friction force on the box?*

N8S.7 A 85-kg crate sits in a 280-kg boat. *Estimate* the upper limit on the thrust that the boat's propeller can exert on the water if we are to ensure that the crate remains motionless relative to the boat.*

N8S.8 A 82-kg worker clings to a lightweight rope going over a lightweight, low friction pulley. The other end of the rope is connected to a 67-kg barrel of bricks. If the worker is initially at rest 15 m above the ground, how fast will he or she be moving when he or she hits the ground?*

N8S.9 Consider the situation shown in Figure N8.6 except assume that the table is inclined at an angle of 22°, with the pulley at the top of the incline. Does block *A* slide up or down the incline? Assume that the table is frictionless, $m_A = 0.75$ kg, $m_B = 0.25$ kg, and that the mass of the pulley and string are negligible.*

N8S.10 Do problem N8S.5 first. Redo problem N8S.5, except assume that the mass of the cable is 1200 kg. Is your answer much different?*

*When you write your solution to these starred problems, be sure to include all of the elements in the problem-solving framework, as outlined in example N8.4.

RICH-CONTEXT

N8R.1 A mule is asked to pull a plow. The mule resists, explaining that "If I tug on the plow, Newton's third law asserts that the plow will tug on me with an equal and opposite force. Since these forces will cancel each other out, it is obvious that we're not going anywhere. Therefore, there is no point in trying." Carefully (but gently) explain to the mule the error in its reasoning, and, using appropriate free-body or free-particle diagrams, explain why it is possible for the mule to accelerate the plow.

N8R.2 A 0.62-kg glider slides frictionlessly on an 3-m air track inclined at an angle of 12°. The glider is connected to a lightweight string which passes over a low-friction pulley at the top of the air track. The other end of the string is connected to a spring scale (whose mass is 0.05 kg), which is connected to a 0.17-kg weight. If you hold the glider at rest, what will the scale read? If you then release the glider, will it accelerate up or down the incline? What will the scale read as the system accelerates?

ADVANCED

N8A.1 Imagine a train consisting of N frictionless cars following a locomotive that is accelerating the whole train forward with acceleration of magnitude a. Assume that the first car behind the locomotive has mass M. If the tension in the coupling at the rear of *each* car is 10% smaller than the tension in the coupling in the front of the car, what is the mass of each successive car, as a fraction of M? What is the total mass of the cars in terms of M? [*Hint*: You might like to know that $1+x+x^2+\ldots+x^n=(1-x^{n+1})/(1-x)$.]

ANSWERS TO EXERCISES

N8X.1 Free-body diagrams for the elevator and the person look like this:

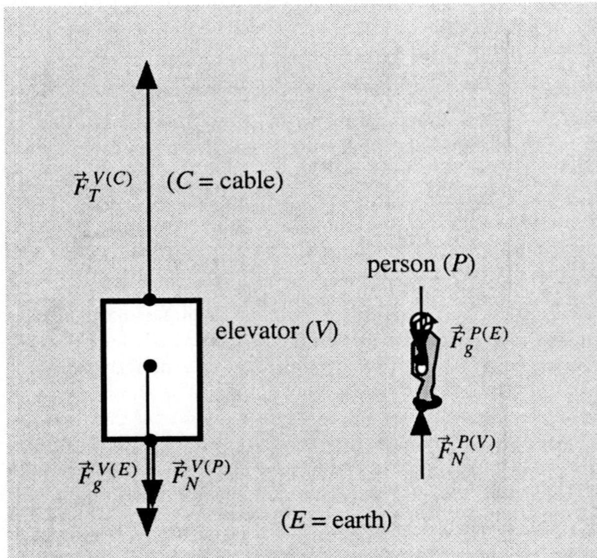

The only third-law partners appearing in this diagram are the normal forces $\vec{F}_N^{V(P)}$ and $\vec{F}_N^{P(V)}$ that the person and elevator floor exert on each other due to their mutual contact interaction. Note that $F_N^{P(V)}$ must be equal to the person's weight if the person is not accelerating vertically; similarly, $F_T^{V(C)}$ must be equal to the sum of $F_g^{V(E)}$ and $F_N^{V(P)}$. The latter force would not be there if there is no person in the elevator, so the tension on the cable has to increase by this amount (which happens to be equal to the person's weight) when the person is in the elevator (and there is no vertical acceleration). So, even though the person's weight acts on the *person*, not the elevator, the effect of this weight is communicated to the elevator (and ultimately to the cable) by the contact interaction between the person and the floor (in a manner constrained by Newton's second and third laws).

N8X.2 Plugging $F_N^{A(H)}=9.0$ N into equation N8.5 we get

$$a_x=\frac{9.0\,\cancel{\text{N}}}{6.0\,\cancel{\text{kg}}}\left(\frac{1\,\cancel{\text{kg}}\cdot\text{m/s}^2}{1\,\cancel{\text{N}}}\right)=1.5\text{ m/s}^2 \quad \text{(N8.17)}$$

Plugging this result then into equation N8.6, we find that $F_N^{A(B)}=F_N^{B(A)}=(2.0\text{ kg})(1.5\text{ m/s}^2)=3.0\text{ N}$.

N8X.3 Solving equation N8.7c for $F_T^{B(S)}$ and using equation N8.7e, we get

$$m_B(a_z+g)=F_T^{B(S)}=F_T^{S(B)} \quad \text{(N8.18)}$$

Plugging this into equation N8.7b, we get:

$$m_Sa_z=-m_Sg+F_T^{S(A)}-m_B(a_z+g) \quad \text{(N8.19)}$$

Solving this for $F_T^{S(A)}$ yields equation N8.8.

N8X.4 Plugging equation N8.8 into N8.7a, we get

$$m_Aa_z=-m_Ag+F_T^{A(X)}-(m_Sa_z+m_Sg+m_Ba_z+m_Bg) \quad \text{(N8.20)}$$

Adding $m_Sa_z+m_Ba_z$ to both sides yields equation N8.9.

N8X.5 Plugging numbers into equation N8.11, we get:

$$F_T^{B(S)}=(2.0\text{ kg})(9.8\text{ m/s}^2+0.98\text{ m/s}^2)=21.6\text{ N} \quad \text{(N8.21)}$$

(Remember that $1\text{ kg}\cdot\text{m/s}^2=1$ N.)

N8X.6 Adding m_Sg to both sides of N8.7b, we get

$$m_S(a_z+g)=+F_T^{S(A)}-F_T^{S(B)} \quad \text{(N8.22)}$$

But $F_T^{S(A)}=F_T^{A(S)}$ and $F_T^{S(B)}=F_T^{B(S)}$ according to Newton's third law. Plugging these into the above we get:

$$m_S(a_z+g)=+F_T^{A(S)}-F_T^{B(S)} \quad \text{(N8.23)}$$

which is equation N8.12.

N8X.7 Equation N8.10 now reads

$$a_z=\frac{54\,\cancel{\text{N}}}{5.001\,\cancel{\text{kg}}}\left(\frac{1\,\cancel{\text{kg}}\cdot\text{m/s}^2}{1\,\cancel{\text{N}}}\right)-9.8\frac{\text{m}}{\text{s}^2}=0.998\frac{\text{m}}{\text{s}^2} \quad \text{(N8.24)}$$

Equation N8.11 now reads

$$F_T^{B(S)}=(2.0\text{ kg})(9.8\text{ m/s}^2+0.998\text{ m/s}^2)=21.6\text{ N} \quad \text{(N8.25)}$$

remembering that $1\text{ kg}\cdot\text{m/s}^2$. Note that this is essentially the same as before), and equation N8.12 reads

$$F_T^{A(S)}-F_T^{B(S)}=(0.001\text{ kg})(10.8\text{ m/s}^2)=0.011\text{ N} \quad \text{(N8.26)}$$

Note that (0.011 N)/(21.6 N) ≈ 0.0005 or 0.05%.

N9

CIRCULARLY CONSTRAINED MOTION

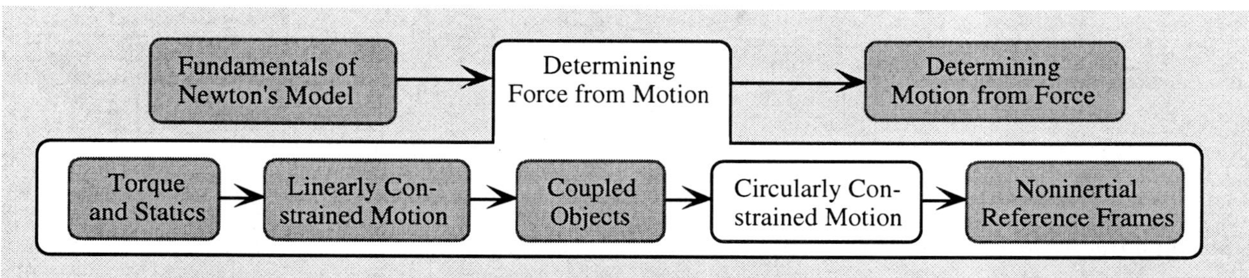

N9.1 OVERVIEW

In chapter N6, we studied how to apply Newton's second law to objects at rest. In chapter N7, we applied this law to objects in linear motion, and in chapter N8, we applied Newton's second and third laws to objects constrained to move together. In this chapter, we continue broadening our repertoire for applications of these laws by studying *circular* motion.

Circular motion is a useful model for an astoundingly wide variety of realistic situations: masses swinging on strings, cars going around corners, satellites orbiting the earth, electrons orbiting nuclei in atoms. An understanding of circular motion is also essential for understanding the behavior and stability of rotating objects from flywheels to planets to galaxies. Learning how to apply Newton's second law to objects in circular motion will therefore greatly increase our ability to interpret the physical world.

Here is an overview of the sections in this chapter.

N9.2 *UNIFORM CIRCULAR MOTION* shows mathematically how we can determine the acceleration of an object in uniform circular motion, illustrating at the same time how we can apply the mathematical ideas of *unit vectors* and the *chain rule* (see the ***Math Skills*** sections), ideas that we will use extensively in the rest of the course.

N9.3 *NON-UNIFORM CIRCULAR MOTION* extends this treatment to the case where an object's speed in its circular path is *not* constant.

N9.4 *BANKING* discusses why planes and bikes moving in a circular path bank into the curve, and why it is advantageous to design roads that do the same for cars.

N9.5 *EXAMPLES* shows how we can use the constrained motion framework to do problems where the object is constrained to move in a circle.

**** *MATH SKILLS: UNIT VECTORS* presents the concept of a *unit vector*, a mathematical tool that we will find very helpful both in the remainder of this unit and also in Unit *E*.

**** *MATH SKILLS: THE CHAIN RULE* discusses the chain rule of differential calculus, which is a very important tool for the serious application of calculus to physical problems.

The ideas and mathematical techniques developed in this chapter provide essential background for *every* chapter in the rest of this unit (and beyond).

N9.2 UNIFORM CIRCULAR MOTION

We introduced the concept of *uniform circular motion* in chapter N2, where we saw that an object moving at a constant speed in a circle was accelerating toward the center of the circle and that the magnitude of this acceleration was plausibly equal to v^2/R, where v is the speed of the object and R is the radius of its circular trajectory. Our task in this section is to put these results on a more firm mathematical foundation, in the process creating tools that we can use to study *nonuniform* circular motion.

General mathematics of circular motion

Consider an object moving in a circle of radius R. If we set up a reference frame so that the circle lies in the xy plane and so that the frame's origin coincides with the circle's center, then we can locate the object on the circle simply by specifying the angle θ that its position vector makes with the x axis (see Figure N9.1). Ignoring the irrelevant z coordinate, its coordinates are:

The object's position

$$\vec{r}(t) \equiv \begin{bmatrix} x(t) \\ y(t) \end{bmatrix} = \begin{bmatrix} R\cos\theta \\ R\sin\theta \end{bmatrix} = R\hat{r}, \quad \text{where } \hat{r} \equiv \begin{bmatrix} \cos\theta \\ \sin\theta \end{bmatrix} \tag{N9.1}$$

where $\hat{r}$ is a **unit vector** (see the ***Math Skills*** section on *Unit Vectors*) that has a magnitude of 1 (since $\sin^2\theta + \cos^2\theta = 1$) and points at any given instant in same direction as the particle's position vector $\vec{r}(t)$. Note that θ is a function of time (since the object is moving around the circle), so $\hat{r}$ is also a function of time. (θ is conventionally understood to increase as the object moves counterclockwise around the circle.)

The object's velocity

To find the velocity, we simply take the time-derivative of $\vec{r}(t)$:

$$\vec{v}(t) = \frac{d\vec{r}}{dt} = R\frac{d\hat{r}}{dt} = R\frac{d}{dt}\begin{bmatrix} \cos\theta \\ \sin\theta \end{bmatrix} = R\begin{bmatrix} -\sin\theta \\ \cos\theta \end{bmatrix}\frac{d\theta}{dt} \tag{N9.2}$$

since by the **chain rule** (see the ***Math Skills*** section on *The Chain Rule*):

$$\frac{d}{dt}\sin[\theta(t)] = \frac{d\sin\theta}{d\theta}\frac{d\theta}{dt} = \cos\theta\frac{d\theta}{dt} \tag{N9.3}$$

$$\frac{d}{dt}\cos[\theta(t)] = \frac{d\cos\theta}{d\theta}\frac{d\theta}{dt} = -\sin\theta\frac{d\theta}{dt} \tag{N9.4}$$

Defining the perpendicular component of the velocity

Now, since $|R\,d\theta|$ is equal to the distance that the object moves around the circle in time dt (see section C9.2), $|R\,d\theta|/dt = v$, the *speed* at which the object moves around the circle. Let us define the symbol $v_\perp \equiv R\,d\theta/dt$: for the case of circular motion, $v_\perp = +v$ if the object moves *counterclockwise* (since θ increases for counterclockwise motion and thus $d\theta/dt$ is positive) and $v_\perp = -v$ if it moves *clockwise*. We can then rewrite equation N9.2 as follows

$$\vec{v}(t) = \begin{bmatrix} v_x(t) \\ v_y(t) \end{bmatrix} = v_\perp\begin{bmatrix} -\sin\theta \\ \cos\theta \end{bmatrix} \quad \text{where } v_\perp \equiv R\frac{d\theta}{dt} \text{ is } \begin{Bmatrix} > 0 \text{ if ccw} \\ < 0 \text{ if cw} \end{Bmatrix} \tag{N9.5}$$

Does this result make sense? Since the velocity of object moving in a circle must be tangent to the circle (and thus perpendicular to the object's position vector $\vec{r}$), inspection of Figure N9.1 shows us that its velocity vector *must* have components $[-v\sin\theta,\ v\cos\theta]$ if it moves counterclockwise and the opposite if it moves clockwise. So equation N9.5 does indeed make sense.

We call $v_\perp \equiv R\,d\theta/dt$ the **perpendicular component** of the object's velocity (meaning the component of $\vec{v}$ that is perpendicular to the object's position vector $\vec{r}$). In the case of circular motion, the object's velocity is always *entirely* perpendicular to $\vec{r}$, so the object's speed is simply $\text{mag}(\vec{v}) = v = |v_\perp|$.

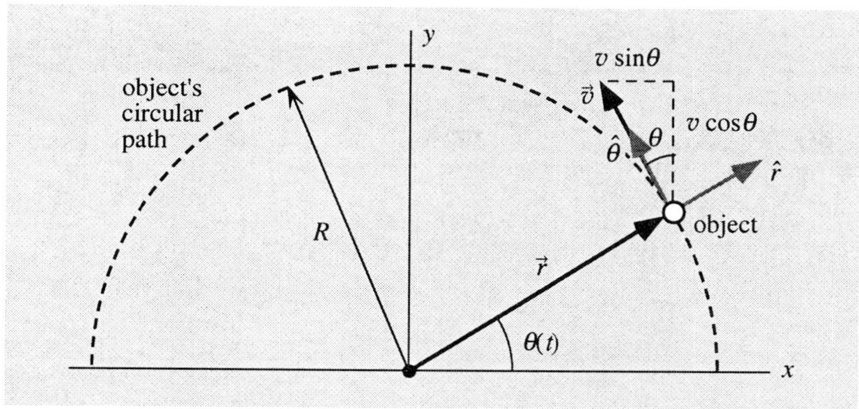

Figure N9.1: An object moving along a circular path. Note that because $\vec{v}$ and $\vec{r}$ are perpendicular, the angle between $\vec{v}$ and the y axis is the same as the angle θ between $\vec{r}$ and the x axis.

The unit vector $[-\sin\theta \;\; \cos\theta]$ always points in the direction that the object would move as θ increases. People commonly give it the symbol $\hat{\theta}$, so

$$\vec{v}(t) = v_\perp \hat{\theta}, \quad \text{where} \quad \hat{\theta} \equiv \begin{bmatrix} -\sin\theta \\ \cos\theta \end{bmatrix} \tag{N9.6}$$

Note that $\hat{\theta}$, like $\hat{r}$, changes direction as time passes and θ changes.

Acceleration for *uniform* circular motion

In *uniform* circular motion, the object's speed in its circular path is a constant. By taking the time derivative of equation N9.6, you can show that

$$\vec{a}(t) \equiv \frac{d\vec{v}}{dt} = v_\perp \frac{d\hat{\theta}}{dt} = v_\perp \frac{d}{dt}\begin{bmatrix} -\sin\theta \\ \cos\theta \end{bmatrix} = v_\perp \begin{bmatrix} -\cos\theta \\ -\sin\theta \end{bmatrix} \frac{d\theta}{dt} \tag{N9.7}$$

Exercise N9X.1: Verify equation N9.7.

Note, however, that the unit vector $[-\cos\theta \;\; -\sin\theta]$ is just $-\hat{r}$ (compare with the definition of $\hat{r}$ given by equation N9.1): instead of pointing away from the origin, $-\hat{r}$ points from the object toward the origin (see Figure N9.1). Note also that by equation N9.5, $d\theta/dt = v_\perp / R$. Therefore the acceleration of an object moving in a circle with a constant speed (**uniform circular motion**) is

$$\vec{a}(t) = -\frac{v_\perp^2}{R}\hat{r} = -\frac{v^2}{R} \quad \text{(for \textit{uniform} circular motion)} \tag{N9.8}$$

In words, this says that the object's acceleration has a magnitude of $a = v^2/R$ and points directly toward the center of the object's circular motion. This is consistent with what we guessed using a motion diagram in Chapter N2.

Useful equations

For the record, note that we have shown in equations N9.2 and N9.7 that

$$\frac{d\hat{r}}{dt} = +\hat{\theta}\frac{d\theta}{dt}, \qquad \frac{d\hat{\theta}}{dt} = -\hat{r}\frac{d\theta}{dt} \tag{N9.9}$$

In uniform circular motion problems, we are often given the time T that it takes to go around the circle once instead of the object's speed. Since the object goes a distance of $2\pi R$ in this time, if its speed is *constant*, its speed is

$$v = 2\pi R/T \tag{N9.10}$$

Exercise N9X.2 An object travels at a constant speed v around a circle whose radius is 3.0 m once every 6.3 s. What is v? The magnitude of its acceleration?

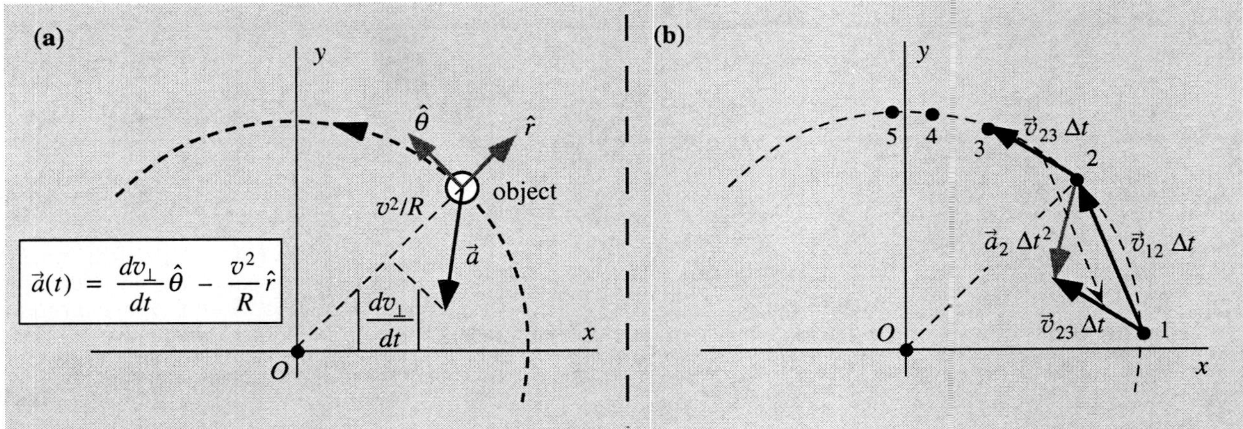

Figure N9.2: An object moving in a circle but slowing down as it goes ($dv/dt < 0$). **(b)** A motion diagram of the same situation, showing that the object's acceleration points a bit backward of directly toward the circle's center

N9.3 NON-UNIFORM CIRCULAR MOTION

Acceleration for nonuniform circular motion

I have introduced rather more mathematical firepower than we strictly need to analyze the case of uniform circular motion problem. However, this makes tackling the case of *nonuniform* circular motion (where R is constant but v is not) much easier. *Everything* that we did in the last section applies whether or not the object's speed v is constant *except* for the very last step of finding the acceleration (equations N9.7 and N9.8) and the practical note in equation N9.10.

If v is *not* constant, then we have to use the product rule to evaluate the time derivative of the velocity (since both $v_\perp$ and $\hat{\theta}$ depend on time):

$$\vec{a}(t) \equiv \frac{d\vec{v}}{dt} = \frac{d}{dt}(v_\perp \hat{\theta}) = \frac{dv_\perp}{dt}\hat{\theta} + v_\perp \frac{d\hat{\theta}}{dt} \qquad \text{(N9.11)}$$

Using equation N9.9 and the definition of $v_\perp$, you can show that

$$\vec{a}(t) = \frac{dv_\perp}{dt}\hat{\theta} - \frac{v^2}{R}\hat{r} \quad \text{and} \quad \text{mag}(\vec{a}) = \sqrt{\left(\frac{dv_\perp}{dt}\right)^2 + \left(\frac{v^2}{R}\right)^2} \qquad \text{(N9.12)}$$

Exercise N9X.3: Verify equation N9.12.

The two parts of the acceleration in this case

This equation says that we can think of the acceleration of an object moving in a circle as the sum of two vectors: the normal acceleration of magnitude v^2/R toward the circle's center due to the way the *direction* of the object's velocity changes with time, and a vector in the direction $\hat{\theta}$ of the object's motion (or opposite to it if $dv_\perp/dt$ is negative) that is due to the way the object's velocity changes. Thus the object's total acceleration vector points somewhat *forward* of directly toward the circle's center when the object is speeding up and somewhat *backward* from directly toward the center if the object is slowing down. Figure N9.2a illustrates this for an object moving counterclockwise.

Exercise N9X.4: Check that the statements "$\vec{a}$ points partly *forward* if the object is speeding up and partly *backward* if the object is slowing down" apply even if the object is moving clockwise around the circle.

Checking this with the help of a motion diagram

Does this all make sense? Figure N9.2b shows a motion diagram of an object moving counterclockwise in a circle and slowing down at the same time. If we use a motion diagram to find the object's acceleration vector, we see that it does indeed point somewhat backward of toward the circle's center.

[We will draw on this material in chapters N12 and N14.]

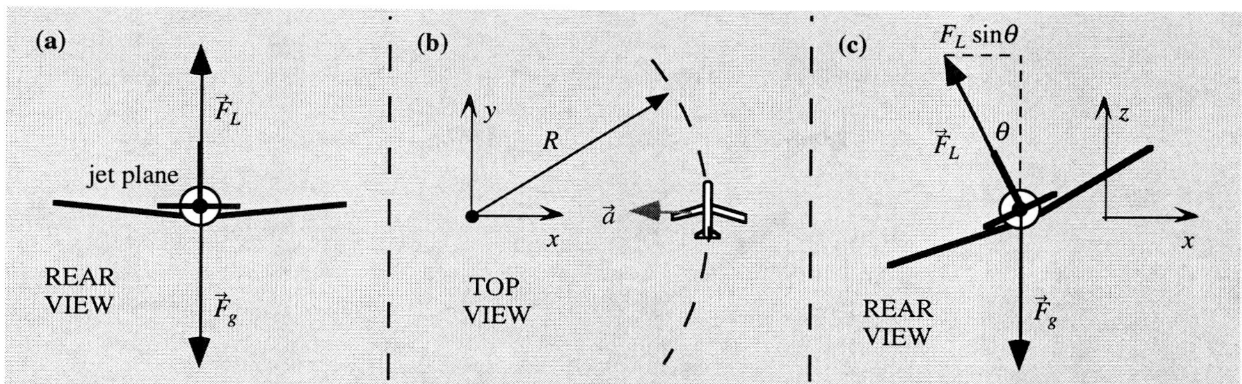

Figure N9.3: (a) A free-particle diagram (rear view) of a jet plane flying straight and level. **(b)** A top view of a plane in a circular turn to the left. **(c)** A free-particle diagram of the plane turning left.

Exercise N9X.5: A car traveling around a circular bend in the road, which has a radius of 450 m. At a certain instant, the car is traveling due west at a speed of 22 m/s and is slowing down at a rate of 1.5 m/s^2. What is the magnitude and direction of the car's acceleration at this instant?

N9.4 BANKING

Why does a plane bank when it turns?

You perhaps know that when a pilot turns a plane, he or she has to bank the plane into the turn, that is, lower one wing and raise the other. Why?

Figure N9.3a shows a rear-view free-particle diagram of the plane. Two forces act vertically on the plane: its weight pulls it downward, and the lift from its wings pushes it upward. When the plane is in straight and level flight, these forces are directly opposite to each other and cancel each other.

We have seen, though, a plane flying in a circular path has an acceleration toward the circle's center, which is *leftward* in Figure N9.3b. Newton's second law tells us that the net force on the plane must also point in that direction. Where does this leftward net force come from? It can't come from the plane's engines, since the thrust force that they exert is always directly forward.

The easiest way to exert a leftward force on a plane is to tilt the wings at an angle θ so that the lift force that they exert (which acts perpendicular to the wings) has both an upward and a leftward component (see Figure N9.3c). If the plane maintains its altitude as it turns, the upward part of the lift force must still balance the downward weight of the plane, leaving the leftward part as the net force on the plane. This net force is what causes the plane to accelerate away from its natural straight-line path into the circular path.

EXAMPLE N9.1

Problem: A jet plane flies at a constant speed of 260 mi/h in a holding pattern that is a horizontal circle of radius 5.0 mi. What is the plane's banking angle?

Solution A sketch of the plane's motion appears in Figure N9.3b and a free-particle diagram for the plane appears as in Figure N9.3c. If the plane is moving in a *horizontal* circle, its acceleration in the z direction is zero: $a_z = 0$. On the other hand, because of its circular motion the plane must have a *nonzero* acceleration toward the circle's center: at the instant shown in Figure N9.3b, $a_x = -v^2/R$ (negative because the acceleration is to the left, that is, in the $-x$ direction as we have defined our coordinates) and $a_y = 0$. Newton's second law then reads:

$$\begin{bmatrix} -mv^2/R \\ 0 \\ 0 \end{bmatrix} = m\vec{a} = \vec{F}_{\text{net}} = \begin{bmatrix} F_{g,x} \\ F_{g,y} \\ F_{g,z} \end{bmatrix} + \begin{bmatrix} F_{L,x} \\ F_{L,y} \\ F_{L,z} \end{bmatrix} = \begin{bmatrix} 0 \\ 0 \\ -mg \end{bmatrix} + \begin{bmatrix} -F_L \sin\theta \\ 0 \\ F_L \cos\theta \end{bmatrix} \quad \text{(N9.13)}$$

EXAMPLE N9.1 (cont.)

If we multiply both sides of the top line of equation N9.13 by –1 and add mg to both sides of the bottom line of equation N9.13, we get

$$\frac{mv^2}{R} = F_L \sin\theta, \quad mg = F_L \cos\theta \tag{N9.14}$$

This problem looks hopeless at first: while we know v and R, we do not know the plane's mass m, the magnitude of the lift force F_L, or the angle θ, meaning that we have more unknowns than equations! But it turns out that we can solve the problem anyway. If we divide the first equation by the second, the unknown m and F_L divide out, and we are left with

$$+\frac{v^2}{Rg} = \frac{\sin\theta}{\cos\theta} = \tan\theta \quad \Rightarrow \quad \theta = \tan^{-1}\left(\frac{v^2}{Rg}\right) \tag{N9.15a}$$

$$\theta = \tan^{-1}\left[\frac{(260\ \cancel{\text{mi}}/\cancel{\text{h}})^2}{(5.0\ \cancel{\text{mi}})(22\ \cancel{\text{mi}}/\cancel{\text{h}}/\cancel{\text{s}})}\left(\frac{1\ \cancel{\text{h}}}{3600\ \cancel{\text{s}}}\right)\right] = 9.7° \tag{N9.15b}$$

Exercise N9X.6: You may notice that the pilot when beginning a sharp turn will power up the engines somewhat if he or she wants to keep the plane from losing altitude. Why would the pilot do this?

A car doesn't need to bank

A car rounding a corner does *not* need to bank into the curve: when you turn the steering wheel to left, the road exerts a static friction force on the tires to the left, which provides the leftward net force necessary to accelerate the car away from its natural straight-line motion (see Figure N9 4a). The car does not need to lean to the left to do this. (In fact, if you consider the torques that have to be exerted on the car, you will find that the car actually leans somewhat to the *right* when turning to the left: see problem N9S.11.)

Roads are sometimes banked for safety's sake

An engineer designing a road will often call for a tight curve to be banked at an angle. This is often easiest to see on freeway overpasses that curve one direction or another. (It is even more obvious in the tracks of roller-coasters, which are often banked at very large angles.) If the banking angle θ is chosen just right for the radius of the curve and the typical speed of a car on that curve, then the tilted *normal* force acting on the car provides the required leftward acceleration (see Figure N9.4b). This is advantageous because the car does not then have to *depend* on static friction to keep it traveling with the curve. Under bad weather conditions, the coefficient of static friction between the roadbed and the tires may become so low that the static friction force cannot keep the car following the curve. The normal force, on the other hand, is not affected by bad conditions. The car cannot move *into* a wet or icy roadbed any more easily than a dry roadbed! Banking the curve thus cuts down on accidents in bad weather. (The advantages in the case of roller coasters are even more obvious!)

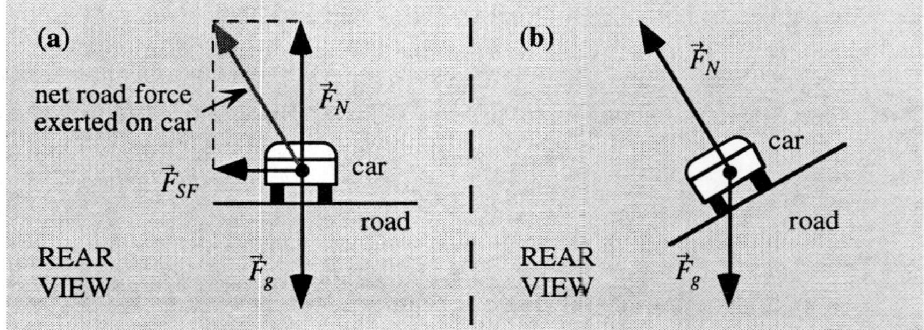

Figure N9.4: (a) A free-particle diagram of a car rounding a curve to the left on a level roadbed. Static friction has to provide the leftward force required to keep the car following the curve. **(b)** A free-particle diagram of a car on an ideally banked roadbed. Here the normal force alone provides the necessary sideward force.

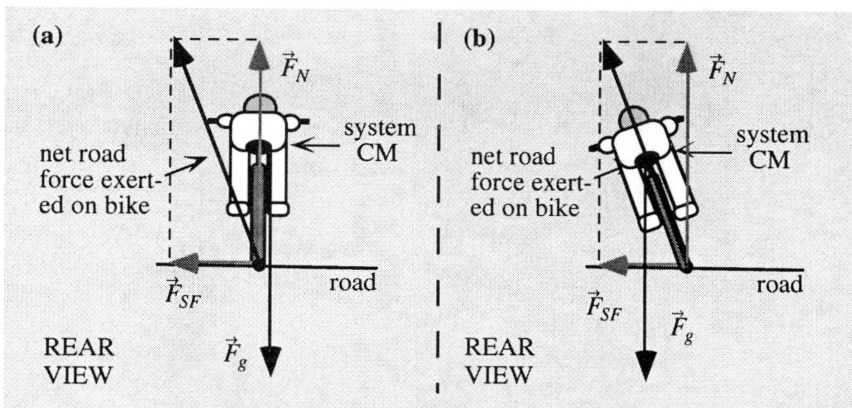

Figure N9.5: (a) Forces acting on a bike turning left while remaining upright. The clockwise torque exerted by the total road force around the system's center of mass will cause it to rotate right and fall over. **(b)** If the bike leans into the curve, then the net road force is directed through the system's center of mass and thus exerts zero torque $\vec{r} \times \vec{F}$ on the system (since the angle between $\vec{r}$ and $\vec{F}$ is 180° implying that the magnitude of the torque from this force is $rF\sin(180°) = 0$).

[If your instructor did not assign chapter N6, skip to the end of the section.]

Why bicyclists lean into a curve

A bicyclist going around a curve will lean into the curve like an airplane does, but for different reasons. As in the case of a car, when the bike's front wheels turn left, the road exerts a leftward force on the bike wheels as well as a vertical normal force. The total force exerted by the road is thus upward and to the left, as shown in Figure N9.5. If the bicyclist does not lean over, this force exerts a net clockwise torque on the bike around the system's center of mass (see Figure N9.5a), that would tend to cause the bike to flip over. On the other hand, if the bicyclist leans into the curve so that the net force exerted by the road points directly at the bicyclist's center of mass, then that force will not exert a torque on the system and the bike will not flip (Figure N9.5b). The first thing that a child learns when riding a bike for the first time is to *keep the torque on the bike zero* (though a child will think of it in terms of "keeping my balance"), so this leaning is automatic for experienced riders. (Note that the gravitational force on the bike does *not* exert a torque on the system around its center of mass because it effectively acts at the center of mass, and thus $\vec{r}$ in $\vec{r} \times \vec{F}$ is zero.)*

N9.5 EXAMPLES

The following examples illustrate how we can use the constrained motion framework to solve various kinds of problems involving circular motion.

How to choose an appropriate reference frame

Circular motion problems, like problems involving linear acceleration, are generally easier to do if you orient your reference frame axes correctly. Usually the best orientation is so that one axis points along the line connecting the object to the center of its circular path. In the case of *uniform* circular motion, the object's acceleration will thus nicely lie along this axis direction and the components of the net force in other axis directions will be zero. In the case of *nonuniform* circular motion, the two axis directions in the plane of the circle will then correspond in a simple way to the $\hat{r}$ and $\hat{\theta}$ directions of equation N9.11. Note that since these directions change as the object moves, the reference frame that you set up will only be useful at one instant. This is usually good enough to solve the problem, since other instants will be analogous.

Banking problems *seem* to have too many unknowns

Most banking angle problems, like the airplane problem discussed in example N9.1, may look at first as though they have too many unknowns to solve. Press ahead anyway and see if you can get some unknowns to divide out.

In most other respects, circular motion problems are like any other constrained motion problem: we take what we know about the object's motion (that it moves in a circle) and use that to determine unknown forces, banking angles, time to complete the circular path, and the like.

*This explanation brushes lightly past some complexities associated with the fact that the system's center of mass is accelerating, but is essentially valid.

EXAMPLE N9.2

OUTLINE OF THE FRAMEWORK for CONSTRAINED-MOTION problems

1. Pictorial Representation

a. Draw a picture of the situation that includes:

① (1) a drawing of the object in the context of its surroundings

② (2) a labeled arrow indicating the direction of the object's acceleration (or say "$\vec{a} = 0$")

③ (3) reference frame axes, with one axis aligned with the object's acceleration (if any)

④ (4) labels defining symbols for relevant quantities

⑤ b. List values for all known labeled quantities and specify which quantities are unknown.

2. Conceptual Representation

The general task is to construct a conceptual model of the situation and link it to an abstract physics model or principle. In constrained motion problems, we do the following:

⑥ a. Draw a free-particle (and/or a free-body) diagram of the object

⑦ b. Describe the constraint on the object's motion and check that the net force on the object is consistent with the acceleration implied by that constraint (use a net-force diagram if necessary)

⑧ c. Describe any approximations or assumptions you have to make to solve the problem

3. Mathematical Representation

⑨ a. Apply Newton's second law *in vector component form*

⑩ b. Solve for any unknown quantities symbolically

⑪ c. Plug in numbers and units and calculate the result and its units.

4. Evaluation

Check that the answer makes sense:

⑫ a. Does it have the correct units?

⑬ b. Does it have the right sign?

⑭ c. Does it seem reasonable?

Problem: A 1500-kg car travels at a constant speed of 22 m/s over the top of a hill whose cross-section near the top is approximately a circle having an effective radius of 150 m. What is the magnitude of the normal force of the car just as it passes the top of the hill and how does this compare to the car's weight?

Solution: A sketch of the situation looks like this:

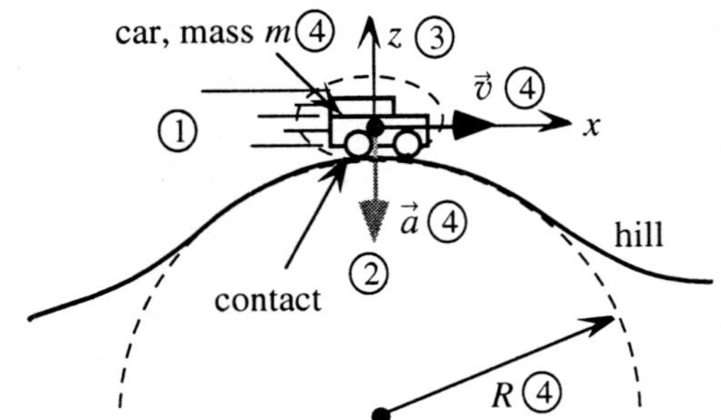

KNOWNS: ⑤

m = 1500 kg
v = 22 m/s
R = 150 m

UNKNOWNS:

we want to ⑤ find F_N / F_g

A free-particle diagram for the car is thus as shown at the right ($\vec{F}_{SF}$ is from the tires' interaction with the road and $\vec{F}_D$ is from the car's interaction with the air). The constraint on the car's motion is that we are told that it moves in a vertical circle (which we have defined to be the xz plane) at a constant speed. This means that the car's acceleration is ⑦ downward (in the $-z$ direction in our reference frame) with a magnitude of $a = v^2/R$. This in turn means that $\vec{F}_{SF}$ from the interaction with the road must be directed *forward* (so that is cancels $\vec{F}_D$) and $\vec{F}_N$ must be smaller than $\vec{F}_g$ (as drawn) to yield a downward net force. The free-particle diagram is then consistent with the observed acceleration. ⑧ We are assuming that the hill's cross-section really *is* circular (it probably isn't quite), and that drag is the only significant opposing force.

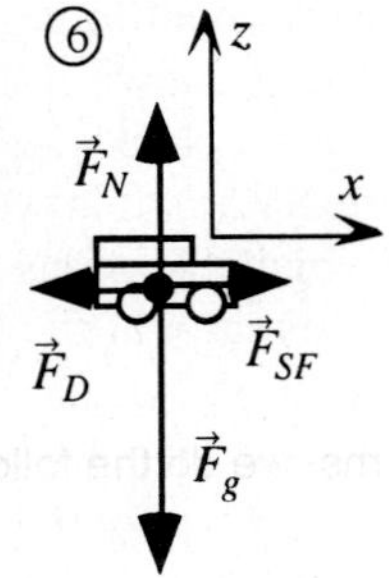

Newton's second law in this case tells us that

⑨
$$\begin{bmatrix} 0 \\ 0 \\ -mv^2/R \end{bmatrix} = m\vec{a} = \vec{F}_{\text{net}} = \begin{bmatrix} F_{g,x} \\ F_{g,y} \\ F_{g,z} \end{bmatrix} + \begin{bmatrix} F_{N,x} \\ F_{N,y} \\ F_{N,z} \end{bmatrix} + \begin{bmatrix} F_{SF,x} \\ F_{SF,y} \\ F_{SF,z} \end{bmatrix} + \begin{bmatrix} F_{D,x} \\ F_{D,y} \\ F_{D,z} \end{bmatrix}$$

$$\Rightarrow \begin{bmatrix} 0 \\ 0 \\ -mv^2/R \end{bmatrix} = \begin{bmatrix} 0 \\ 0 \\ -mg \end{bmatrix} + \begin{bmatrix} 0 \\ 0 \\ +F_N \end{bmatrix} + \begin{bmatrix} +F_{SF} \\ 0 \\ 0 \end{bmatrix} + \begin{bmatrix} -F_D \\ 0 \\ 0 \end{bmatrix} \quad (1)$$

The top line of this equation tells us that $F_{SF} = F_D$, which we knew intutively. Solving the bottom line for F_N, we get:

⑩
$$F_N = mg - \frac{mv^2}{R} = (1500\text{ kg})(9.8\text{ m/s}^2) - (1500\text{ kg})\frac{(22\text{ m/s})^2}{150\text{ m}}$$

⑪
$$= (14{,}700 - 4800)\,\cancel{\text{kg·m/s}^2}\left(\frac{1\text{ N}}{1\,\cancel{\text{kg·m/s}^2}}\right) = 9900\text{ N} \quad (2)$$

Since we can read from this calculation that mg = 14,700 N, F_N / F_g =
⑪ 9900 N / 14700 N = 0.67, so the normal force is 2/3 of the weight. The
⑫⑬ units check out and magnitudes look reasonable here ($F_N < F_g$ is a good
⑭ sign). The sign of F_N is also positive, which is appropriate for a vector magnitude (a negative result would indicate something seriously awry!).

EXAMPLE N9.3

OUTLINE OF THE FRAMEWORK for CONSTRAINED-MOTION problems

1. Pictorial Representation

a. Draw a picture of the situation that includes:

① (1) a drawing of the object in the context of its surroundings

② (2) a labeled arrow indicating the direction of the object's acceleration (or say "$\vec{a} = 0$")

③ (3) reference frame axes, with one axis aligned with the object's acceleration (if any)

④ (4) labels defining symbols for relevant quantities

⑤ b. List values for all known labeled quantities and specify which quantities are unknown.

2. Conceptual Representation

The general task is to construct a conceptual model of the situation and link it to an abstract physics model or principle. In constrained motion problems, we do the following:

⑥ a. Draw a free-particle (and/or a free-body) diagram of the object

⑦ b. Describe the constraint on the object's motion and check that the net force on the object is consistent with the acceleration implied by that constraint (use a net-force diagram if necessary)

⑧ c. Describe any approximations or assumptions you have to make to solve the problem

3. Mathematical Representation

⑨ a. Apply Newton's second law *in vector component form*

⑩ b. Solve for any unknown quantities symbolically

⑪ c. Plug in numbers and units and calculate the result and its units.

4. Evaluation

Check that the answer makes sense:

⑫ a. Does it have the correct units?

⑬ b. Does it have the right sign?

⑭ c. Does it seem reasonable?

Problem: What is the banking angle needed to keeep a 1500-kg car following a circular bend in the road of radius 330 m when it is going 25 m/s (≈ 58 mi/h) without requiring that *any* static friction force be exerted on the tires?

Solution: Rear and top-view diagrams of the situation look like this:

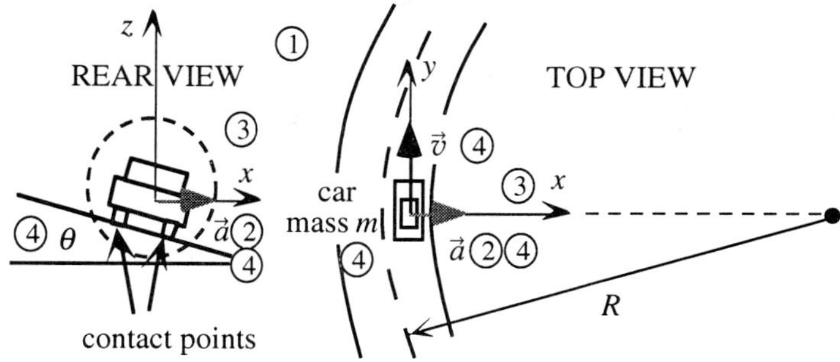

⑤ **KNOWNS:** $v = 25$ m/s, $R = 330$ m, $m = 1500$ kg

UNKNOWNS: θ

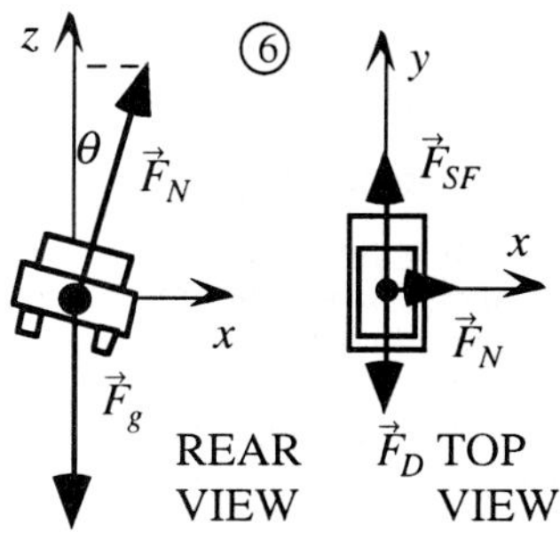

Rear and top-view free-particle diagrams for the car are shown to the right. ⑦ The constraint here is that the car is moving in a horizontal circle at a constant speed. Its acceleration therefore points directly inward, which (as we have defined the frame axes) is in the $+x$ direction at the instant shown. This is consistent with the free-particle diagrams, which show the net force ⑧ to point in that direction (due to the tilting of the normal force). We are assuming that the bend in the road is really circular, that it is level, and that the car's speed is constant. If this is so, then there is no component ⑧ of acceleration in either the y direction or the z direction. I am also assuming that the force opposing the car's forward motion is drag.

Newton's second law in this case implies that

⑨
$$\begin{bmatrix} +mv^2/R \\ 0 \\ 0 \end{bmatrix} = m\vec{a} = \vec{F}_{\text{net}} = \begin{bmatrix} F_{g,x} \\ F_{g,y} \\ F_{g,z} \end{bmatrix} + \begin{bmatrix} F_{N,x} \\ F_{N,y} \\ F_{N,z} \end{bmatrix} + \begin{bmatrix} F_{SF,x} \\ F_{SF,y} \\ F_{SF,z} \end{bmatrix} + \begin{bmatrix} F_{D,x} \\ F_{D,y} \\ F_{D,z} \end{bmatrix}$$

$$\Rightarrow \begin{bmatrix} +mv^2/R \\ 0 \\ 0 \end{bmatrix} = \begin{bmatrix} 0 \\ 0 \\ -mg \end{bmatrix} + \begin{bmatrix} +F_N \sin\theta \\ 0 \\ +F_N \cos\theta \end{bmatrix} + \begin{bmatrix} 0 \\ +F_{SF} \\ 0 \end{bmatrix} + \begin{bmatrix} 0 \\ -F_D \\ 0 \end{bmatrix}$$

The middle line of this equation tells us that $F_D = F_{SF}$, which is not particularly relevant to what we want to find. The bottom line, on the other hand, tells us that $mg = F_N \cos\theta$, whereas the first line tells us that $mv^2/R = F_N \sin\theta$. If we divide the second by the first, we get

⑩ ⑪
$$\frac{mv^2}{mRg} = \frac{F_N \sin\theta}{F_N \cos\theta} \Rightarrow \tan\theta = \frac{v^2}{Rg} = \frac{(25\,\text{m/s})^2}{(330\,\text{m})(9.8\,\text{m/s}^2)} = 0.19$$

⑪ So $\theta = \tan^{-1}(0.19)$, implying that the road should be banked toward the ⑫ center at an angle of $\theta = 11°$. Note how the units work out so that ⑬ $\tan\theta$ is correctly unitless, and θ is positive (consistent with the pic- ⑭ tures) and its magnitude is reasonable (> 45° would be unreasonable).

(Note that the result does not depend on m but does depend on v. A car not going at exactly the right speed will need some x component of static friction to hold it on its path.)

EXAMPLE N9.4

OUTLINE OF THE FRAMEWORK for CONSTRAINED-MOTION problems

1. Pictorial Representation

a. Draw a picture of the situation that includes:

① (1) a drawing of the object in the context of its surroundings

② (2) a labeled arrow indicating the direction of the object's acceleration (or say "$\vec{a} = 0$")

③ (3) reference frame axes, with one axis aligned with the object's acceleration (if any)

④ (4) labels defining symbols for relevant quantities

⑤ b. List values for all known labeled quantities and specify which quantities are unknown.

2. Conceptual Representation

The general task is to construct a conceptual model of the situation and link it to an abstract physics model or principle. In constrained motion problems, we do the following:

⑥ a. Draw a free-particle (and/or a free-body) diagram of the object

⑦ b. Describe the constraint on the object's motion and check that the net force on the object is consistent with the acceleration implied by that constraint (use a net-force diagram if necessary)

⑧ c. Describe any approximations or assumptions you have to make to solve the problem

3. Mathematical Representation

⑨ a. Apply Newton's second law *in vector component form*

⑩ b. Solve for any unknown quantities symbolically

⑪ c. Plug in numbers and units and calculate the result and its units.

4. Evaluation

Check that the answer makes sense:

⑫ a. Does it have the correct units?

⑬ b. Does it have the right sign?

⑭ c. Does it seem reasonable?

Problem: A satellite in a circular orbit is a freely-falling object that happens to be moving at just the right speed so that the gravitational force acting on it provides exactly the right acceleration to hold it in its circular path. How fast would an object have to travel to be a circular orbit just above the earth's surface? How long would it take the object to go around the earth at this speed? Ignore air friction (?!?).

Solution: A picture of the situation looks like this:

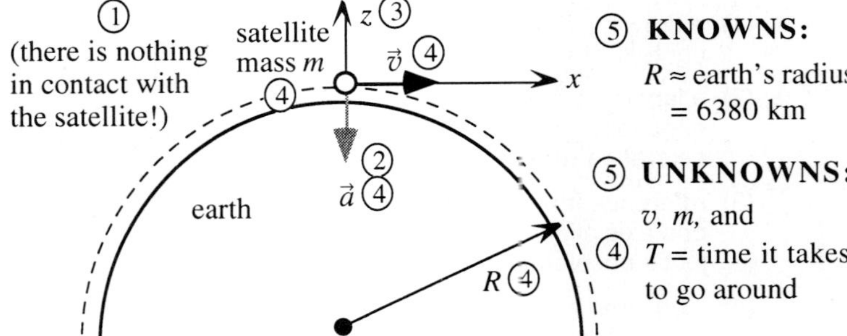

⑧ A free-particle diagram of the object here is almost too trivial to draw: if we ignore air friction, the only force that can possibly act on the satellite is the force of gravity. ⑦ A satellite in a circular orbit is (by definition) moving in a circular path. We will assume that its speed v is constant: if this is so, then the satellite's acceleration will have a magnitude of $a = v^2/R$ and will point toward the center of the earth. This is consistent with the net force drawn on the diagram, since it must also point toward the center of the earth. At the representative instant shown, then, both the net force and the satellite's acceleration have components only in the $-z$ direction (according to our frame). At that instant, Newton's second law implies

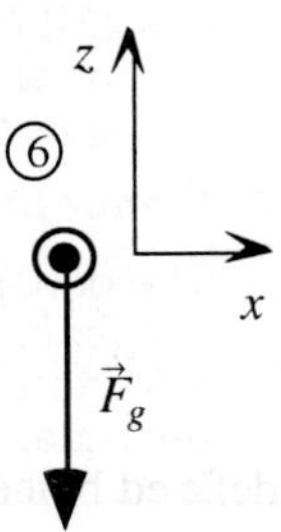

$$⑨ \quad \begin{bmatrix} 0 \\ 0 \\ -mv^2/R \end{bmatrix} = m\vec{a} = \vec{F}_{\text{net}} = \begin{bmatrix} F_{g,x} \\ F_{g,y} \\ F_{g,z} \end{bmatrix} = \begin{bmatrix} 0 \\ 0 \\ -mg \end{bmatrix}$$

Solving the last line of this equation for v^2 and then v, we get:

$$v^2 = Rg \;\Rightarrow\; \overset{⑩}{v = \sqrt{Rg}} = \sqrt{(6{,}380{,}000\text{ m})(9.8\text{ m/s}^2)} = \overset{⑪}{7900\text{ m/s}}$$

This is pretty fast! (Note that the mass of the satellite is irrelevant.) Since the distance around the earth is $2\pi R$, the time T required for a satellite to cover this distance is

$$T = \frac{\text{distance}}{\text{speed}} \overset{⑩}{=} \frac{2\pi R}{v} = \frac{2\pi(6380\,\cancel{\text{km}})}{7.9\,\cancel{\text{km}}/\cancel{\text{s}}}\left(\frac{1\text{ min}}{60\,\cancel{\text{s}}}\right) \overset{⑪}{=} 85\text{ min } (= 1.4\text{ h}) \quad ⑫$$

⑬ Note how the units all work out in both cases, and that v comes out positive (as a magnitude should). ⑭ The speed is pretty fast, but this is correct (one of the reasons that space travel is difficult and expensive).

The problem with a satellite orbiting "just above" the earth's surface (in addition to the hazard that it might pose to airplanes and mountains) is that drag is *not* going to be negligible at 7.9 km/s. (Indeed, the space shuttle's wings glow red-hot from friction when it reenters the atmosphere at a speed comparable to this!)

MATH SKILLS: Unit Vectors

A *vector* is a quantity having both a magnitude and a direction. We have already defined a symbol for the *magnitude* of a vector: given a vector $\vec{w}$, its magnitude is written w (the vector symbol without the vector arrow). This magnitude is simply a numerical value without any direction: the information about its parent vector's direction has been stripped away.

The **unit vector** $\hat{w}$ corresponding to an arbitrary vector $\vec{w}$ expresses just the opposite idea: it represents the *direction* of the original vector with all information about the magnitude stripped away. (Note that $\vec{w}$ here is meant to represent any *arbitrary* vector, not just acceleration.) Technically speaking, the unit vector $\hat{w}$ that corresponds to $\vec{w}$ is a vector whose *magnitude* is equal to 1 (with *no* units) and whose *direction* is the same as that of its parent vector.

The unit vector corresponding to a vector

By convention, the symbol for the unit vector corresponding to $\vec{w}$ is the same letter with the vector arrow replaced by a symbol called a *caret* (sometimes informally called a *hat*). We can calculate the unit vector for any $\vec{w}$ as follows:

$$\hat{w} \equiv \frac{\vec{w}}{w} \tag{N9.16}$$

Note that the magnitude of the unit vector so defined is simply

$$\text{mag}(\hat{w}) = \frac{\text{mag}(\vec{w})}{w} = \frac{w}{w} = 1 \tag{N9.17}$$

Since $\vec{w}$ and w have the same units, their ratio not only has the numerical value of 1 but the units cancel, leaving the ratio unitless. Since $\hat{w}$ is simply $\vec{w}$ multiplied by the scalar w^{-1}, though, $\hat{w}$ has the same direction as $\vec{w}$. The quantity defined by equation N9.16 is thus consistent with the technical description of $\hat{w}$ given in the previous paragraph.

According to equation N9.16, the components of $\hat{w}$ are

$$\hat{w} = \begin{bmatrix} \frac{w_x}{w}, \frac{w_y}{w}, \frac{w_z}{w} \end{bmatrix} \tag{N9.18}$$

For example, if $\vec{r}$ is the position of some object relative to the origin, then the unit vector indicating the direction of that object relative to the origin is:

$$\hat{r} = \begin{bmatrix} \frac{x}{r}, \frac{y}{r}, \frac{z}{r} \end{bmatrix} \tag{N9.19}$$

Unit vectors as directional indicators

I find it most helpful to *think* of a unit vector as being a pure direction, conveying exactly the same kind of information as a hand pointing "thataway". In this spirit, we can define $\hat{x}$, $\hat{y}$, and $\hat{z}$ to be vectors with magnitude 1 pointing in the x, y, and z directions respectively (even though these vectors do not have obvious parents vectors). For example, I would read the vector equation

$$\vec{g} = -(9.8\text{ m/s}^2)\hat{z} \tag{N9.20}$$

as saying that "the gravitational field vector $\vec{g}$ has a magnitude of 9.8 m/s^2 and points in the $-\hat{z}$ direction (that is, *downward*)". The $-\hat{z}$ in this kind of equation serves purely as a directional indicator. The $\hat{\theta}$ vector defined in section N9.2 is an indicator of this type: it indicates the direction perpendicular to the $\vec{r}$ vector in the direction that θ increases.

Even so (and in spite of the way the hat in the unit vector symbol replaces the vector arrow) you should *always* keep in mind that $\hat{a}$ is really a *vector* that just happens to have a magnitude of 1. Whenever a unit vector symbol appears in a mathematical equation, you should treat it and interpret it *exactly* as you would *any other vector*.

MATH SKILLS: The Chain Rule

The **chain rule** is one of the most important and useful theorems of differential calculus. Assume that we have a function $f(u)$ is a function of u, which in turn is a function $u(t)$ of t. This means that f is also a function of t and thus has a meaningful t derivative. The chain rule claims that

Statement of the chain rule

$$\frac{df}{dt} = \frac{df}{du}\frac{du}{dt} \tag{N9.21}$$

This theorem looks relatively pedestrian, but in fact is *very* useful. For example, imagine that we want to find the derivative of the function

$$f(t) = (at^2 + bt + c)^5 \tag{N9.22}$$

We *could* evaluate this derivative by multiplying out the polynomial and then taking the derivative of each term, using the constant rule and

$$\frac{d}{dt}t^n = nt^{n-1} \tag{N9.23}$$

This would be very tedious. It is much simpler to define $u \equiv at^2 + bt + c$, which means that $f(u) = u^5$. The chain rule then tells us that

$$\frac{df}{dt} = \frac{df}{du}\frac{du}{dt} = \left(\frac{d}{du}u^5\right)\left(\frac{d}{dt}[at^2 + bt + c]\right)$$
$$= \left(5u^4\right)(2at + b) = 5(at^2 + bt + c)^4(2at + b) \tag{N9.24}$$

Exercise N9X.7: Use this technique to find the derivative of $f(t) = 1/(t^2 - b)^3$.

The "physicist's proof" as a mnemonic device

The physicist's "proof" of this theorem, by the way, is that we simply multiply top and bottom of df/dx by du and rearrange things a bit (physicists tend to think of df, dt, and du simply as small numbers). Real mathematicians worry more about the definitions of the derivatives, the limit-taking process, and all the things that might go wrong with this cross-multiplication, but the basic result is the same. At the very least, you can think of "multiplying top and bottom by du" as a way of *remembering* the chain-rule formula.

Applying the chain rule to trigonometric functions

We will be using the chain rule fairly regularly from now on. We will be especially interested in evaluating time derivatives of the trigonometric functions $\sin\theta$ and $\cos\theta$. The derivatives of these functions with respect to the angle θ are

$$\frac{d}{d\theta}\sin\theta = \cos\theta, \qquad \frac{d}{d\theta}\cos\theta = -\sin\theta \tag{N9.25}$$

These functions become dependent on time if the angle θ depends on time. If this is so, then the chain rule tells us that the time derivatives of these functions are

$$\frac{d}{dt}\sin\theta = \frac{d\sin\theta}{d\theta}\frac{d\theta}{dt} = \cos\theta\frac{d\theta}{dt}, \qquad \frac{d}{dt}\cos\theta = -\sin\theta\frac{d\theta}{dt} \tag{N9.26}$$

In particular, if θ increases linearly with time, so that $\theta(t) = \omega t$ (where ω is a constant), then $d\theta/dt = \omega$ and

$$\frac{d}{dt}\sin\omega t = \omega\cos\omega t, \qquad \frac{d}{dt}\cos\omega t = -\omega\sin\omega t \tag{N9.27}$$

We will use these formulas extensively in this chapter and those that follow.

SUMMARY

I. CALCULUS OF CIRCULAR MOTION
   A. The general mathematics of circular motion
      1. Consider an object moving in a circle of radius R with speed v [$\text{mag}(\vec{r}) = R$ is constant but v is not necessarily constant]
      2. Orient a reference frame so the circle is in xy plane
      3. The object's location is specified by angle θ it makes with x axis
         a) this angle will change as the object moves around the circle
         b) therefore it is a function of time: $\theta = \theta(t)$.
      4. We can define the following unit vectors
         a) $\hat{r} \equiv [\cos\theta, \sin\theta]$ (points radially outward from the origin)
         b) $\hat{\theta} \equiv [-\sin\theta, \cos\theta]$ (points in direction of increasing θ)
      5. The chain rule then implies that

$$\frac{d\hat{r}}{dt} = +\hat{\theta}\frac{d\theta}{dt}, \qquad \frac{d\hat{\theta}}{dt} = -\hat{r}\frac{d\theta}{dt} \qquad \text{(N9.9)}$$

      6. Define the *perpendicular component* of $\vec{v}$ to be $v_\perp = R\,d\theta/dt$ (for ccw circular motion, $v_\perp = +v$; for cw motion, $v_\perp = -v$)
      7. According to these definitions, the object's position and velocity
         a) $\vec{r}(t) = R\hat{r}$ (N9.1)
         b) $\vec{v}(t) \equiv d\vec{r}/dt = R(d\theta/dt)\hat{\theta} = v_\perp\hat{\theta}$ (N9.5)
   B. The case of *uniform circular motion*: if v = constant, then
      1. Then $\vec{a}(t) = d\vec{v}/dt = v_\perp(d\theta/dt)(-\hat{r}) = -(v^2/R)\hat{r}$ (N9.9)
         a) $\vec{a}$ thus points toward the circle's center; $\text{mag}(\vec{a}) = v^2/R$
         b) this is consistent with what we found in chapter N2
      2. $v = 2\pi R/T$ where T is time to complete one circle (N9.10)
   C. The case of *nonuniform circular motion*: if $v \neq$ constant, then
      1. $\vec{a}(t) = (dv_\perp/dt)\hat{\theta} - (v^2/R)\hat{r}$ (N9.12)
      2. in addition to inward component, $\vec{a}$ has component $\perp$ to $\vec{r}$
      3. this is qualitatively consistent with the results of motion diagram

II. BANKING
   A. A turning plane banks, lowering wing closer to center of turn. Why?
      1. When plane follows a circular path, it must accelerate toward center
      2. Therefore there must be a net horizontal force on the plane
      3. banking plane lets lift force provide this horizontal force
      4. banking angle does not depend on plane's mass or lift force
   B. Banking roads
      1. A turning car does not need to bank, because
         a) static friction on the tires provides the needed horizontal force
         b) this force can be exerted on tires without the car banking
      2. Nonetheless, if curved roadbed is tipped at the appropriate angle
         a) then the normal force on car can provide horizontal acceleration
         b) this is advantageous because bad conditions do not affect the normal force whereas water or ice can reduce $F_{SF,\max}$
   C. A bicycle banking into a curve
      1. If bicycle doesn't bank, the total road force exerts a torque on bike (this would cause the bike to rotate and thus to fall over)
      2. To prevent this, one leans so the net road force goes through CM (bike riders learn to do this automatically)

III. EXAMPLES
   A. Setting up the right reference frame
      1. set it up so that one axis is parallel to object's $\vec{r}$ vector (this makes things easier in both uniform and nonuniform cases)
      2. this frame only works for one instant, but this is usually enough
   B. A satellite freely falls around the earth!
      (it is moving forward so fast, the earth curves away as fast as it falls)

GLOSSARY

unit vector: a vector with magnitude equal to 1 (no units) that we can use to indicate a pure direction.

chain rule: a mathematical theorem that says that if f is a function of u which is a function of t, then

$$\frac{df}{dt} = \frac{df}{du}\frac{du}{dt} \tag{N9.21}$$

uniform circular motion: the motion of an object that travels along a circular trajectory at a constant speed.

perpendicular component $v_\perp$ (of the velocity of an object moving in a plane around some origin): the component of $\vec{v}$ along the direction perpendicular to the object's position vector $\vec{r}$ in the direction of increasing θ, where θ is the angle in the plane between $\vec{r}$ and a given direction (usually taken to be the x axis).

nonuniform circular motion: the motion of an object that travels along a circular path with a varying speed.

banking (into a turn): when an airplane or bicycle rotates around an axis parallel to its motion as it goes around a curve. Both will lean toward the center of the curve (for different physical reasons). A roadbed also may be banked into a curve in the same way. If the speed of a car is just right, the banking will allow it to negotiate the curve without the road exerting a sideways static friction force.

TWO-MINUTE PROBLEMS

N9T.1 When a speeding roller coaster car is at the bottom of a loop, the magnitude of the normal force exerted on the car's wheels due to its interaction with the track is
A. greater than the weight of the car and its passengers
B. equal to the weight of the car and its passengers
C. less than the weight of the car and its passengers

N9T.2 A child grips tightly the outer edge of a playground merry-go-round as other kids push on it to give it a dizzying rotational velocity. When the other kids let go, the net force on the child points most nearly
A. inward toward the center of the merry-go-round
B. outward away from the center of the merry-go-round
C. in the direction of rotation
D. nowhere: the net force is zero
E. other (specify)

N9T.3 A car is traveling counterclockwise along a circular bend in the road whose effective radius is 100 m. At a certain instant of time, the car is traveling due *north*, has a speed of 10 m/s and is in the process of *increasing* that speed at a rate of 1 m/s^2. The direction of the car's acceleration at that instant is most nearly
A. north
B. northeast
C. east
D. northwest
E. west
F. southwest
T. zero

N9T.4 A plane flying in a certain circular holding pattern banks at an angle of 8° when flying at a speed of 150 mi/h. If a second plane flies in the same circle at 300 mi/h, what is its banking angle?
A. a bit less than 16°
B. exactly 16°
C. a bit more than 16°
D. a bit less than 32°
E. exactly 32°
F. a bit more than 32°
T. answer depends on the planes' masses

N9T.5 Car 1 with mass m rounds a curve of radius R traveling at a constant speed v. Car 2 with mass $2m$ rounds a curve of radius $2R$ traveling at a constant speed $2v$. How does the magnitude F_2 of the sideward static friction force acting on car 2 compare with the magnitude F_1 of the sideward static friction force acting on Car 1?
A. $F_1 = 4F_2$
B. $F_1 = 2F_2$
C. $F_1 = F_2$
D. $F_2 = 2F_1$
E. $F_2 = 4F_1$
F. other (specify)

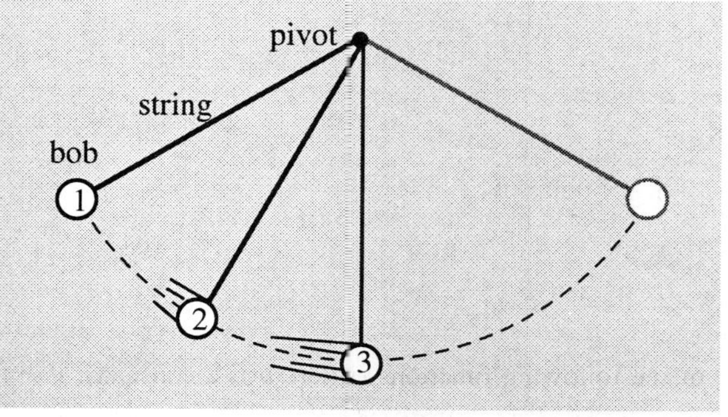

Figure N9.6: (For problems N9T.6 through N9T.8.) A mass (bob) swings from the end of a string. At point 1, the bob is at the extreme point of the swing and thus is instantaneously at rest. At point 3, the bob is directly below its suspension point and has its maximum speed.

N9T.6 (See Figure N9.6.) Which one of the arrows to the right most closely indicates the direction of the bob's acceleration when it is at the point 1?

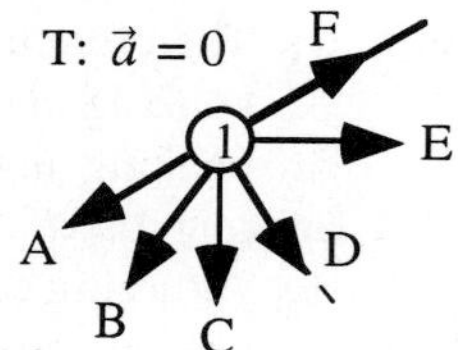

N9T.7 (See Figure N9.6.) Which one of the arrows to the right most closely indicates the direction of the bob's acceleration when it is at the point 2?

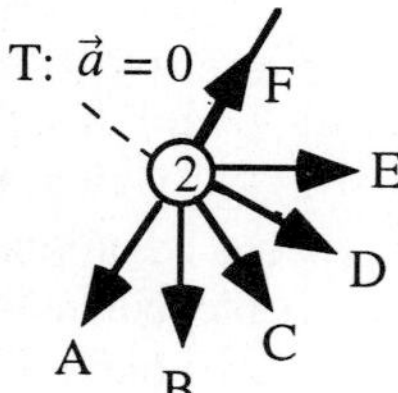

N9T.8 (See Figure N9.6.) Which one of the arrows to the right most closely indicates the direction of the bob's acceleration when it is at the point 3?

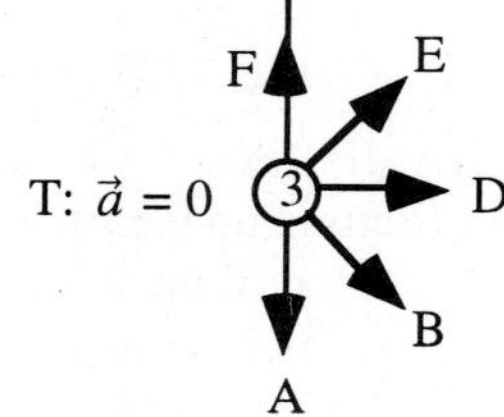

HOMEWORK PROBLEMS

BASIC SKILLS

N9B.1 A car traveling at a constant speed of 50 mi/h travels around a curve. An accelerometer in the car measures its sideward acceleration to be $0.1g$. What is the effective radius of the curve? (*Hint*: g = 22 [mi/h]/s.)

N9B.2 A plane is traveling in a circular path 32 km in diameter. It is banking at an angle of 12°, which indicates that its sideward acceleration is 2.1 m/s^2. What is its speed?

N9B.3 An airplane banks at an angle of 11° while flying in a level circle at 320 mi/h. What is the radius of its circular path?

N9B.4 A car is traveling counterclockwise along a circular bend in the road with an effective radius of 200 m. At a certain instant, the car's speed is 20 m/s, but it is slowing down at a rate of 1.0 m/s^2. What is the magnitude of the car's acceleration?

N9B.5 The position $\vec{r}$ of a certain object relative to the origin has components [3.0 m, 5.0 m, −2.0 m]. What are the components of the unit vector $\hat{r}$ that points in the same direction?

N9B.6 The position $\vec{r}_{B(A)}$ of object B relative to object A has the components [7 m, −12 m, 0 m]. What are the components of the unit vector $\hat{r}_{B(A)}$ that points in the same direction?

N9B.7 Use the chain rule to evaluate the time derivatives of the following functions (where b is a constant):

A. $3(t^3+b^3)^4$ B. $\sqrt{t^2-b^2}$ C. $\cos(bt^2)$

N9B.8 Use the chain rule to evaluate the time derivatives of the following functions (where b is a constant):
A. $5b/(bt-1)^4$ B. $(bt+1)^{3/2}$ C. $\sin(bt+\pi)$

SYNTHETIC

N9S.1 A bicycle and rider with a total mass of 85 kg traveling at 13 m/s (≈ 30 mi/hr) attempts to round an unbanked curve with a radius of 60 m (≈ 200 ft). **(a)** What is the magnitude of the force required to keep the bike on its circular path? **(b)** At what angle does the bicycle lean? **(c)** What must the minimum coefficient of static friction be between the tires and the road if the bike is not to skid?*

N9S.2 A stunt driver drives a car over the top of a hill having a cross section that can be approximated by a circle of radius 250 m. What is the greatest speed the car can have before it leaves the road at the top of the hill?*

N9S.3 A car travels at a constant speed of 23 m/s through a small valley whose cross section is like a circle of radius 310 m. What is the magnitude of the normal force on the car, expressed as a multiple of the car's weight?*

N9S.4 A child places a lunch-box on the rim of a playground merry-go-round that has a radius of 2.0 m (≈ 6 ft). If the merry-go-round goes once around every 6.0 seconds, what is the speed of the box? What must the coefficient of static friction between the box and the merry-go-round be if the box is to stay on?*

N9S.5 The acceleration of gravity near the surface of the Moon has about 1/6 the magnitude that it does on Earth. The radius of the Moon is 1740 km. How long would a satellite orbiting just above the Moon's surface take to go once around the Moon? (The Moon has no atmosphere!)*

N9S.6 Imagine that you are designing a circular curve in a highway that must have a radius of 330 ft and will carry traffic moving at 60 mi/hr. **(a)** At what angle should the roadbed be banked for maximum safety? **(b)** If the roadway is *not* banked, what would the necessary coefficient of static friction between the tires and the road have to be to keep a car on the road? Is such a coefficient reasonable?* (*Hint*: 1 m = 3.3 ft, 1 m/s = 2.24 mi/h.)

N9S.7 A 150-lb student rides a ferris wheel that rotates at a constant rate. At the highest point, the seat exerts a normal force of magnitude 110 lb on the student. What would the magnitude of this normal force be at the lowest point?* (*Hint*: An object with a mass of 1 kg weighs 2.2 lbs.)

N9S.8 A ball of mass m is tied to one end of a string of length L, the other end of which is fixed to the ceiling. The ball is then set in motion at a constant speed in a horizontal circle of radius $R < L$ around the axis that the string would make if the ball were to hang at rest (the string thus makes a constant angle θ with the vertical direction). Determine how long it takes the ball to go around the circle once in terms of m, g, L, and θ.*

N9S.9 As you are riding in a 1650-kg car, you approach a hairpin curve in the road whose radius is 50 m. The roadbed is banked inwards at an angle of 10°. **(a)** Suppose the road is very icy, so that the coefficient of static friction is essentially zero. What is the maximum speed with which you can go around the curve? **(b)** Now suppose that the road is dry and the static friction coefficient between the tires and the road is 0.6. Now what is the maximum speed with which you can safely go around the curve?*

*As you write your solution to the starred problems, please follow the outline of the constrained-motion framework presented in examples N9.2 through N9.4.

N9S.10 At a certain instant of time, a 1200-kg car traveling along a curve 250 m in radius is moving at a speed of 10 m/s (22 mi/hr) but is slowing down at a rate of 2 m/s^2. Ignoring air friction, what is the total static friction force on the car as a fraction of its weight at that instant?

N9S.11 Consider a car turning a corner to the left. Using the methods of Chapter N6, show that in order to balance the torques on the car that seek to rotate the car around an axis along the direction of its motion, the roadbed has to exert more normal force on the right wheels of the car than on the left wheels. (Since this normal force will be transmitted to the car's body through the springs of the car's suspension, this means that the suspension springs on the right will have to compress more than on the left, and thus the car will lean to the right.)

N9S.12 An unpowered roller-coaster car starts at rest at the top of a hill of height H, rolls down the hill, and then goes around a vertical loop of radius R. Determine the minimum value for H required if the car is to stay on the track at the top of the loop. Please use the worksheet. (*Hints*: At the top of the loop, the car is upside down. If it is in contact with the track, though, the contact interaction will exert a normal force on the car perpendicular to the track and *away* from the track, since the normal force is a compression force. You may find it helpful to use conservation of energy here.)

RICH-CONTEXT

N9R.1 "Rotor" is a ride found in many amusement parks that consists of a hollow cylindrical room (roughly 8 ft in radius) that rotates around a central vertical axis. Riders enter the room and stand against the canvas-covered wall. The room begins to rotate, and when a certain speed is reached, the floor of the room drops away, revealing a deep pit. The riders do not fall, though: they are supported by a static friction force exerted by the person's contact interaction with the wall. Estimate the rate at which the room should rotate (in revolutions per minute) to safely pin the riders to the wall. (You may have to make some estimates).

N9R.2 Imagine that you are driving on a large, dark parking lot in the middle of the night. Your headlights suddenly illuminate a wall directly ahead of you perpendicular to your direction of travel and stretching away on both sides as far as the eye can see. To avoid hitting the wall is it better to turn your car to the right or left without braking or brake as hard as you can while moving straight toward the wall?

N9R.3 You are the technical consultant for a car-chase sequence in an action movie. In a certain part of the scene, the director wants a car to round a certain curve while braking from 66 mi/h to rest before it travels 290 ft along the curve. You measure the radius of the curve to be 590 ft. Is this scene possible? Defend your response and suggest an alternative scene if not.

ADVANCED

N9A.1 The position of an object in the xy plane can be expressed in terms of its distance r from the origin and the angle θ that its position vector makes with the x axis. As the object moves, r and θ change with time. **(a)** Show that if r is *not* necessarily constant (the object's motion is not circular), the object's velocity can be written

$$\vec{v}(t) = \frac{dr}{dt}\hat{r} + r\frac{d\theta}{dt}\hat{\theta} \qquad \text{(N9.28)}$$

We call the quantity $v_r \equiv dr/dt$ and $v_\perp \equiv r d\theta/dt$ the *radial* and *perpendicular* components of the object's velocity. **(b)** Find the object's acceleration $\vec{a}$ and express it as the vector sum of sometimes $\hat{r}$ plus something times $\hat{\theta}$.

ANSWERS TO EXERCISES

N9X.1 The components of $\vec{v}$ are $[-v\sin\theta,\ v\cos\theta]$. Taking the time derivative of each component and remembering that v = constant, we have

$$a_x = \frac{d}{dt}(-v\sin\theta) = -v\frac{d\sin\theta}{dt} = -v\cos\theta\frac{d\theta}{dt} \qquad \text{(N9.29a)}$$

$$a_y = \frac{d}{dt}(v\cos\theta) = v\frac{d\cos\theta}{dt} = -v\sin\theta\frac{d\theta}{dt} \qquad \text{(N9.29b)}$$

where I have used equations N9.3 to evaluate the derivatives of the trigonometric functions. These equations say the same thing as equation N9.7 does.

N9X.2 By equation N9.10, we have

$$v = \frac{2\pi R}{T} = \frac{2\pi(3.0\text{ m})}{6.3\text{ s}} = 3.0\text{ m/s} \qquad \text{(N9.30)}$$

So the magnitude of its acceleration is

$$a = \frac{v^2}{R} = \frac{(3.0\text{ m/s})^2}{(3.0\text{ m})} = 3.0\text{ m/s}^2 \qquad \text{(N9.31)}$$

N9X.3 Equation N9.11 says that

$$\vec{a}(t) = \frac{dv_\perp}{dt}\hat{\theta} + v_\perp\frac{d\hat{\theta}}{dt} \qquad \text{(N9.32)}$$

According to equation N9.9, though

$$\frac{d\hat{\theta}}{dt} = -\hat{r}\frac{d\theta}{dt} \qquad \text{(N9.33)}$$

Also, equation N9.5 implies that $d\theta/dt = v_\perp/R$. Plugging these things into equation N9.32, we get:

$$\vec{a}(t) = \frac{dv_\perp}{dt}\hat{\theta} + v_\perp\frac{d\theta}{dt}(-\hat{r}) = \frac{dv_\perp}{dt}\hat{\theta} - \frac{v_\perp^2}{R}\hat{r} \qquad \text{(N9.34)}$$

Since $v_\perp^2 = v^2$ in the case of circular motion, this is equivalent to the first equality in equation N9.12. Since the quantities $|dv_\perp/dt|$ and v^2/R form two legs of a right triangle whose hypotenuse is mag($\vec{a}$), the second equality follows from the pythagorean theorem.

N9X.4 If the object is moving clockwise, then its $v_\perp$ is negative. If it is speeding up, then $v_\perp$ is becoming more negative, so $dv_\perp/dt$ is negative. This means that the vector $(dv_\perp/dt)\hat{\theta}$ points in the $-\hat{\theta}$ direction, which is *forward* as far as the object's motion is concerned. Similarly, if the object is slowing down, then $v_\perp$ is becoming more positive, so $(dv_\perp/dt)\hat{\theta}$ points in the $+\hat{\theta}$ direction, which is *backward* relative to the object's motion.

N9X.5 The radial part of this car's acceleration has a magnitude of v^2/R = (22 m/s)^2/(450 m) = 1.1 m/s^2. We are told that the car is slowing down at the rate of 1.5 m/s^2, so the component of the acceleration in the direction of motion is $dv_\perp/dt$ = −1.5 m/s^2. The magnitude of the total acceleration (according to the pythagorean theorem) is

$$a = \sqrt{(1.1\text{ m/s}^2)^2 + (-1.5\text{ m/s}^2)^2} = 1.85\text{ m/s}^2 \qquad \text{(N9.35)}$$

Note that $\vec{a}$ points somewhat backward here. The angle ϕ that $\vec{a}$ makes with the backward direction is

$$\phi = \tan^{-1}\left|\frac{v^2/R}{dv_\perp/dt}\right| = \tan^{-1}\left|\frac{1.1\ \cancel{\text{m/s}^2}}{1.5\ \cancel{\text{m/s}^2}}\right| = 36° \qquad \text{(N9.36)}$$

N9X.6 Notice that in Figure N9.3a, the whole lift force goes to supporting the plane $mg = F_L$. In Figure N9.3c, though, we see that the plane is supported by only the vertical component of the lift force: $mg = F_L\cos\theta$. This means that the lift force must increase in magnitude as the plane banks to keep its vertical component equal in magnitude to the plane's weight. The easiest way to make the wings exert more lift is to increase the speed of the plane through the air.

N9X.7 Define $u \equiv t^2 - b$: then $f(u) = u^{-3}$. Then

$$\frac{df}{dt} = \frac{df}{du}\frac{du}{dt} = \frac{d(u^{-3})}{du}\frac{d(t^2-b)}{dt}$$

$$= -3u^{-4}(2t) = \frac{-6t}{(t^2-b)^4} \qquad \text{(N9.37)}$$

N10

NONINERTIAL REFERENCE FRAMES

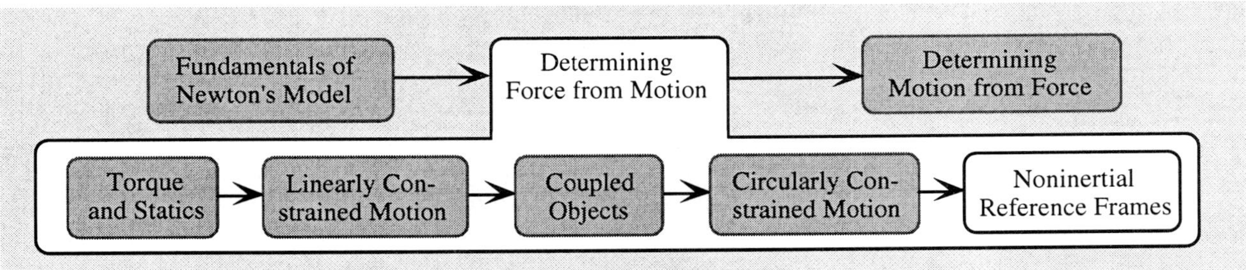

N10.1 OVERVIEW

In the last four chapters, we have been studying how to infer the existence and magnitude of forces from observations of an object's motion. However, this process can lead us *incorrectly* to infer the existence of forces if we are not using an appropriate (inertial) reference frame to analyze the situation.

One point that I have repeatedly made since the beginning of Unit *C* is that in the newtonian model, *forces arise from interactions*. Yet there are forces that we seem to experience on a daily basis that do *not* arise from one's interaction with something else but rather from one's motion (for example, the *centrifugal force* that seems to pull things away from the center of a rotating object). These forces have no place in newtonian physics, and yet they *seem* real.

This problem is the last and greatest challenge that everyone's aristotelian intuition about physics presents to our budding newtonian intuition. The successful explanation of these apparent forces *within* the newtonian model is our goal in this chapter. Here is an overview of this chapter's sections.

N10.2 *FICTITIOUS FORCES* reviews the evidence that suggests the existence of forces that seem to arise from motion.

N10.3 *THE GALILEAN TRANSFORMATION* how an object's motion in one reference frame appears in another frame moving relative to it.

N10.4 *INERTIAL REFERENCE FRAMES* shows how we can divide reference frames into two classes: *inertial* frames, where Newton's model applies, and *noninertial* frames, where fictitious forces appear to arise.

N10.5 *LINEARLY ACCELERATING FRAMES* argues that we can explain all the *effects* of the fictitious force that seems to arise in an accelerating frame *without* postulating the existence of the force if we analyze the situation using an inertial reference frame.

N10.6 *CIRCULARLY ACCELERATING FRAMES* argues the same thing for the effects of the centrifugal force.

N10.7 *GRAVITY AND ACCELERATING FRAMES* discusses why gravity seems to be erased in freely-falling reference frames and the general curious similarity between gravitational and fictitious forces.

This chapter provides essential background for unit *R* and (particularly the last section) for our discussion of planetary motion. Understanding this material is also essential for anyone wanting to develop a fully newtonian world-view.

N10.2 FICTITIOUS FORCES

According to the newtonian model, *all* physical forces express the interaction of two objects and all are either long-range forces or contact forces. The latter are easy to recognize since you can easily see when two objects are touching each other, and the only long-range forces that operate over macroscopic distances are gravitational and electromagnetic forces.

Examples of *fictitious forces* in daily life

Yet there are certain situations where forces appear to act that do *not* fit these categories. For example, imagine yourself in a jet that is accelerating for takeoff. When the pilot opens up the throttle, a magical force seems to appear from nowhere that pushes you backward in your seat. The more extreme the plane's acceleration, the stronger that this apparent force seems to be.

When the plane lands, the reverse happens. After the wheels touch the ground, the pilot applies the brakes and reverses the thrust on the engines, and when this happens, a mysterious force appears that tugs you forward. Again, the more extreme the acceleration, the stronger this force appears to be. If you are in an automobile and the driver suddenly applies brakes sharply, you feel the same kind of force pulling you forward.

Imagine yourself now in a car that is going at a high speed around a sharp turn. As the car turns, a magical force seems to press you against the side of the car away from the turn. Again, this force seems to be associated with the car's acceleration (we know that a car going around a curve is accelerating toward the center of the curve). The more extreme the car's acceleration (that is, the tighter the curve or the higher the car's speed) the stronger this apparent force becomes.

These alleged forces do not fit in any of the categories described in chapter N3. They are obviously not contact forces (nothing that you are touching presses you back into the plane seat during takeoff). Such forces feel like gravitational forces, but they cannot *be* gravitational (an extremely large and massive object does not magically appear behind the plane when the pilot opens the throttle). Nor are they electrostatic or magnetic (you do not become electrically charged or magnetized when your car turns a sharp corner). These alleged forces do not in fact seem to be a consequence of the presence of *any* external object.

Our task: to explain the effects without the forces

Even though these forces "feel" quite real, *I claim that these alleged forces are in fact inventions of your imagination*. Ladies and gentlemen of the jury, I will show you beyond a shadow of a doubt that it is possible within the context of the newtonian model to explain *all* of the described *effects* of these forces *without* assuming that these forces really exist. Since the argument for the existence of these alleged forces is based *entirely* on the circumstantial evidence of their effects, an alternative explanation of those effects makes the case for the existence of these **fictitious forces** disappear. Let me begin my case.

N10.3 THE GALILEAN TRANSFORMATION

Fictitious forces are associated with *accelerating* reference frames

The first step in understanding this situation is to recognize that these forces only seem to arise when you are riding in something (like a plane or car) that is *accelerating* relative to the earth's surface. While being pushed back into your chair is something you might expect during an airplane takeoff, you would be very surprised (even terrified) if a mysterious force were to spontaneously push you back into you chair while you were sitting at home reading a book!

When you are riding in a plane, car, or even a playground merry-go-round, you automatically and unconsciously use your surroundings (the cabin of the plane, the frame of the car, or the structure of the merry-go-round) as your frame of reference, and you judge your motion and the motion of objects around you in terms of that frame. When your jet or car is cruising at a constant speed, no magical forces appear: only when the jet or car changes its velocity do strange things seem to happen. *The presence of these forces has something to do with observing motion from within an accelerating reference frame.*

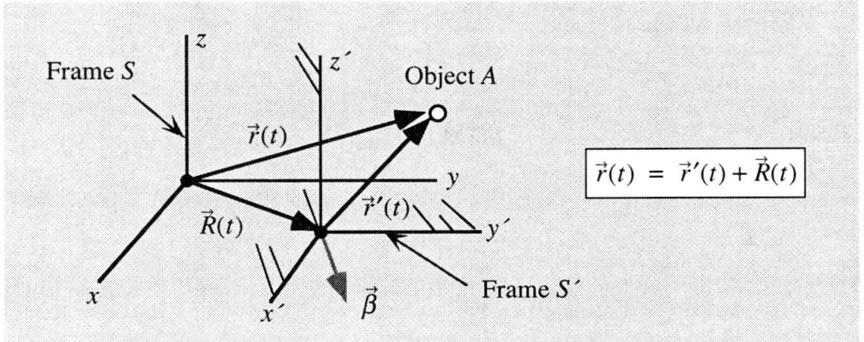

Figure N10.1: Two reference frames S and S'. Frame S' moves relative to S at a velocity $\vec{\beta}$, as shown. The positions of object A as measured in the two reference frames are related by the vector equation in the box.

Connecting measurements made in one frame with those made in another

The next step is to find a means of mathematically connecting observations made in one reference frame with observations made in another. Figure N10.1 shows two abstract reference frames, S and S' (the latter is read *S-prime*). For the sake of argument, let us consider frame S to be at rest on the surface of the earth, while S' is moving with respect to S at some (possibly time-dependent) relative velocity $\vec{\beta}$ (the Greek letter *beta* here refers to the "boost" in velocity required to take you from frame S to frame S'). Let's also take advantage of our ability to choose the orientation of reference frames to set them up so that their axes point in the same directions (this makes the analysis easier). Assume that we measure the position of object A as a function of time in both reference frames: let $\vec{r}(t)$ be its position vector as measured in frame S at time t, and let $\vec{r}'(t)$ be its position as measured in S' at the same instant, and let the position of S' relative to S at that instant be $\vec{R}(t)$. As shown in Figure N10.1, the definition of vector addition means that the mathematical relationship between $\vec{r}(t)$, $\vec{r}'(t)$, and $\vec{R}(t)$ at any time (no matter how S' is moving) is given by:

Transformation equation for positions

$$\vec{r}(t) = \vec{r}'(t) + \vec{R}(t) \tag{N10.1}$$

In words, this equation tells us that the object's position in frame S is the vector sum of its position in frame S' and the position of S' relative to S.

If we take the time-derivative of both sides of this equation, we get:

$$\vec{v}(t) = \vec{v}'(t) + \vec{\beta}(t) \tag{N10.2}$$

where $\vec{v}(t)$ is the object's velocity in frame S, $\vec{v}'(t)$ is its velocity in frame S', and $\vec{\beta}(t)$ is the velocity of frame S' with respect to S. This equation allows you to *transform* an object's velocity measured in the S' frame to that object's velocity measured in the S frame. If instead we know the object's velocity in S and want to find it in S', we can of course solve equation N10.2 for $\vec{v}'(t)$, getting:

(Galilean) transformation equation for velocities

$$\vec{v}'(t) = \vec{v}(t) - \vec{\beta}(t) \tag{N10.3}$$

We call equation N10.3 the **galilean velocity transformation equation.** In component form, equations N10.2 and N10.3 say that

$$\begin{bmatrix} v_x(t) \\ v_y(t) \\ v_z(t) \end{bmatrix} = \begin{bmatrix} v_x'(t) \\ v_y'(t) \\ v_z'(t) \end{bmatrix} + \begin{bmatrix} \beta_x(t) \\ \beta_y(t) \\ \beta_z(t) \end{bmatrix} \qquad \begin{matrix} \text{(N10.4a)} \\ \text{(N10.4b)} \\ \text{(N10.4c)} \end{matrix}$$

$$\begin{bmatrix} v_x'(t) \\ v_y'(t) \\ v_z'(t) \end{bmatrix} = \begin{bmatrix} v_x(t) \\ v_y(t) \\ v_z(t) \end{bmatrix} - \begin{bmatrix} \beta_x(t) \\ \beta_y(t) \\ \beta_z(t) \end{bmatrix} \qquad \begin{matrix} \text{(N10.5a)} \\ \text{(N10.5b)} \\ \text{(N10.5c)} \end{matrix}$$

Does this make any sense? Let's consider some examples.

EXAMPLE N10.1

Problem: Imagine that you are on a train traveling in the x direction at a speed of 25 m/s with respect to the ground. If you throw a baseball in the $+x$ direction at a speed of +12 m/s relative to the train, what is the ball's x-velocity relative to the ground? *Intuitively*, you might say that the ball's velocity with respect to the ground should be the *sum* of the train's velocity with respect to the ground and the ball's velocity with respect to the train, that is, 37 m/s. Is this correct?

Solution In order to apply any one of equations N10.2 through N10.5 correctly, we have to do two things: (1) determine what the reference frames are and which is S and which is S', and (2) determine which of the stated velocities correspond to which of the symbolic quantities $\vec{v}$, $\vec{v}'$, and $\vec{\beta}$.

The first of these steps is actually a fairly arbitrary decision: as long as we keep everything straight, we would get the same ultimate answer no matter which frame (ground or train) we take to be frame S. But note that $\vec{\beta}$ is defined to specify the velocity of frame S' *relative* to S. In the problem description, the train's velocity is specified relative to the ground, so it is *convenient* to let S' be the train frame and S be the ground frame (because then we know right away from the problem description that $\beta_x = +25$ m/s). When one frame in the problem is attached to the earth's surface we also conventionally take that frame to be frame S (in situations where both the frames are attached to objects floating in deep space, though, this convention is no help!).

So let's decide that the train frame is frame S' and the ground frame is frame S. We are told that the ball moves at a speed of 12 m/s in the x direction with respect to the train, so this velocity must be $\vec{v}'$, since the train is the "primed" frame (that is, $v_x' = +12$ m/s). We are trying to find the velocity of the ball with respect to the ground, so $\vec{v}$ is the unknown quantity. Equation N10.4a gives us v_x in terms of v_x' and β_x:

$$v_x = v_x' + \beta_x = +12 \text{ m/s} + 25 \text{ m/s} = 37 \text{ m/s} \text{ (as expected)} \tag{N10.6}$$

EXAMPLE N10.2

Problem: Imagine that you are in a train traveling in the $+x$ direction with a speed of 35 m/s relative to the ground and you observe a car traveling in the same direction at a speed of 29 m/s relative to the ground. What is the car's speed relative to you?

Solution Again, we are *given* that the train's velocity with respect to the ground is 35 m/s in the x direction. If we take the train to be frame S' and the ground to be frame S, then this means that $\beta_x = +35$ m/s. We are given that the velocity of the car in the ground frame is 29 m/s in the $+x$ direction, so $v_x = +29$ m/s. The x velocity of the car in the train frame is the unknown quantity v_x'. Equation N10.5a allows us to find this quantity in terms of the others:

$$v_x' = v_x - \beta_x = 29 \text{ m/s} - 35 \text{ m/s} = -6 \text{ m/s} \tag{N10.7}$$

meaning that the car will appear to drift backward at a speed of 6 m/s relative to the train. If you visualize the situation, perhaps you will agree that this is right.

EXAMPLE N10.3

Problem: An airplane is flying with a velocity of 75 m/s due north relative to the air. If the wind is blowing 12 m/s west relative to the ground, what is the plane's velocity (magnitude and direction) relative to the ground?

Solution Again, we will take the ground to be frame S and the air to be frame S'. According to the description of the situation, the air is moving relative to the ground at 12 m/s west. If both our frames are oriented in the usual way relative to the earth's surface, this means that $\vec{\beta} = [-12 \text{ m/s}, 0, 0]$. The plane's velocity with respect to the air, on the other hand, is $\vec{v}' = [0, +75 \text{ m/s}, 0]$. We want to find the plane's velocity with respect to the ground, that is, $\vec{v}$. According to equation N10.4, we have

$$\begin{bmatrix} v_x \\ v_y \\ v_z \end{bmatrix} = \begin{bmatrix} v_x' \\ v_y' \\ v_z' \end{bmatrix} + \begin{bmatrix} \beta_x \\ \beta_y \\ \beta_z \end{bmatrix} = \begin{bmatrix} 0 \\ +75 \text{ m/s} \\ 0 \end{bmatrix} + \begin{bmatrix} -12 \text{ m/s} \\ 0 \\ 0 \end{bmatrix} = \begin{bmatrix} -12 \text{ m/s} \\ +75 \text{ m/s} \\ 0 \end{bmatrix} \quad \text{(N10.8)}$$

The magnitude of this velocity is

$$v = \sqrt{v_x^2 + v_y^2 + v_z^2} = \sqrt{(-12 \text{ m/s})^2 + (75 \text{ m/s})^2 + 0} = 76 \text{ m/s} \quad \text{(N10.9a)}$$

The angle that this velocity makes with the y (north) axis is

$$\theta = \tan^{-1}\left(\frac{12 \text{ m/s}}{75 \text{ m/s}}\right) = 9° \quad \text{(N10.9b)}$$

Since v_x is negative, the plane is flying at 76 m/s, 9° *west* of north.

Now, if we take the time derivative of both sides of equation N10.3, we get

Transformation equation for accelerations

$$\vec{a}'(t) = \vec{a}(t) - \vec{A}(t) \quad (\vec{A} \text{ is the acceleration of } S' \text{ relative to } S) \quad \text{(N10.10)}$$

This equation tells us how to calculate an object's *acceleration* $\vec{a}'$ in the S' frame if we know its acceleration $\vec{a}$ in the S frame and the acceleration $\vec{A}$ of frame S' with respect to S. We can apply this equation to a given situation in much the same way as we applied equation N10.2 or N10.3 in the examples.

Note that if frame S' moves at a *constant velocity* with respect to S (so that $\vec{A} = 0$) the object's acceleration is the same in both reference frames:

$$\vec{a}'(t) = \vec{a}(t) \quad (\text{if } S' \text{ moves at a constant velocity relative to } S) \quad \text{(N10.11)}$$

We will build our case regarding fictitious forces on an analysis of equations N10.10 and N10.11.

Exercise N10X.1: A train is moving at a constant velocity of 23 m/s in the $+x$ direction. A child throws a ball out the back end of the train with a velocity of 3 m/s in the $-x$ direction relative to the train. What is the velocity of the ball with respect to the ground?

Exercise N10X.2: A ship is moving at a constant velocity of 8 m/s due north. A sailboat nearby has a velocity of 6 m/s due west. What is the velocity of the sailboat relative to the ship?

Exercise N10X.3: Imagine that you are in an elevator whose vertical acceleration is 3.0 m/s^2 upward. You drop a ball that falls with a downward acceleration of g = 9.8 m/s^2 in the frame of the earth. What is the ball's acceleration in the frame of the elevator?

N10.4 INERTIAL REFERENCE FRAMES

Let's consider the implications of equation N10.10 for Newton's laws. Imagine that an object is completely isolated from external interactions. This means that $\vec{F}_{\text{net}} = 0$, and by Newton's second law, the object should have *zero* acceleration. Newton's first law says this even more directly: *an isolated object moves at a constant velocity.*

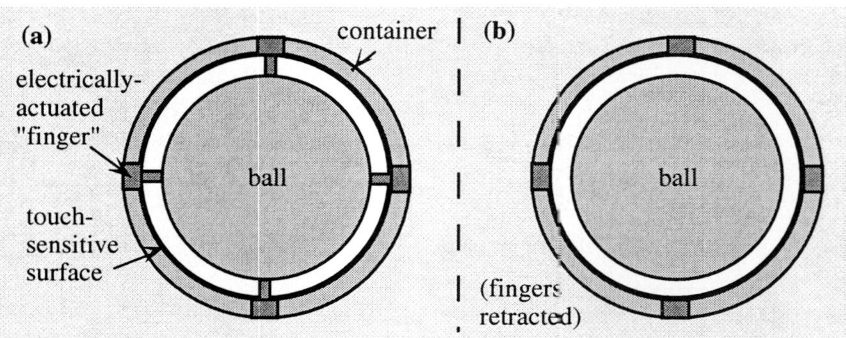

Figure N10.2: (a) A cross-sectional view of a floating-ball first-law detector. Electrically-actuated "fingers" hold the ball initially at rest in the spherical container. **(b)** After the fingers are retracted, the ball should continue to float at rest in the container if the container frame is inertial.

Now, an object is either isolated from external interactions or it is not: you do not have to use a reference frame to determine whether an object is massive or whether it is electrically charged or whether it is touching something else. Everyone, no matter what reference frame they might be using, should therefore agree as to whether a given object is isolated or not.

An isolated object doesn't obey Newton's laws in an noninertial reference frame

Imagine that a universally-accepted isolated object *is* observed to move at a constant velocity (zero acceleration) in some frame S. Equation N10.10 implies that all *other* frames can be divided into two categories. In those frames that move at a constant velocity with respect to S (so that their acceleration $\vec{A}$ relative to S is zero) the isolated object will *still* be measured to have zero acceleration and thus be observed to obey Newton's first and second laws. Such frames are called **inertial reference frames** (because Newton's first law is sometimes called the *law of inertia*). In those frames that are accelerating relative to frame S, the isolated object will be measured to have a nonzero acceleration $\vec{a}' = -\vec{A}$, meaning that Newton's first and second laws *fail* to work in such reference frames. Such frames are called **noninertial reference frames**.

Let me emphasize again that *Newton's first and second laws do not apply in noninertial reference frames*: such frames are "bad" for analyzing motion in the newtonian model. If we want to apply Newton's laws to analyze an object's motion, we must do the analysis in an *inertial* reference frame.

Constructing an idealized first-law detector

We can distinguish an inertial frame from an noninertial frame *without* determining the frame's motion with respect to something else using a **first law detector**. Figure N10.2 shows such a detector. Electrically-actuated "fingers" hold an electrically uncharged and non-magnetic ball in the center of a spherical container from which the air has been removed (Figure N10.2a). When the fingers are retracted, the ball is (at least momentarily) at rest and completely isolated from contact and electromagnetic interactions (Figure N10.2b). An isolated object at rest will *remain* at rest if the frame is inertial. If the frame to which the container is attached is non-inertial, though, the ball will accelerate relative to that frame and thus the container (see equation N10.10). This can easily be detected, because if the ball drifts away from rest in any direction it will eventually hit the container wall, where touch-sensitive sensors can register the violation of the first law (and trigger the "fingers" to reset the ball).

Actually, the detector as just described will only distinguish inertial from noninertial frames in deep space (far away from any gravitating objects) because there is no other way to isolate the internal ball from gravity. We might adapt our detector for operation in a gravitational field by changing the spherical container to a cylindrical one and allowing the ball to drop from rest. Since the ball's acceleration should be perfectly vertical by definition, any sideward deviation that would cause the ball to strike a container wall will indicate that the frame of the container is noninertial. Also, since all objects fall with the same acceleration, the ball *should* take a well-defined amount of time to fall the length of the cylinder. If it arrives at the bottom early or late, the frame is not inertial.

A floating-puck first-law detector

A practical way to make a first-law detector for use in a gravitational field would be to replace the ball with a puck that floats on a cushion of air above a flat, level air table. This puck would be isolated from external interactions that might affect its horizontal motion, so if we place it at rest at the center of the table, it should *remain* at rest. If we observe the puck to accelerate away from the center, the frame in which the table sits is not inertial. This detector will not register violations of the first law due to the acceleration of the table frame in the vertical direction, but it is more practical than a floating-ball or falling-ball detector for applications near the earth's surface

Formal definition of *inertial* and *noninertial* frames

Imagine now attaching floating-ball detectors in a number of places throughout a reference frame. A frame is defined to be *inertial* if *no* detector in the frame registers any violation of Newton's first law; the frame is *noninertial* otherwise.

If we apply this definition to a frame attached to the surface of the earth, we find that such a frame is *not* perfectly inertial (because the earth is rotating). The earth rotates fairly slowly though, so the deviations from perfection are *usually* negligibly small. Unless otherwise stated, we will assume that a reference frame attached to the earth's surface is sufficiently inertial for our purposes.

N10.5 LINEARLY ACCELERATING FRAMES

What you see in a linearly accelerated frame

Consider again our example of an airplane accelerating for take-off. Imagine that you hold a floating-puck first-law detector in your lap. Then the pilot turns on the engines and the plane begins to accelerate down the runway. You feel a force pushing you back into your seat. You will also see the puck accelerate toward the rear of the plane. In the plane frame, then, it certainly looks like some magical force is acting on you and the puck (see Figure N10.3a), pulling everything backward. Yet the puck is horizontally *isolated* and therefore there is no way for an external force to act on them. How can we resolve this paradox?

Explanation from point of view of ground frame

The problem is that we are trying to make Newton's second law work in an accelerating frame, when we have just argued that the second law *fails* in a frame accelerating with respect to the ground. Let's look at the motion of the puck from the perspective of a reference frame attached to the ground. When the pilot turns up the engines, and the plane begins to accelerate. The puck is horizontally isolated, so it will continue to move in a horizontal straight line relative to the ground. As the plane accelerates forward relative to the ground, the air table, on the other hand, is carried with the plane forward with respect to the ground and thus with respect to the puck, making the puck appear to accelerate backward with respect to the table. In the inertial frame of the ground, no "magical forces" are needed to explain the puck's behavior (see Figure N10.3b).

We can also see this from equation N10.10. Let the ground frame be frame S and the plane frame be frame S'. Since the puck has zero net physical force on it, its acceleration will be zero in the inertial ground frame S ($\vec{a} = 0$). If the plane

Figure N10.3: (a) In a plane accelerating for takeoff, a puck in a first-law detector deflects to the rear of the plane relative to the box. **(b)** When viewed in the ground frame, the puck remains at rest. It only deflects relative to the air table because the *table* is accelerating forward with you and the plane. (Note that I have displaced successive images of the situation to the right for clarity's sake. The plane is accelerating straight forward and is not moving to the right.)

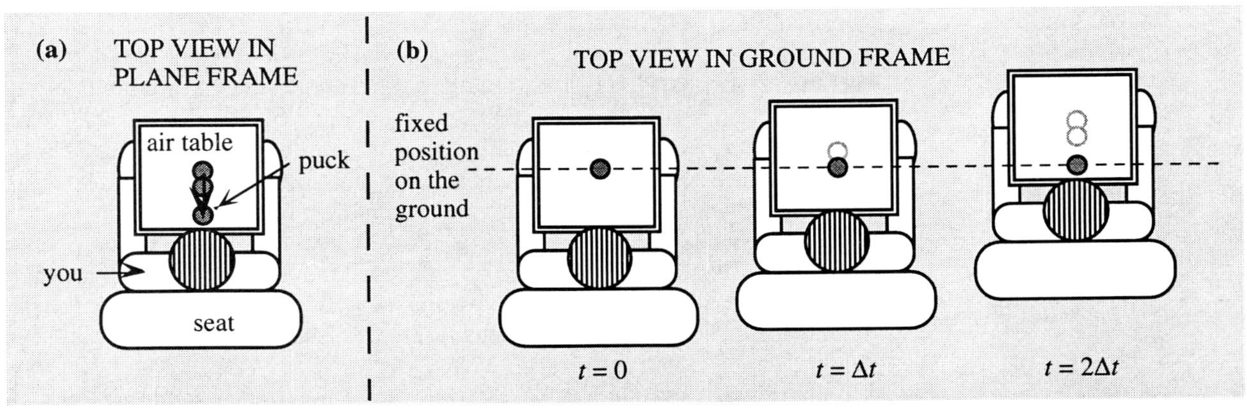

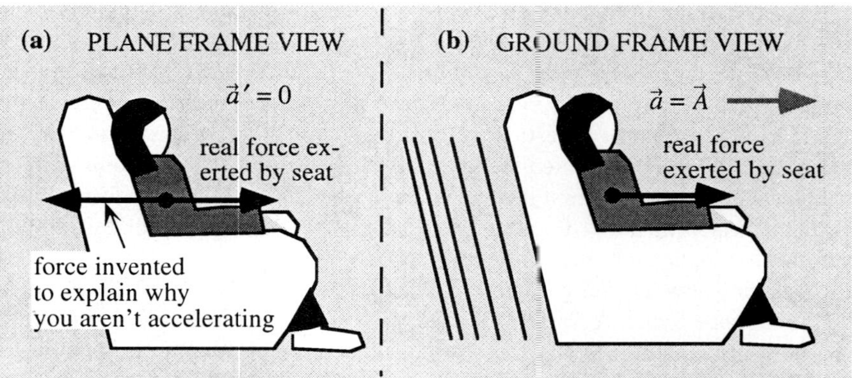

Figure N10.4: (a) In an accelerating plane, you seem to be at rest and thus have zero acceleration. You feel the chair pressing forward on you, so you assume that there must be a force pressing back on you to make the net force on you zero. **(b)** But the net force on you really is not zero: the chair has to push forward on you to accelerate you along with the plane. When you analyze things in the ground frame, there is no need for a backward force.

has a non-zero acceleration with respect to the ground frame ($\vec{A} \neq 0$), then the puck's acceleration measured in the plane frame is:

$$\vec{a}' = \vec{a} - \vec{A} = 0 - \vec{A} = -\vec{A} \qquad \text{(N10.12)}$$

(This applies to *any* object not accelerating in the S frame.) This tells us that the puck's acceleration as measured in the plane is equal in magnitude and opposite in direction to the plane's acceleration relative to the ground. This is what we would expect if the puck has no horizontal acceleration relative to the ground.

The point is that we do not need a "magical rearward force" to explain the behavior of the puck: we can explain its behavior just fine in the ground frame *without* assuming such a force exists. This alleged force thus vanishes if we analyze this force from a proper inertial reference frame: *it is not real.*

Resolving intuitive objections to this explanation

"Wait," you say, "If this force is not real, what causes the puck to drift backwards?" Think about it. The air table is being accelerated forward along with the plane, as observed in the ground frame. If the puck is to be at rest with respect to the table, it would have to accelerate forward relative to the ground frame also. But there is no force acting on the puck that can accelerate it forward. Therefore, it is unable to keep up with the table as the latter accelerates forward.

"But," I hear you cry, "we can *feel* this backwards force. Therefore, it *must* be real!" Not so, I say. What you *feel* is the airplane seat pressing you forward. In an *inertial* reference frame, if your seat presses you forward but you don't accelerate forward, it must be because some other force is pressing you backward. But in the noninertial plane frame, this logic does not apply, because Newton's laws do not apply. In the inertial frame of the ground, the seat presses you forward not because something else is pressing you backward, but because the seat *has* to press you forward to accelerate you along with the plane (Figure N10.4).

Part of the reason that you seem to really *feel* this force is that parts of your body have to flex and stretch to accelerate other parts of your body along with the plane. For example, if you hold your hand in front of you, it feels as if something were pushing it toward you. But what is really going on is that you have to flex the muscles in your arm to push your hand forward if it is going to accelerate along with the plane. You feel this muscular action and intuitively interpret it as a response to something pushing your hand toward you.

In short, when you feel the seat pressing you forward, you intuitively *invent* a force pressing you backward, because otherwise your lack of acceleration in the plane frame and the way that your body feels would be inexplicable. In doing this, though, you are illegally applying Newton's second law in an accelerated reference frame (an offense punishable by nonsensical answers throughout the universe!). In the inertial reference frame of the ground, such a force (1) cannot be made consistent with the idea that *forces express interactions*, and (2) is entirely *unnecessary* to explain what we observe if we use an inertial frame. Therefore, from the point of view of the newtonian model, this force is *not real.*

Exercise N10X.4: If you are in a car that suddenly slows down, you seem to feel a magical force acting to draw you forward. What is *really* going on? In a serious accident, people without seat belts can go through the windshield. If they are not being thrown forward by this magical force, how is this possible?

N10.6 CIRCULARLY ACCELERATING FRAMES

Now imagine being in a car that turning a corner or riding a playground merry-go-round. The outward force that you seem to feel under such circumstances is such a vivid and common part of daily experience that it even has a colloquial English name: **centrifugal force**. This force is so deeply imbedded in our pre-Newtonian intuition about the world that even science writers and (sadly) high-school science texts use this term as if this force were real.

We can analyze what you experience in the car turning a corner in a manner entirely similar to our work in the last section. When viewed in the inertial reference frame of the ground, as the car turns left (say), a floating puck would continue to move forward in the direction of the car's original velocity. In the car's frame, the air table is accelerating toward the left, so the puck seems to move right relative to the table. Viewed in the car frame, it looks like the puck is pushed outward (to the right) by centrifugal force. What is *really* happening, though, is the table (along with the car) is accelerating left relative to the puck.

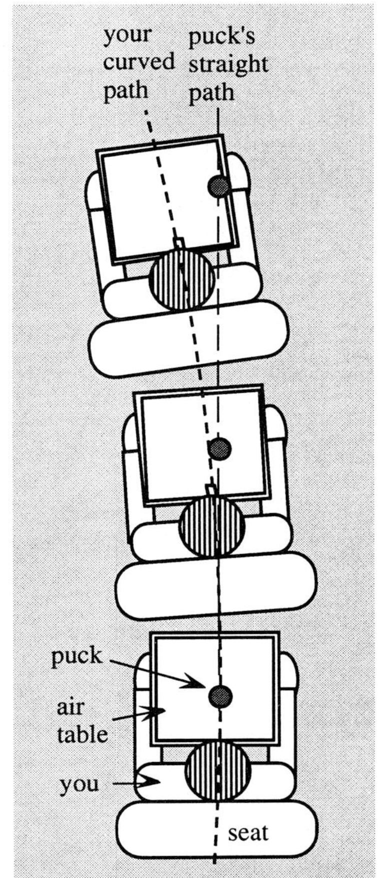

Figure N10.5: The puck in a floating-puck first-law detector seems to accelerate to the right as you ride in a car turning left because the car and you accelerate to the left under the puck while it moves straight forward.

Your body would do the same thing as the puck (that is, move forward in a straight line) if the seat belt or the side of the car did not exert a leftward force on you to accelerate you along with the car. In the car's frame, you do not *seem* to be accelerating, you therefore might interpret the leftward force exerted by the belt or side of the car as being due to a rightward force pulling you outward. But one is not *allowed* to use Newton's second law in the accelerating frame of the car, so this inference is *false*.

Really, we have two choices in both of these situations. We could in principle *insist* that Newton's laws apply in *all* reference frames. This would require us to invent forces in some reference frames that are not due to contact or any other interaction with an external object. Moreover, these forces seem to exist at some times and not at others, and exist in one reference frame (for example, the plane) but not exist in another (for example, the ground). It seems to me that forces that either exist or don't exist depending on one's arbitrary choice of reference frame have an extremely poor claim on being "real".

It is better, simpler, and entirely sufficient to simply assert that Newton's laws *don't apply* in noninertial reference frames and insist that we use only *inertial* reference frames when analyzing motion using Newton's laws. We have no physical evidence for these alleged forces except for the effects that they seem to produce. We can entirely explain these effects *without* the use of these alleged forces by using an inertial reference frame. These alleged forces are, therefore, not real in any useful sense of the word, and we call them *fictitious* forces.

Closing argument

Ladies and gentlemen of the jury, I have shown beyond a shadow of a doubt that motion can be easily interpreted and analyzed in an inertial reference frame without invoking the magical forces others claim to exist. Are these forces real? Only one verdict is logical, sensible, and right. Do your duty. I rest my case.

Exercise N10X.5: When you are on a rapidly rotating playground merry-go-round, you have to hold on tight, using the full strength of your arms to keep your body from flying off. If centrifugal forces are not real, why do you have to exert so much strength to keep yourself at rest relative to the merry-go-round?

N10.7 GRAVITY AND ACCELERATING FRAMES

Gravity seems to be erased in freely falling frame

Equation N10.10 has a very curious consequence. Imagine an object that is completely isolated *except* for the force of gravity. The net force on that object in the ground frame will thus be $m\vec{a} = \vec{F}_{\text{net}} = m\vec{g}$. Dividing through by m, we get $\vec{a} = \vec{g}$, independent of the object's mass. Thus *all* such otherwise isolated objects fall with this acceleration.

So imagine viewing these objects in a reference frame that is itself freely falling. Its acceleration relative to the ground will be $\vec{A} = \vec{g}$ too. Applying equation N10.10, we find that in this frame any falling object has acceleration

$$\vec{a}' = \vec{a} - \vec{A} = \vec{g} - \vec{g} = 0 \qquad \text{(N10.13)}$$

and will thus behave entirely like an isolated object that is *not* affected by gravity. If you place an object at rest relative to a freely falling reference frame, it will remain at rest (since it falls along with the frame). In particular, if you put a floating-ball detector in such a frame, it will behave as if the frame were inertial in deep space. It is as if the force of gravity has been completely turned off for all objects in this frame.

More generally, assume that the net force on a certain object of mass m is $\vec{F}_{\text{net}} = m\vec{g} + \vec{F}_1 + \vec{F}_2 + \ldots$, where $m\vec{g}$ is the gravitational force on the object and $\vec{F}_1, \vec{F}_2, \ldots$ are other forces acting on the object. Newton's second law tells us that in an inertial frame, this object's acceleration is

$$m\vec{a} = \vec{F}_{\text{net}} = m\vec{g} + \vec{F}_1, \vec{F}_2, \ldots \qquad \text{(N10.14)}$$

However, in the freely falling frame its acceleration is $\vec{a}' = \vec{a} - \vec{A} = \vec{a} - \vec{g}$, so

$$m\vec{a}' = m\vec{a} - m\vec{g} = (m\vec{g} + \vec{F}_1 + \vec{F}_2 + \ldots) - m\vec{g} = \vec{F}_1 + \vec{F}_2 + \ldots \qquad \text{(N10.15)}$$

This looks the way Newton's second law *would* look if the gravitational force were not present. Thus even when the objects are *not* isolated, *objects analyzed in a freely falling frame behave as if they were in an inertial frame where no external gravitational forces act.*

Applications

For example, objects appear to be weightless in the orbiting space shuttle not because there is no gravity up where the shuttle flies ($\vec{g}$ is not much smaller a couple of hundred kilometers above the earth than it is on the ground) but because the shuttle is freely falling around the earth as it orbits. Astronauts train for the apparently weightless conditions in orbit by riding on a plane that is put in a freely falling dive for several minutes.

If you think about it, in the newtonian scheme of things, the earth *cannot* be an inertial frame: it is accelerating in its circular orbit around the sun, which in turn is accelerating in its orbit around the galaxy, which in turn is accelerating in response to other galaxies and so on. Newton's laws *seem* to operate in reference frames attached to the earth because the earth is *freely falling* in the gravitational field of the sun and the rest of the universe. Even though these gravitational forces are *real* in the newtonian model, objects analyzed in a non-rotating frame fixed to the center of the earth behave almost exactly as they would in a really inertial frame where these forces were not present. (A frame attached to the surface of the earth is not quite so good because the earth is rotating, but the rotation is slow enough that it doesn't make much of a difference in most cases.)

Similarly, if we want to analyze the motion of the earth and other planets in the solar system, we can use a non-rotating reference frame attached to the center of mass of the solar system (which essentially coincides with the center of the sun). Even though the whole solar system is attracted by and orbits the center of our galaxy, its center of mass freely falls in that gravitational field, so we can ignore the effects of the galaxy (and indeed the rest of the universe).

This makes physics much simpler than it otherwise might be! Imagine how difficult it would be if we had to know about the gravitational effects of the sun and moon and distant galaxies to analyze the motion of an object on the earth.

This makes physics much simpler!

Note that all of the reference frames we have been using in the course so far are *freely-falling* frames, which are not really inertial in newtonian physics but rather "functionally inertial" (in that external gravitational interactions are in effect canceled by the frame's acceleration). Here is a list of such frames that we commonly use in newtonian physics, in decreasing order of how well they mimic a truly inertial reference frame

A list of practical frames that are functionally inertial

1. a non-rotating frame whose origin is the solar system's center of mass
2. a non-rotating frame falling in space near the earth (like the GPS)
3. a non-rotating frame whose origin is fixed to the earth's center of mass
4. a frame fixed to the earth's rotating surface
5. a frame moving with constant velocity relative to the earth's surface

Note that gravity, even though it *is* linked to a real and tangible interaction between objects, can be erased by using the "correct" (in this case, freely-falling) frame, just like a ficticious force! The idea that gravity is in many ways *equivalent* to the fictitious forces that one feels in an accelerated frame lies at the foundation of Einstein's theory of *general* relativity, our current theory of gravity.

There is a reverse effect that is interesting as well. Consider an object m acted on by a net force $\vec{F}_{\text{net}} = m\vec{g} + \vec{F}_1 + \vec{F}_2 + \ldots$. Again, in an inertial frame, equation N10.14 correctly states Newton's second law. In a noninertial frame whose acceleration relative to the inertial frame is $\vec{A}$, we have

Fictitious forces due to translational accelerations behave like gravitational forces

$$\begin{aligned} m\vec{a}' &= m\vec{a} - m\vec{A} = (m\vec{g} + \vec{F}_1 + \vec{F}_2 + \ldots) - m\vec{A} \\ &= m(\vec{g} - \vec{A}) + \vec{F}_1 + \vec{F}_2 + \ldots \end{aligned} \tag{N10.16}$$

This is what Newton's second law *would* look like in this frame if the acceleration of gravity were $\vec{g} - \vec{A}$ instead of $\vec{g}$. Therefore, we can *pretend* that Newton's second law works in an accelerated frame if we imagine that there is a fictitious gravitational force $-m\vec{A}$ acting on an object of mass m in that frame. This provides a practical means of doing newtonian physics in frames (like the surface of the rotating earth) that are measurably noninertial but for which more inertial alternatives are hard to use. The application of this idea is an important topic in upper-level courses in newtonian mechanics.

EXAMPLE N10.4

Problem: Astronauts train for the high accelerations they experience during launch using a centrifuge that consists of a little room suspended at the end of a long boom that is rotated in a horizontal plane. If the apparent acceleration of gravity in the room is $5g$, and the distance between the room's center and the center of rotation is 8.0 m, how long does it take the boom to rotate once?

Solution The acceleration $\vec{A}$ of the room relative to the ground is directed horizontally and toward the center of rotation and has a magnitude of v^2/R, where v is the speed of the room in its circular path of radius R = 8.0 m. Since $\vec{g}$ is downward (perpendicular to $\vec{A}$), saying that the mag$(\vec{g} - \vec{A}) = 5g$ means that

$$5g = \sqrt{g^2 + A^2} \quad \Rightarrow \quad 25g^2 - g^2 = A^2 \quad \Rightarrow \quad A = \sqrt{24}g \tag{N10.17}$$

Since $v = 2\pi R/T$, where T is the time that it takes the boom to rotate once, then

$$\sqrt{24}g = A = \frac{v^2}{R} = \frac{(2\pi R/T)^2}{R} = \frac{4\pi^2 R}{T^2} \quad \Rightarrow \quad T = 2\pi\sqrt{\frac{R}{\sqrt{24}g}} \tag{N10.18}$$

Plugging in the numbers, we get T = 2.6 s (pretty fast for a large thing!)

SUMMARY

I. TRANSFORMATION EQUATIONS BETWEEN REFERENCE FRAMES
  A. Consider two arbitrary reference frames S and S'. Let:
    1. their corresponding axes be oriented in the same direction
    2. $\vec{r}(t)$, $\vec{r}\,'(t)$ be an object's position relative to the origin of S, S'
    3. $\vec{R}(t)$ be the position of origin of S' relative to the origin of S
    4. $\vec{\beta}(t)$ and $\vec{A}(t)$ be the velocity and acceleration of S' relative to S
  B. Given frames as described above, we have
    1. $\vec{r}(t) = \vec{r}\,'(t) + \vec{R}(t)$ or $\vec{r}\,'(t) = \vec{r}(t) - \vec{R}(t)$ (N10.1)
    2. time derivative $\Rightarrow$ $\vec{v}\,'(t) = \vec{v}(t) - \vec{\beta}(t)$ (N10.3)
    3. time derivative $\Rightarrow$ $\vec{a}\,'(t) = \vec{a}(t) - \vec{A}(t)$ (N10.10)
  C. To apply these formulas in a given problem
    1. decide which frame in problem is S and which is S'
       (Whatever is easy, but we often choose the *ground* frame to be S)
    2. link problem quantities with symbols in the equations

II. INERTIAL AND NON-INERTIAL FRAMES
  A. An *inertial* frame is one in which Newton's *first* law is true
    1. that is, an isolated object moves with constant velocity ($\vec{a} = 0$)
    2. this is required for Newton's second law to be true in that frame
    3. a frame in which this is not true is called a *noninertial frame*
  B. Equation N10.10 implies that if a frame S is inertial, then
    1. any frame S' moving with a constant velocity relative to S
    2. any frame S' accelerating relative to S is *noninertial*.
  C. We can use a first-law detector to distinguish inertial frames
    1. a *floating-ball* detector uses an isolated ball initially at rest at the center of a container: any drift signals a violation of the first law
    2. a *floating-puck* detector uses a puck floating on a flat, level air table (more practical than a floating-ball detector near the earth)

III. FICTITIOUS FORCES
  A. Newton's second law does not apply in noninertial reference frames
    1. to analyze motion using this law, we must use an inertial frame.
    2. a frame on the earth's surface is fairly good (see below)
  B. Fictitious forces are *imaginary* forces that we *invent* to make the second law work in a noninertial reference frame
    1. we do this automatically and intuitively
    2. but we *can* explain everything in an *inertial* frame w/o such forces
    3. forces that disappear when one changes reference frame are not real.
  C. An example: accelerating for takeoff
    1. the seat pushes you *forward* to keep you accelerating with plane
    2. the seat wouldn't do this in a normal (inertial) frame unless something pushes you back into the seat
      a) So you imagine that such a force *is* pushing back on you
      b) this also explains why the push from the seat does not accelerate you forward
    3. but you can't apply second law in noninertial plane frame!
    4. we can explain everything in the ground frame w/o this force
  D. We can explain apparent "centrifugal forces" similarly

IV. GRAVITY AND ACCELERATING FRAMES
  A. Gravity appears to vanish in a freely falling frame
    1. since everything falls with same acceleration along with frame
    2. we can ignore external gravitational forces if we use a falling frame
    3. for example, in frames based on the space shuttle, earth, sun, etc.
    4. makes physics much simpler!
  B. All the frames we use are really freely-falling frames in some sense!
  C. The apparent force $-m\vec{A}$ in an accelerating frame acts like gravity

GLOSSARY

inertial reference frame: a reference frame in which isolated object are observed to move with constant velocity (that is, zero acceleration), in accordance with Newton's first law. Inertial frames move at constant velocities with respect to each other.

noninertial reference frame: a reference frame in which isolated objects are observed to move with nonzero acceleration, in violation of both Newton's first and second laws, making it impossible to use the latter to link the true forces acting on the object with its motion. Noninertial frames *accelerate* relative to inertial frames.

first-law detector: any device which examines the paths of isolated objects for deviations from nonaccelerated motion. Such devices include the *floating-ball* and *floating-puck* detectors described in section N10.4.

fictitous force: a force that we infer to exist in a noninertial reference frame when we try (either intuitively or consciously) to apply Newton's second law in that frame (even though it *doesn't* apply in a noninertial frame).

centrifugal force: a fictitious force that seems to pull objects away from the center of a rotating reference frame.

functionally inertial: a freely-falling frame is "functionally inertial" because even though it is accelerating, objects in such a frame behave *as if* they were inertial if we ignore all external gravitational fields.

TWO-MINUTE PROBLEMS

N10T.1 In a Western movie, a person shoots an arrow backward from a fleeing horse. If the velocity of the horse relative to the ground is 13 m/s west, and the arrow's velocity relative to the horse is 38 m/s east, what is the arrow's velocity with respect to the ground?
A. 41 m/s east C. 25 m/s east
B. 41 m/s west D. 25 m/s west

N10T.2 A blimp has a velocity of 8.2 m/s due west relative to the air. There is a wind blowing at 3.5 m/s due north. The speed of the balloon relative to the ground is:
A. 11.7 m/s C. 7.4 m/s E. other (specify)
B. 8.9 m/s D. 4.7 m/s

N10T.3 An elevator moves downward with an acceleration of 6.2 m/s^2. A ball dropped from rest by a passenger will have what *downward* acceleration relative to the elevator?
A. 3.6 m/s^2 C. 9.8 m/s^2 E. other (specify)
B. 6.2 m/s^2 D. 16.0 m/s^2

N10T.4 When you go over the crest of a hill in a roller-coaster, a force appears to lift you up out of your seat. This is a fictitious force. (T or F)

N10T.5 If your car is hit from behind, you are suddenly pressed back into your seat. The normal force that the seat exerts on you is a fictitious force (T or F).

N10T.6 You are pressed downward toward the floor as an elevator begins to move upward. The force pressing you down is a fictitious force (T or F).

N10T.7 A beetle (black dot on the top view shown at the right) sits on a rapidly rotating turntable. The table rotates faster and faster and eventually the beetle loses its grip. What is its subsequent trajectory relative to the ground (assuming it cannot fly)?

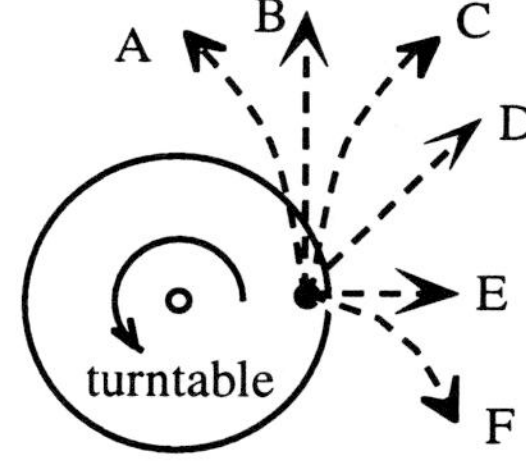

N10T.8 When is an elevator a reasonably good inertial reference frame?
A. Never, under any circumstances
B. Always, since it is attached to the surface of the earth
C. It is except when it is changing speed
D. Other conditions (describe)

N10T.9 Is a freely falling elevator an accurate inertial reference frame?
A. No -- such a frame is accelerating toward the Earth
B. It is not a *real* inertial frame, but we can treat it as one if we ignore the gravitational forces on objects inside
C. Yes D. It depends (explain)

HOMEWORK PROBLEMS

BASIC SKILLS

N10B.1 A person shoots a tranquilizer dart at a fleeing bear. If the dart's speed is 33 m/s relative to the ground, and the bear is fleeing at 9 m/s relative to the ground, what is the dart's speed relative to the bear?

N10B.2 In a certain baseball game a well-hit line drive moves almost horizontally westward at 125 mi/h. The center fielder runs toward the ball at a speed of 15 mi/h. What is the ball's speed relative to the fielder?

N10B.3 An airplane flies due north at speed of 145 km/hr relative to the ground. If there is a wind blowing east at 15 km/hr, what is the plane's speed relative to the air?

N10B.4 Two cars approach an intersection, one traveling north at 18 m/s and the other traveling west at 14 m/s. What is the cars' speed relative to each other?

N10B.5 A boat originally traveling at a speed of 5.2 m/s due east hits a sandbar and comes to rest in 1.3 s. What is the approximate average acceleration (magnitude and direction) of a piece of slippery ice sitting on a table in the ship's galley **(a)** relative to the ground **(b)** relative to the ship? Explain your answers.

N10B.6 Two cars in a race start from rest. Both cars accelerate forward, car *A* at 12 mi/h/s and car *B* at 9 mi/h/s. What is the acceleration of car *B* relative to car *A* (magnitude and direction)?

N10B.7 Imagine that you toss a baseball to a friend while you are both in the back of a bus moving due west at a constant speed 15 m/s. While the baseball is in the air, its acceleration with respect to the ground is 9.8 m/s^2 downward (since only the force of gravity acts on it once it leaves your hand). What is the ball's acceleration relative to the bus? Explain.

N10B.8 A car originally moving at 25 mi/hr hits a brick wall and comes to rest in 0.12 seconds. As the the car comes to rest, about how fast will an unbelted passenger accelerate **(a)** relative to the ground and **(b)** relative to the windshield? Explain your response. (Ignore friction.)

SYNTHETIC

N10S.1 A boat moves directly toward a dock 2.0 km due west across a river flowing 1.0 m/s due south. How fast does the boat have to travel relative to the water to make it across in 20 minutes?

N10S.2 An airplane flies due north at 250 mi/h relative to the ground in air that is moving east at 25 mi/h. At what angle is the plane pointed relative to north? What is its speed relative to the air?

N10S.3 An airplane accelerates for takeoff, reaching a speed of 150 mi/h in 30 s. A 65-kg person feels pressed back into his or her seat during this time. **(a)** What is the person's acceleration relative to the ground? **(b)** What is the magnitude of the forward force that the seat must exert on the person during this time? **(c)** What is the person's acceleration in the plane frame? **(d)** If you were to naively apply Newton's second law in this frame, what is the magnitude of the fictitious force that you infer must be pushing the person backward?

N10S.4 A 35-kg child is holding on for dear life on a playground merry-go-round that has a radius of 2.5 m and is turning at a rate of one turn every 5.0 s. **(a)** What is the child's acceleration relative to the ground? **(b)** What is the magnitude of the force acting on the child through the child's arms? **(c)** What is the child's acceleration relative to the place on the merry-go-round where the child is standing? **(d)** If you were to naively apply Newton's second law to this situation, what is the magnitude of the apparent centrifugal force acting on the child?

N10S.5 An elevator is accelerating upward at 1.8 m/s^2. Someone is standing on a bathroom scale in the elevator and it reads 180 lbs. What is the person's real weight?

N10S.6 A test tube is placed in a centrifuge. When the centrifuge rotates at full speed, the test tube's bottom is about 38 cm from the axis of rotation. The material near the bottom of the test tube experiences an effective gravitational force 18 times larger than normal. How many times does the centrifuge rotate per second?

N10S.7 Imagine that the ball in a floating-ball detector can travel only 1.0 cm before hitting the side of its container. Imagine also that we have figured out a way to cancel exactly the true effects of gravity. If we place our detector on the earth's equator, how long will it take for the ball (after it is released from rest) to reach the side, and in which direction will it drift? Compare this to the time that it would take in a frame that accelerates from rest to walking speed (about 1 m/s) in one minute.

RICH-CONTEXT

N10R.1 You are kidnapped and put blindfolded in an elevator at the ground floor of a building. As the elevator starts, you notice that your weight seems to increase by 10% for 3 s, remain normal for 24 s, then decrease by 10% for 3 s. You are then taken out and put into a locked room. What is the approximate floor number your room is on?

N10R.2 A freely-falling frame does not behave *precisely* like an inertial frame because $\vec{g}$ is not precisely constant near a gravitating object. For example, $\vec{g}$ near the earth always points directly toward the center of the earth, which is in different directions at different points. Consider a boxcar 10 m long that is freely falling toward the earth (assume the boxcar is oriented horizontally). Two marbles float initially at rest at opposite ends of the boxcar. Estimate the marbles' acceleration relative to each other. (Hint: $\sin\theta \approx \tan\theta \approx \theta$ if θ is measured in radians and $\theta \ll 1$.)

ADVANCED

N10A.1 A jet flies a distance L and back first against and then with a wind with speed u. One might think that since the wind will slow the jet one way and speed it equally the other way, the effects would cancel and the trip time would be $2L/v$, where v is the jet's speed relative to the air. Show that the round-trip time is *actually* $(2L/v)(1 - u^2/v^2)^{-1}$.

ANSWERS TO EXERCISES

N10X.1 Here the train is the S' frame and the ground is the S frame, and we are given that $\beta_x = +23$ m/s and $v'_x = -3$ m/s, so by equation N10.4a, $v_x = \beta_x + v'_x = +20$ m/s,

N10X.2 Implicitly, the speeds of the boats given in the problem are both relative to the ground. We choose the ground to be the S frame and the ship to be the S' frame, and the sailboat to be the object whose motion is examined in both frames. Thus we are given that $\vec{v} = 6$ m/s west, while $\vec{\beta} = 8$ m/s north, and we want to find $\vec{v}'$. Using equation N10.5, we find that $\vec{v}' = [-6 \text{ m/s}, -8 \text{ m/s}, 0]$. This vector has a magnitude of 10 m/s and a direction 37° west of south.

N10X.3 Let the elevator and the ground be the S' and S frames respectively. We are given $A_z = +3.0$ m/s^2 and that the ball's acceleration in S is $a_z = -9.8$ m/s^2; we want to find its acceleration a'_z in S'. The z component of equation N10.10 tells us that $a'_z = a_z - A_z = -12.8$ m/s^2.

N10X.4 The "magical force" that seems to draw a person forward is simply the tendency of the person's body to continue moving at a constant velocity while the car accelerates backward. If the person is not held to the car by a seat belt, the person will continue moving through the windshield as the car comes to rest. Thus the person is not "thrown through the windshield" so much as the windshield is thrown backward by the collision past the person's body as the latter moves naturally forward.

N10X.5 You have to hold on tight in order to exert the inward force on your body required to keep it accelerating in the circular path of the rim of the merry-go-round!

N11

PROJECTILE MOTION

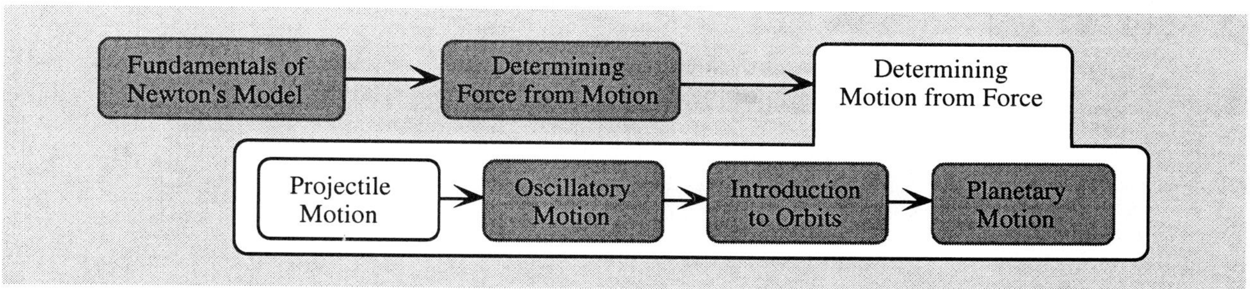

N11.1 OVERVIEW

In chapter N5, we saw that Newton's second law enables us (in principle) to predict that object's motion if we know (1) how the net force on the object varies with time, and (2) the object's initial velocity and position. The problems discussed in that chapter were, however, somewhat artificial. With this chapter, we begin a four-chapter subunit exploring practical situations where we use our knowledge about forces to determine an object's motion. The three situations that we will examine in depth are projectile motion (in this chapter), oscillatory motion (in the next chapter), and planetary motion (in the last two chapters).

This chapter extends what we did in chapter N5 by applying the mathematics there to determine the motion of falling objects (both with and without drag). Studying this important application will prepare us for the analysis of still more complicated situations in future chapters.

Here is an overview of the sections in this chapter.

N11.2 *WEIGHT AND PROJECTILE MOTION* briefly summarizes things that we have already learned about *weight* (the gravitational force on an object), and defines some terms.

N11.3 *SIMPLE PROJECTILE MOTION* develops the mathematics we need to predict motion from forces in three dimensions and applies this mathematics to the problem of projectile motion.

N11.4 *SOME BASIC IMPLICATIONS* discusses some of the implications of the mathematical analysis of projectile motion presented in the previous section.

N11.5 *A PROJECTILE MOTION FRAMEWORK* describes an adaptation of our problem-solving framework that guides you through a process for solving practical problems involving simple projectile motion.

N11.6 *DRAG AND TERMINAL SPEED* looks at how drag modifies the motion of projectiles.

N11.2 WEIGHT AND PROJECTILE MOTION

One of the fundamental forces of nature that we experience on a daily basis is the *force of gravity*, the force that incessantly and insistently tugs us toward the center of the earth. Since this force is such a common part of our experience, it is worth looking at it in some detail.

As we've said before, in physics we define an object's *weight* $\vec{F}_g$ to be *the force that gravity exerts on that object*. Note that this force expresses the interaction between the object and whatever nearby body (usually the earth) is massive enough to create an appreciable gravitational field. In section C8.7, we saw that near earth's surface, the weight force on an object points downward and has a magnitude of mg, where g is empirically measured to have the value 9.80 m/s^2. In section N5, we summarized all information in the single equation

Defining the gravitational field vector $\vec{g}$

$$\vec{F}_g = m\vec{g} \tag{N11.1}$$

where $\vec{g}$ is *vector* that points toward the center of the earth and whose magnitude is $g = 9.80$ m/s^2. We can take this equation as the *definition* of $\vec{g}$. Since $\vec{g}$ expresses the effect of the gravitational interaction on an object of given mass at a certain point in space, and has the same value for all objects at that point, it expresses something basic about the character of the gravitational field the earth creates at that point. We therefore call $\vec{g}$ the earth's **gravitational field vector** at that point. (We can define $\vec{g}$ near other celestial objects in a similar way.)

If the *only* force on an object is its weight $\vec{F}_g$, we say that it is **freely falling**. As we saw in chapter N5.5, Newton's second law tells us that since the object's weight is strictly proportional to its mass, its acceleration is

The acceleration of a freely-falling object is $\vec{g}$

$$\vec{a} = \frac{\vec{F}_{\text{net}}}{m} = \frac{\vec{F}_g}{m} = \frac{m\vec{g}}{m} = \vec{g} \tag{N11.2}$$

implying that *every* object at a given point in space (independent of its mass, composition or other characteristics) will fall with the *same* acceleration equal to the value of $\vec{g}$ at that point. Since the value of $\vec{g}$ at a point characterizes the common acceleration of all falling objects at that point, people sometimes call it the **acceleration of gravity** (even though it obviously isn't *gravity* that is accelerating!). I will use the term *gravitational field vector* not only because it more accurately describes the physical meaning of $\vec{g}$, but it also underlines its analogy with the *electric field vector* $\vec{E}$ that we will discuss in Unit *E*.

The simple projectile motion model

As long as (1) an object remains "sufficiently close" to the surface of the earth, and (2) its trajectory is "sufficiently short" so that the curvature of the earth is not significant, the magnitude and direction of $\vec{g}$ (and thus of the object's weight $\vec{F}_g$) will be approximately constant. If in addition (3) the object is "not significantly affected" by other forces except possibly air drag, then, we call it a **projectile** and its motion **projectile motion**. When (4) we *also* neglect drag, we say that the object's motion is **simple projectile motion.** Newton's second law implies that such an object will have constant acceleration $\vec{a} = \vec{g}$.

When is drag important?

What do these restrictions mean in practice? Let's look at each in turn, starting with the last. A *freely falling* object is strictly defined to be an object that is not influenced by anything *except* its own weight. In practice it is hard to isolate falling objects from the effects of air friction (drag). In some cases, these effects can be substantial: it is clear that a feather does not fall in the same way that a steel ball does. The statement that an object is "freely falling" thus could only strictly apply to objects in a vacuum. However, the description is *approximately* true in any case where the drag force is small compared to its weight.

Whether this approximation is good or not depends both on the object's *mass*, *shape*, *size* and *speed*. For example, we would expect two objects of the same size and shape and moving at the same speed to experience the same drag. However, if one object is more massive than the other, the same drag force will change its velocity *less* in a given time interval (according to Newton's second law) and thus affect its motion less. Thus a steel ball will behave more like a freely falling object than a ping-pong ball of the same size. An object's shape is also important: for example, a sheet of paper behaves more like a freely falling object if it is compressed into a tight ball than if it is left flat, because more air would interact with the flat sheet than would interact with the ball. Finally, the speed of an object is important: higher speeds produce higher drag forces. So the approximation works best if the object is small, massive, and moving slowly.

When is an object "sufficiently close" to the earth?

What does it mean for an object to be "sufficiently close" to the earth's surface? We have seen that the effect of the earth's gravity diminishes with distance from the earth's center. However, you have to travel a significant distance vertically before variations in $\vec{g}$ become noticeable. Empirically, the variation in the value of g is smaller than $\pm 0.1\%$ within a range of ± 3 km relative to sea level. Even at altitudes of several hundred kilometers (roughly the altitude where the space shuttle flies) the value of g is only 2% to 3% smaller than it is at sea level. So depending on what we would consider a significant variation in the value of g to be, "sufficiently close" could range from within 3 km to within hundreds of kilometers of the earth's surface.

When is its trajectory "sufficiently short"?

The other restriction is that the object's trajectory be "sufficiently short" compared to the earth's curvature. This is because $\vec{g}$ points toward the earth's *center*, so if the object moves far enough horizontally, the direction of $\vec{g}$ could change significantly (relative to the distant stars, say). The earth is pretty large, though: since the earth's circumference is about 40,000 km, and the change in the direction of $\vec{g}$ for an object going around the earth will be 360°, we see that we have to travel more than 100 km for the direction of $\vec{g}$ to change by more than 1°. So we can consider the direction of $\vec{g}$ to be constant as long as the object's trajectory is not longer than 100 km or so.

Exercise N11X.1: Imagine that the drag force on a 1.5-inch sphere moving through air at a speed of 5 m/s is 0.009 N. What fraction is this of the weight of a 5-g ping-pong ball of this size? How about a 220-g steel ball?

Exercise N11X.2: A rocket is launched near the equator and travels about 1200 km due east. By about how much does the direction of $\vec{g}$ at the rocket's position change (relative to the distant stars) during the rocket's flight?

N11.3 SIMPLE PROJECTILE MOTION

Integrating $\vec{a}(t)$ to find $\vec{v}(t)$ and $\vec{r}(t)$

So, when the drag on an object is negligible and $\vec{g}$ is relatively constant in magnitude and direction, we can model the object as a "simple projectile" and take its acceleration to be $\vec{a} = \vec{g}$ = constant. Since this *is* a constant, we can easily integrate this using the methods of section N5.4 to find the falling object's velocity and position as a function of time. Integrating the acceleration to find the velocity (see equation N5.16), we find that:

$$\vec{v}(t) - \vec{v}(0) = \int_0^t \vec{a}(t)\,dt = \int_0^t \vec{g}\,dt = \vec{g}\int_0^t dt = \vec{g}(t-0) = \vec{g}t$$

$$\Rightarrow \quad \vec{v}(t) = \vec{g}t + \vec{v}(0) \qquad \text{(N11.3)}$$

Similarly, integrating the velocity to find the position (see N5.19) we get:

$$\vec{r}(t) - \vec{r}(0) = \int_0^t \vec{v}(t)\,dt = \int_0^t [\vec{g}t + \vec{v}(0)]\,dt = \vec{g}\int_0^t t\,dt + \vec{v}(0)\int_0^t dt$$

$$= \vec{g}(\tfrac{1}{2}t^2 - \tfrac{1}{2}0^2) + \vec{v}(0)(t-0) = \tfrac{1}{2}\vec{g}t^2 + \vec{v}(0)t$$

$$\Rightarrow \quad \vec{r}(t) = \tfrac{1}{2}\vec{g}t^2 + \vec{v}(0)t + \vec{r}(0) \qquad \text{(N11.4)}$$

Note how I have treated both $\vec{g}$ and $\vec{v}(0)$ as simple constants in the integration process. In practical component language, if we define the z axis to be vertically upward and the x and y axes to be horizontal, then $\vec{g} = [0, 0, -g]$. (The z component of $\vec{g}$ is negative because $\vec{g}$ points vertically downward. Note that $g \equiv \text{mag}(\vec{g})$ is always a positive number, no matter how $\vec{g}$ points.) If we also define

$$\vec{v}(0) \equiv \begin{bmatrix} v_{0x} \\ v_{0y} \\ v_{0z} \end{bmatrix} \quad \text{and} \quad \vec{r}(0) \equiv \begin{bmatrix} x_C \\ y_C \\ z_C \end{bmatrix} \tag{N11.5}$$

then we can express equations N11.3 and N11.4 in component form as follows:

The basic equations describing simple projectile motion

$$\begin{bmatrix} v_x(t) \\ v_y(t) \\ v_z(t) \end{bmatrix} = \begin{bmatrix} 0t \\ 0t \\ -gt \end{bmatrix} + \begin{bmatrix} v_{0x} \\ v_{0y} \\ v_{0z} \end{bmatrix} \Rightarrow \begin{bmatrix} v_x(t) \\ v_y(t) \\ v_z(t) \end{bmatrix} = \begin{bmatrix} v_{0x} \\ v_{0y} \\ -gt + v_{0z} \end{bmatrix} \tag{N11.6}$$

$$\begin{bmatrix} x(t) \\ y(t) \\ z(t) \end{bmatrix} = \begin{bmatrix} \frac{1}{2}0t^2 \\ \frac{1}{2}0t^2 \\ -\frac{1}{2}gt^2 \end{bmatrix} + \begin{bmatrix} v_{0x}t \\ v_{0y}t \\ v_{0z}t \end{bmatrix} + \begin{bmatrix} x_0 \\ y_0 \\ z_0 \end{bmatrix} \Rightarrow \begin{bmatrix} x(t) \\ y(t) \\ z(t) \end{bmatrix} = \begin{bmatrix} v_{0x}t + x_0 \\ v_{0y}t + y_0 \\ -\frac{1}{2}gt^2 + v_{0z}t + z_0 \end{bmatrix} \tag{N11.7}$$

We will use equations N11.6 and N11.7 very often in what follows.

Exercise N11X.3: According to equation N11.7, how long would it take for an object to fall a vertical distance of 4.9 m from rest? How would this result change if the object initially had a horizontal x-velocity of 5.0 m/s?

N11.4 SOME BASIC IMPLICATIONS

Gravity only affects the vertical component of an object's motion

Equations N11.6 and N11.7 imply three basic but important things about simple projectile motion that are worth underlining. First, notice that *gravity only affects the z component of a simple projectile's motion*. In particular, equation N11.6 tells us that the horizontal part of the projectile's original velocity is completely unaffected by the presence of gravity: as it falls, a simple projectile will maintain whatever original horizontal velocity it had. For example, a package dropped from a plane will continue to move horizontally at the same rate as the plane (see Figure N11.1a).

The component equations are independent

Secondly, *the component equations* in either equations N11.6 or N11.7 are *completely independent of each other*. In particular, note that the projectile's vertical motion is completely unaffected by its horizontal velocity. This means, for example, that an object dropped from rest will have the same vertical motion as an object with a large initial horizontal velocity, as shown in Figure N11.1b.

We can always define the horizontal motion to be along the x axis

Thirdly, since a projectile's horizontal motion is unaffected by gravity, the projection of a projectile's motion on the horizontal (xy) plane will always be a straight line. It is thus generally possible to orient our reference frame so that its x axis lies in the direction of motion (that is, so that $v_y(t) = v_{0y} = 0$). If we do

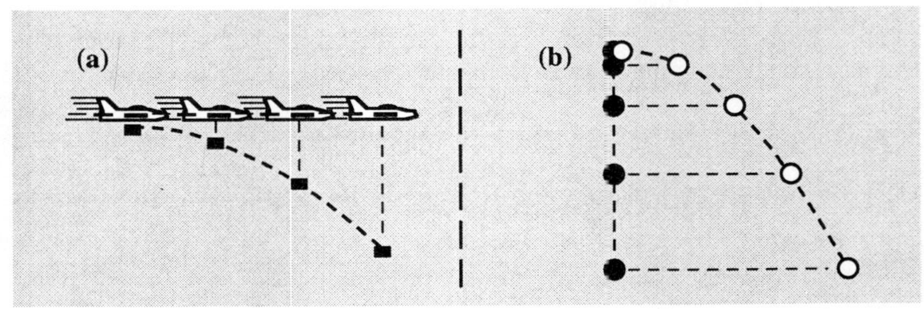

Figure N11.1: (a) A projectile's horizontal motion is unaffected by gravity: as an object falls, it will continue to move forward at whatever horizontal velocity it had originally. **(b)** A projectile's vertical motion is independent of its horizontal motion: objects fall vertically at the same rate no matter how they are moving horizontally

this, the projectile's motion lies entirely in the xz plane, which simplifies things somewhat. I will commonly do this in the examples that follow.

When does a (simple) projectile reach its peak?

Most actual projectiles are launched from near the ground and follow a trajectory that looks like an arc, going first upward and then curving back downward to the ground. How do we determine when the object has reached its maximum altitude and when it reaches the ground?

If the initial velocity of a projectile is at all upward, then $v_{0z} > 0$. The bottom line of equation N11.6 tells us that subsequently $v_z(t) = -gt + v_{0z}$, so the projectile's z-velocity starts positive, but decreases linearly through zero and eventually becomes negative as shown in Figure N11.2. I claim that the projectile is at its peak when its z-velocity passes through zero. Why? At an instant when $v_z(t)$ is positive, the projectile is still going upward, and so will reach a still higher position later. At an instant where $v_z(t)$ is negative, it is moving downward and thus was at a higher position earlier. Only at the exact instant where $v_z(t) = 0$ is it neither climbing to or coming down from a higher position: this must be the peak. Thus to find t_p (≡ time of the peak), we solve

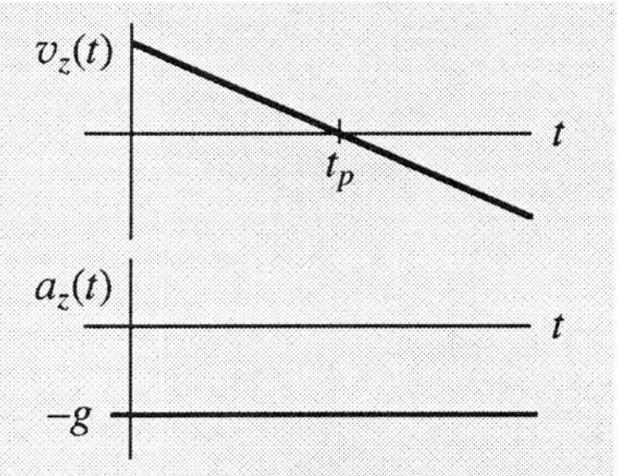

Figure N11.2: Graphs of a projectile's z-velocity and z-acceleration near the peak of its trajectory.

$$0 = v_z(t_p) = -gt_p + v_{0z} \quad \Rightarrow \quad t_p = \frac{v_{0z}}{g} \tag{N11.8}$$

(Note, by the way, that even though the projectile's z-velocity is passing through zero at this point, its z-acceleration remains nonzero and equal to $-g$ always. This can be seen clearly in Figure N11.2: even as the value of $v_z(t)$ passes through zero, its slope is always constant and negative.)

When does a (simple) projectile hit the ground?

If we have set up our reference frame so that the ground is at $z = 0$, then finding the time t_h (≡ time when the projectile hits the ground) is simply a matter of solving equation $z(t_h) = 0$ for t_h using the quadratic equation

$$0 = z(t_h) = -\tfrac{1}{2}gt_h^2 + v_{0z}t_h + z_0 \quad \Rightarrow \quad t_h = \frac{v_{0z} \pm \sqrt{v_{0z}^2 + 2gz_0}}{g} \tag{N11.9}$$

Exercise N11X.4: Check that the last result in equation N11.9 is correct.

Note that this equation yields *two* solutions. This is because if the projectile had been freely falling for all time and has sufficient energy to get up to $z = 0$ at all, then its parabolic trajectory would cross $z = 0$ *twice*, once going up and once going down. Usually we will be more interested in the later solution. Note in particular that since the projectile is *not* freely falling before it is launched, neither equation N11.8 nor N11.9 apply for times before the launch time, and so if either yield such a solution, this solution is not physically meaningful.

Exercise N11X.5: Are there values of v_{0z} and z_0 for which equation N11.9 yields *no* solution? If so, explain in physical terms *why* there is no solution.

Other implications

Other implications of equations N11.6 and N11.7 include the *parabolic* shape of a projectile's trajectory in simple projectile motion (see problem N11S.9) and the fact that the horizontal distance a projectile launched from level ground will go before returning to the ground is greatest if it is launched at a 45° angle (see problem N11S.10).

N11.5 A PROJECTILE-MOTION FRAMEWORK

As usual, we can adapt our general problem-solving framework to help us solve problems involving simple projectile motion. Examples N11.1 and N11.2 on the next two pages illustrate the use of such an adapted framework (see the outline on the left side of each example), as well as some other useful techniques for solving simple projectile motion problems.

EXAMPLE N11.1

OUTLINE OF THE FRAMEWORK for PROJECTILE-MOTION problems

1. Pictorial Representation

a. Draw a picture of the situation that includes:

① (1) a drawing of the object indicating its initial position and subsequent trajectory

② (2) a labeled arrow indicating the direction of the object's initial velocity (or say "$\vec{v}_0 = 0$")

③ (3) reference frame axes (conventionally chosen so that $\vec{v}_0$ lies in the *xz* plane) with a clearly specified origin

④ (4) a description of when you are defining $t = 0$ to be, and a list of other time symbols that you will use in the problem

⑤ (5) labels defining other symbols for relevant quantities

⑥ b. List values for all known labeled quantities and specify which quantities are unknown.

2. Conceptual Representation

The general task is to construct a conceptual model of the situation and link it to an abstract physics model or principle. In projectile problems, we have to do the following:

⑦ a. Describe the approximations and assumptions you are making in order to apply the simple projectile model to this situation

⑧ b. Specify the range of time during which the object can be considered a simple projectile

3. Mathematical Representation

⑨ a. Apply equations N11.6 and/or N11.7 *in vector component form*

⑩ b. Solve for any unknown quantities symbolically

⑪ c. Plug in numbers and units and calculate the result and its units.

4. Evaluation

Check that the answer makes sense:

⑫ a. Does it have the correct units?

⑬ b. Does it have the right sign?

⑭ c. Does it seem reasonable?

Problem: Your friend, who is an unemployed actor, has a chance at a role in a Western movie if he can learn to slide a mug precisely down the length of a saloon bar. After practicing a lot, he thinks he can do it. But when asked to perform in front of the producer, he is a bit nervous, and the mug goes off the end of the bar (which is 1.4 m high) at a speed of 2.5 m/s. Does the mug compound your friend's problem by hitting the producer's foot, which is 2.0 m beyond the end of the bar?

Solution: A sketch of the situation appears below. Note that $t = 0$ is ④ the instant that the mug leaves the edge of the bar; t_h is the instant when the mug hits the floor.

KNOWNS:⑥

$$\vec{v}_0 = \begin{bmatrix} v_{0x} \\ v_{0y} \\ v_{0z} \end{bmatrix} = \begin{bmatrix} +2.5 \text{ m/s} \\ 0 \\ 0 \end{bmatrix} \qquad z_0 = 1.4 \text{ m},\quad x_0 = y_0 = 0,\quad z_h \equiv z(t_h) = 0,\quad D = 2.0 \text{ m}$$

UNKNOWNS:⑥ $t_h = ?$ $x_h = ?$ Is $x_h < D$?

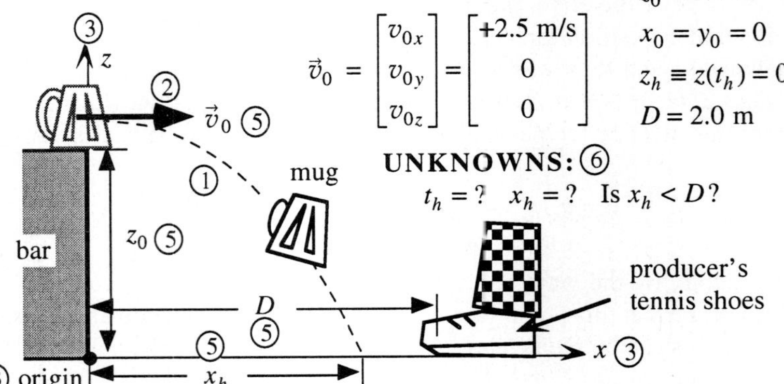

I am assuming that the effects of air friction are negligible (a mug is typically pretty small and massive, and this one is not moving very fast). If this is so, the mug moves under the influence of gravity alone after it leaves the bar at $t = 0$ until it hits the floor (or the producer's ⑦ leg) at $t = t_h$. ⑧ I am also treating the mug as a point particle, essentially ignoring its size. If x_h turns out to be very close to D, I may need to do the calculation again, making some guesses as to the actual size of the mug. I'm also assuming that the mug's initial velocity is exactly horizontal. Equation N11.7 tells us that

$$⑨ \begin{bmatrix} x(t) \\ y(t) \\ z(t) \end{bmatrix} = \begin{bmatrix} v_{0x}t + x_0 \\ v_{0y}t + y_0 \\ -\frac{1}{2}gt^2 + v_{0z} + z_0 \end{bmatrix} = \begin{bmatrix} v_{0x}t \\ 0 \\ -\frac{1}{2}gt^2 + z_0 \end{bmatrix} \qquad (1)$$

where in the last step, I have used $v_{0y} = v_{0z} = 0$ and $x_0 = y_0 = 0$ to simplify the expression. Now, what we want to find is $x_h \equiv x(t_h)$. The first line of equation (1) implies that $x_h \equiv x(t_h) = v_{0x}t_h$, so if we just knew what t_h was, we could calculate x_h. How can we find t_h? We know that $z(t_h) = 0$ (since the mug hits the floor at $t = t_h$ by definition), so if we evaluate the *third* line of equation (1) at $t = t_h$ we get:

$$0 = z(t_h) = -\tfrac{1}{2}gt_h^2 + z_0 \;\Rightarrow\; \tfrac{1}{2}gt_h^2 = z_0 \;\Rightarrow\; t_h = \pm\sqrt{\frac{2z_0}{g}} \;⑩ \qquad (2)$$

Since the mug is not a projectile before $t = 0$, the negative solution is not relevant. Plugging the positive result into the first line of equation (1), we get

$$x_h = v_{0x}t_h = v_{0x}\sqrt{\frac{2z_0}{g}} \;⑩ = (2.5 \text{ m/}\not{s})\sqrt{\frac{2(1.4 \not{m})}{(9.8 \not{m}/\not{s}^2)}} = 1.3 \text{ m} \;⑪ \qquad (3)$$

⑫ This has the right units, the right sign (the mug should hit to the right ⑭⑬ of $x = 0$), and seems plausible. So the mug misses hitting the producer (we can just hope that it also does not splash the producer's shoes).

EXAMPLE N11.2

OUTLINE OF THE FRAMEWORK for PROJECTILE-MOTION problems

1. Pictorial Representation

a. Draw a picture of the situation that includes:

(1) (1) a drawing of the object indicating its initial position and subsequent trajectory

(2) (2) a labeled arrow indicating the direction of the object's initial velocity (or say "$\vec{v}_0 = 0$")

(3) (3) reference frame axes (conventionally chosen so that $\vec{v}_0$ lies in the *xz* plane) with a clearly specified origin

(4) (4) a description of when you are defining $t = 0$ to be, and a list of other time symbols that you will use in the problem

(5) (5) labels defining other symbols for relevant quantities

(6) b. List values for all known labeled quantities and specify which quantities are unknown.

2. Conceptual Representation

The general task is to construct a conceptual model of the situation and link it to an abstract physics model or principle. In projectile problems, we have to do the following:

(7) a. Describe the approximations and assumptions you are making in order to apply the simple projectile model to this situation

(8) b. Specify the range of time during which the object can be considered a simple-projectile

3. Mathematical Representation

(9) a. Apply equations N11.6 and/or N11.7 *in vector component form*

(10) b. Solve for any unknown quantities symbolically

(11) c. Plug in numbers and units and calculate the result and its units.

4. Evaluation

Check that the answer makes sense:

(12) a. Does it have the correct units?

(13) b. Does it have the right sign?

(14) c. Does it seem reasonable?

Problem: During a soccer game, you kick the ball so that it leaves the ground at an angle of 33° traveling at a speed of 15 m/s. The ball reaches its maximum height just as it passes the goalie, who is standing well in front of the net. Is the ball high enough at this point to be out of the goalie's reach?

(4) *Solution:* A drawing of the situation appears below. Note that $t = 0$ is the instant the ball leaves the ground and the foot that it kicking it. Let t_p be the time when the ball reaches its peak.

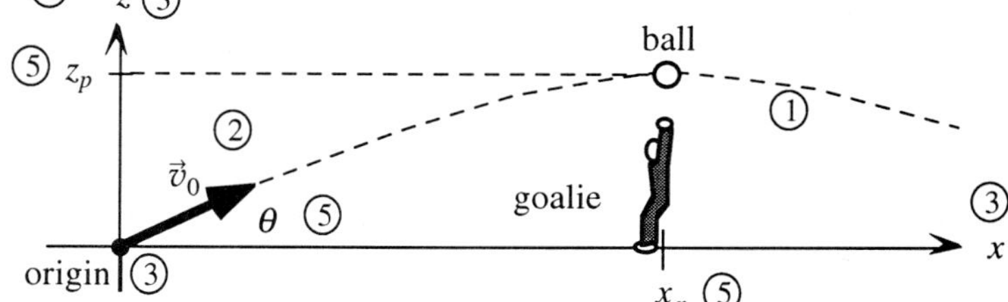

(6) **KNOWNS:** $v_0 = 15$ m/s, $\theta = 33°$

$$\begin{bmatrix} x_0 \\ y_0 \\ z_0 \end{bmatrix} = \begin{bmatrix} 0 \\ 0 \\ 0 \end{bmatrix} \qquad \vec{v}_0 = \begin{bmatrix} v_{0x} \\ v_{0y} \\ v_{0z} \end{bmatrix} = \begin{bmatrix} +v_0 \cos\theta \\ 0 \\ +v_0 \sin\theta \end{bmatrix}$$

(6) **UNKNOWNS:**

$t_p = ?$

$z_p \equiv z(t_p) = ?$

$x_p \equiv x(t_p) = ?$

(7) If the effects of air resistance are negligible (maybe not a very good assumption for this fairly lightweight ball which is moving pretty fast), then after the ball leaves the ground at $t = 0$, it moves under the influence of gravity alone until it hits the ground or is caught: we will (8) assume this is after $t = t_p$. I also am going to assume that the goalie can catch the ball if z_p is less than about 3.0 m (which is ≈ 9.5 ft): this is about how high I think I could reach (with a bit of a jump). Equations N11.6 and N11.7 (with known values plugged in) read:

$$\begin{bmatrix} v_x(t) \\ v_y(t) \\ v_z(t) \end{bmatrix} = \begin{bmatrix} v_0 \cos\theta \\ 0 \\ -gt + v_0 \sin\theta \end{bmatrix} \quad (1) \qquad \begin{bmatrix} x(t) \\ y(t) \\ z(t) \end{bmatrix} = \begin{bmatrix} (v_0 \cos\theta)t \\ 0 \\ -\frac{1}{2}gt^2 + (v_0 \sin\theta)t \end{bmatrix} \quad (2) \quad (9)$$

What we want to find is $z_p \equiv z(t_p)$. We could use the third line of equation (2) to find this if only we knew what t_p is. What other equation might we use to find t_p? Well, as discussed in the previous section, the peak in the trajectory occurs when $v_z(t) = 0$, so setting $t = t_p$ in the third line of equation (1), we get:

(10)

$$0 = v_z(t_p) = -gt_p + v_0 \sin\theta \quad \Rightarrow \quad t_p = \frac{v_0 \sin\theta}{g} \qquad (3)$$

Plugging this into the third line of equation (2), we then get

$$z_p = z(t_p) = -\tfrac{1}{2}gt_p^2 + v_0 \sin\theta\, t_p \;=\; -\tfrac{1}{2}g\left(\frac{v_0 \sin\theta}{g}\right)^2 + \frac{(v_0 \sin\theta)^2}{g}$$

$$= \left(-\frac{1}{2} + 1\right)\frac{v_0^2 \sin^2\theta}{g} = \frac{v_0^2 \sin^2\theta}{2g} \quad (10)$$

$$= \frac{(15\text{ m/}\not{s})^2 \sin^2(33°)}{2(9.8\text{ m/}\not{s}^2)} = 3.4\text{ m} \quad (11) \qquad (4)$$

(12)(13) This result has the right units, the right sign (the peak is above the (14) ground!), and a reasonable magnitude. The kick is almost certainly out of the goalie's reach, though, so it's a GOOAAL!!

The pictorial representation part of the framework

Establishing an appropriate reference frame is at least as important for simple projectile motion problems as it was for constrained motion problems, since the equations that describe simple projectile motion (equations N11.6 and N11.7) are expressed in terms of components that have meaning only in the context of a well-defined reference frame. However, there are some other things that are important in projectile motion problems that are not so important in constrained-motion problems. For example, time t is an important variable in equations N11.6 and N11.7 but not particularly so in constrained motion problems. It is *crucial* in simple projectile motion problems to specify clearly the origin of time (that is, describe how we define the instant when $t = 0$) and important to define symbols for other important instants of time. Knowing where the origin of the spatial coordinate system also is more important in projectile motion problems than in most other kinds problems. Clearly specifying initial conditions is also very important. The extra steps in the *pictorial representation* part of the framework outline helps you pay appropriate attention to these issues.

(In contrast, an object's mass, which is very important in many kinds of problems, is irrelevant in projectile motion problems. Therefore, there is no particular need to specify the mass in projectile motion problems.)

The conceptual representation part

The point of the *conceptual representation* section of the framework in all kinds of physics problems is to assess the applicability and limitations of the model that we are applying. With simple projectile motion problems, the main issues that we need to address are (1) exactly *when* does the model apply (this helps us recognize whether a mathematically possible answer is in fact physically absurd) and (2) what *approximations* do we have to make to make the situation fit the simple projectile model.

The mathematical representation part

In the mathematical part of the framework, it is worth remembering that the component equations in equations N11.6 and N11.7 are independent and must be satisfied simultaneously (like a conservation of momentum problem). Note also how in both problems, we solved one of these independent equations for the time, and then plugged this into a different equation to compute another quantity of interest. This is a very common pattern for projectile motion problems.

Because of this, you might find it helpful in complicated problems, after extracting the important component equations from the equations N11.6 and/or N11.7, to divide your paper into columns, putting each important component equation at the head of a column. As you continue to solve the problem, you can work down each column as far as you can toward getting something useful, then work down the next column, and so on. This can often help you discover the links between columns that you need to make to solve the whole problem.

N11.6 DRAG AND TERMINAL SPEED

Up to now, we have been considering *simple* projectile motion, where the only significant force acting on an object is its weight. Realistically, though, projectiles moving through air experience some drag. How would including drag affect our predictions about a projectile's motion?

In section N7.5, we saw that the drag force on a large object with cross-sectional area A moving at a reasonably large speed v through air has a magnitude

$$F_D = \tfrac{1}{2} C\rho A v^2 \tag{N11.10}$$

where ρ is the density of air (which is about 1.2 kg/m^3 at room temperature and standard pressure), and C is a unitless constant (called the *drag coefficient*) that depends on an object's shape.

The terminal speed of a falling object

What is the effect of such a drag force on the motion of a projectile? Let us begin by considering the simplest possible situation: an object falling vertically from rest. Without a great deal of mathematical analysis, we can pretty easily determine at least qualitatively what will happen. At first (because v is very small) the magnitude of the drag force is negligible, and the object falls essentially free-

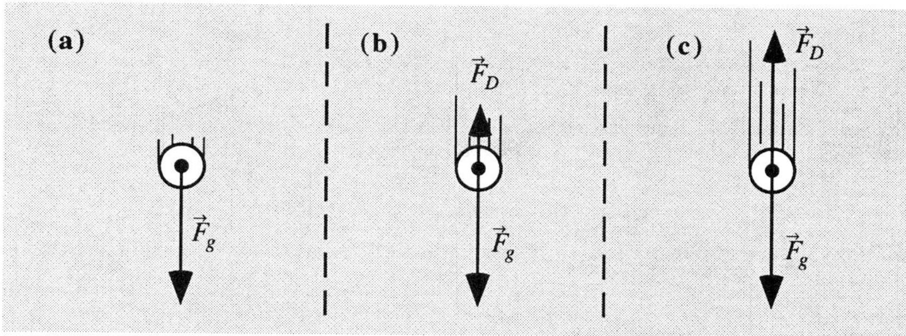

Figure N11.3: (a) Just after an object begins to fall, the magnitude of the drag force on it is very small. **(b)** As it begins to pick up downward speed, the drag force grows in magnitude. **(c)** When the magnitude of the drag force equals the magnitude of the object's weight, the object does not continue to accelerate, but instead maintains a *constant* downward velocity.

ly downward (see Figure N11.3a). But as its downward speed increases, the drag force also increases, decreasing the net force on the object and thus its downward acceleration (Figure N11.3b). The downward speed of the object will continue to increase, however, until the drag force becomes essentially equal to the object's weight (Figure N11.3c). As the object approaches the speed where this happens, the net force on it approaches zero. When the net force is essentially zero, the object no longer accelerates, but continues to fall with constant downward velocity. We call the magnitude of this final constant downward velocity the object's **terminal speed** v_T. You can easily show that its value must be

$$v_T = \sqrt{\frac{2mg}{C\rho A}} \tag{N11.11}$$

Exercise N11X.6: Verify equation N11.11. (*Hint:* when is $F_D = F_g$?)

Exercise N11X.7: A person's terminal speed in air is typically about 60 m/s. If so, what is the value for CA for a falling person? (Assume that $m \approx 60$ kg.)

Verifying this predicted behavior mathematically

Let us see if we can show mathematically that this qualitative description is correct. If we define the z axis to be positive upward, then the z component of Newton's second law then implies that

$$ma_z = F_{\text{net},z} = -mg + \tfrac{1}{2}C\rho A v_z^2 = -mg + mg\left(\frac{C\rho A}{2mg}\right)v_z^2 = -mg\left(1 - \frac{v_z^2}{v_T^2}\right)$$

$$\Rightarrow \quad a_z = -g[1 - v_z^2 / v_T^2] \tag{N11.12}$$

(note that the drag force term is positive because it acts *upward* on an object falling downward). You can see directly from this equation that the vertical acceleration becomes smaller and smaller in magnitude until it becomes essentially zero as $|v_z|$ approaches the terminal speed v_T.

The next step would be to integrate both sides of this equation with respect to time to find v_z as a function of t. But here we run into a roadblock. In order to *find* v_z as a function of t, we have to integrate the right side of equation N11.12, and to do that, we already have to *know* how v_z depends on t!

This illustrates a typical problem that arises when we attempt to use forces to determine motion. *In principle*, once we know the forces acting on an object, we can find its acceleration $\vec{a}(t)$ using Newton's second law and then integrate with respect to time to find the object's velocity $\vec{v}(t)$ and position $\vec{r}(t)$ as functions of time. *In practice*, however, the expression for an object's acceleration almost always refers to the object's unknown velocity and/or position, making a straightforward integration of the acceleration impossible. For realistic problems, we always have to resort to some kind of trick to perform the integration.

One trick for doing the integral in this case

In this particular case, the trick is as follows. Since $a_z \equiv dv_z/dt$, we can rewrite equation N11.12 as $dv_z/dt = -g[1 - v_z^2 / v_T^2]$. If we multiply both sides of this equation by $dt / (1 - v_z^2 / v_T^2)$ and then integrate both sides, we get

$$\frac{dv_z}{1-v_z^2/v_T^2} = -g\,dt \quad\Rightarrow\quad \int_{v_z(0)}^{v_z(t)} \frac{dv_z}{1-v_z^2/v_T^2} = -g\int_0^t dt \qquad \text{(N11.13)}$$

The integral on each side of this expression involves a *single* variable (v_z on the left, t on the right: note that v_T is just a constant), so it is possible *in principle* to find the antiderivative. Note that whenever we integrate both sides of an expression, the lower limit of each integral is the initial value of the integration variable and the upper limit is the final value of that variable.

Checking that this trick yields the right answer when $b = 0$

We can test that this trick yields the right answer if there is no drag (that is, if the v_z^2/v_T^2 term is zero): using the fact that $\int du = u$ for *any* variable u, (and treating v_z as an ordinary variable), you can show that $v_z(t) = -gt + v_z(0)$, which we have already seen to be true for a freely falling object.

Exercise N11X.8: Check this. [*Hint:* Argue that $\int_{v_z(0)}^{v_z(t)} dv_z = v_z(t) - v_z(0)$.]

Using a table of integrals

When there is drag, integrating the left side of equation N11.13 involves finding the antiderivative of the function $f(v_z) = [1 - v_z^2/v_T^2]^{-1}$, which is something that I, at least, don't know off the top of my head. This is exactly why we have *tables of integrals* (some selfless mathematician has worked this out for us already, so why do it again?). My table of integrals says that

$$\int \frac{dx}{a^2 - b^2x^2} = \frac{1}{ab}\tanh^{-1}\frac{bx}{a} \qquad \text{(as long as } a^2 > b^2x^2\text{)} \qquad \text{(N11.14)}$$

where $\tanh^{-1}()$ refers to the *inverse hyperbolic tangent* function (which is found on many scientific calculators). If we identify $a^2 = 1$ and $b^2 = 1/v_T^2$, then we can see that equation N11.13 becomes

$$v_T \tanh^{-1}\frac{v_z}{v_T} - v_T \tanh^{-1} 0 = -g(t-0) \quad\Rightarrow\quad v_T \tanh^{-1}\frac{v_z}{v_T} = -gt \qquad \text{(N11.15)}$$

since if we start from rest, $v_z(0) = 0$ and also $\tanh^{-1} 0 = 0$ (you can check it with your calculator!). Now, by definition, $\tanh(\tanh^{-1} u) = u$, where tanh() is the hyperbolic tangent function. So if we divide both sides of equation N11.15 by v_T and take the hyperbolic tangent of both sides, we find that

$$\frac{v_z}{v_T} = \tanh\left(\frac{-gt}{v_T}\right) \qquad \text{(N11.16)}$$

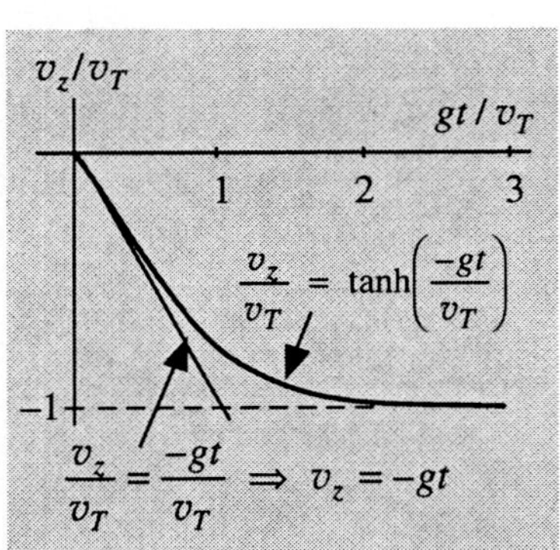

Figure N11.4: A graph of equation N11.18. The straight line shows that at very early times, the object's downward speed increases at the normal free-fall rate.

This, at last, expresses v_z as a function of time. A graph of this function appears in Figure N11.4 ($v_z < 0$ because the object is falling in the $-z$ direction). Note that $gt/v_T = 1$ when enough time has passed for an object *freely* falling from rest to reach a downward speed of whatever v_T is. You can see from the graph that by this time, an object falling *with* drag has reached a speed of only about $\frac{3}{4}v_T$. After twice this time has passed, the object has essentially reached the terminal speed. For early times (times so that $gt/v_T < 0.2$), the object's downward speed increases at essentially the free-fall rate of $v_z = -gt$.

Exercise N11X.9: For a human being, whose $v_T \approx 60$ m/s, about how long would a person have to fall to get to the point where $v_z \approx v_T$?

If the object is not falling from rest, or has an initial horizontal component of velocity the problem becomes even *more* difficult. The most straightforward way to find the trajectory then is to write a computer program to calculate it using the inverse-motion-diagram technique described in chapter N5 (see problem N11A.2). The important issue to understand here is that even in very straightforward circumstances, it may not be easy to solve Newton's second law for the trajectory of an object (a problem we will struggle with for the rest of the unit).

SUMMARY

I. WEIGHT AND SIMPLE PROJECTILE MOTION

A. An object's *weight* ≡ the gravitational force $\vec{F}_g$ acting on it

1. weight ≠ mass, but at a given position in a gravitational field an object's weight is *proportional* to its mass: $\vec{F}_g = m\vec{g}$ (N5.1)

   a) the constant of proportionality $\vec{g}$ at that location is called the *gravitational field vector* at that location

   b) $\vec{g}$ expresses the character of the gravitational field there

   c) $\vec{g} = 9.8\ \text{m/s}^2 = (22\ \text{mi/hr})/\text{s}$ downward near the earth's surface

2. an object is *freely falling* if its weight is the *only* force on it: then Newton's 2nd law ⇒ object's acceleration is $\vec{a} = \vec{g}$

B. Requirements for *simple projectile motion*

1. object remains "sufficiently close" to earth (so that $g \approx$ constant) (g varies by less than ±1% within 30 km of earth's surface)

2. trajectory is "sufficiently short" (so direction of $\vec{g} \approx$ constant) (direction of $\vec{g}$ varies by less that 1° during 100-km trajectory)

3. drag forces are "sufficiently small" compared to weight

II. MATHEMATICAL CONSTRUCTION OF TRAJECTORIES

A. Consider the case of *simple* projectile motion ($\vec{a} = \vec{g}$)

B. Integrating both sides of $d\vec{v}/dt \equiv \vec{a}(t) = \vec{g}$ and $d\vec{r}/dt \equiv \vec{v}(t)$, we get

$$\vec{v}(t) = \vec{g}t + \vec{v}(0) \quad \text{and} \quad \vec{r}(t) = \tfrac{1}{2}\vec{g}t^2 + \vec{v}(0)t + \vec{r}(0) \qquad \text{(N11.3,4)}$$

[note that we can treat $\vec{g}$ and $\vec{v}(0)$ like constant scalars]

C. each equation stands for three analogous component equations (see equations N11.6 and N11.7, the most important equations in chapter N11)

III. IMPORTANT IMPLICATIONS

A. Important implications of the simple projectile motion equations:

1. the horizontal components of motion are unaffected by gravity

2. an object's vertical motion is independent of its horizontal motion

3. a simple projectile's motion is confined to a plane (which we can choose to be the xz plane)

B. When a projectile reaches a peak and when it returns to the ground

1. to find t for the peak: solve $v_z(t) = 0$ for t

2. to find t for the return: solve $z(t) = 0$ for t (the resulting quadratic has two solutions, but one may not be physically reasonable)

IV. A FRAMEWORK FOR SIMPLE PROJECTILE MOTION PROBLEMS

A. In simple projectile motion problems, the following are important:

1. in the *pictorial representation* part: clear definitions of the reference frame, the origins of both space and time measurements, initial conditions, definitions of time as well as other symbols

2. in the *conceptual representation* part

   a) establish *when* the simple projectile model applies

   b) state approximations and assumptions

3. in the *mathematical representation part*: solve the *component* versions (N11.6 and N11.7) of the equations of motion (remember that each component equation is independent).

B. See Examples N11.1 and N11.2 for a complete outline

V. DRAG AND TERMINAL VELOCITY

A. Drag forces in air are usually accurately modeled by $F_D = \frac{1}{2}C\rho A v^2$ (ρ = density of air, A = cross-sectional area, C is a unitless coefficient)

B. The simplest possible case: an object falling vertically from rest

1. as time passes, drag force will grow until it essentially cancels $\vec{F}_g$

2. at this point the object will no longer accelerate, so its speed does not increase beyond the *terminal speed* $v_T = [2mg / C\rho A]^{1/2}$

3. the math solution shows that this happens at $t \approx 2v_T/g$

C. The solution here also illustrates that finding $\vec{v}(t)$ and $\vec{r}(t)$ from acceleration can be difficult and require mathematical tricks

GLOSSARY

weight: the gravitational force $\vec{F}_g$ acting on an object. At a given location in space and time, this force is proportional to the object's mass: $\vec{F}_g = m\vec{g}$.

gravitational field vector $\vec{g}$: the quantity that links an object's mass to its weight at a given place in a gravitational field, characterizing the strength and direction of the gravitational field at that point. This quantity also specifies the acceleration that all freely-falling objects will have at that point, and thus is sometimes called the **acceleration of gravity**. Near the surface of the earth, g has the magnitude 9.8 m/s^2 = (22 mi/h)/s.

projectile: a falling object whose trajectory is "sufficiently short" and "sufficiently near" to the earth's surface so that $\vec{g}$ is essentially constant. If air drag is also negligible, then we say that the object is **freely falling** and its motion is **simple projectile motion.**

terminal speed v_T: the speed where the drag force on an object is equal in magnitude to its weight. At such a speed, the net force on an object falling vertically downward is zero, and therefore the object will no longer accelerate, but rather remain at that speed. Under normal circumstances, $v_T = [2mg / C\rho A]^{1/2}$, where ρ is the density of air, A is the cross-sectional area the object presents to the air going by, and C is a unitless coefficient ($\approx$ 1) that depends on the object's shape and surface qualities. An object falling from rest reaches this speed after $t \approx 2v_T/g$.

TWO-MINUTE PROBLEMS

N11T.1 You are driving 5 feet or so behind a pickup truck (don't actually try such a stupid tailgating stunt, of course!). A crate slips of the back of the truck. The crate will not hit your car until after it hits the road, regardless of your speed (T or F). (Ignore air resistance.)

N11T.2 A person standing in the cabin of a jet plane drops a coin. This coin hits the floor of the cabin at a point directly below where it was dropped (as seen in the cabin) no matter how fast the plane is moving (T or F).

N11T.3 A tennis ball is dropped from rest at the exact same instant and height that a bullet is fired horizontally. Which hits the ground first (ignoring air resistance)?
A. the bullet
B. the ball
C. both hit at the same time

N11T.4 As a projectile moves along its parabolic trajectory, which of the following remain constant (ignoring air resistance, and defining z axis to point upward)?
A. its speed
B. its velocity
C. its x-velocity and y-velocity
D. its z-velocity
E. its acceleration
F. C and E
T. other combination

N11T.5 Imagine that we throw baseball with an initial speed of 12 m/s in a direction 60° upward from the horizontal. What is the baseball's speed at the peak of its trajectory? (*Hint*: You do not need to do a *lot* of calculating here.)
A. 12 m/s
B. 10.4 m/s
D. 6 m/s
E. 3 m/s
E. 0 m/s
F. other (specify)

N11T.6 Imagine that you serve a tennis ball with an initial speed of 10 m/s in a direction 10° below the horizontal. What is its speed at the peak of its trajectory?
A. 10 m/s
B. 9.8 m/s
C. 1.7 m/s
D. 0 m/s
E. there is no "peak" to this tennis ball's trajectory
F. other (specify)

N11T.7 Imagine that you throw a tennis ball vertically into the air. At the exact top of its trajectory it is at rest. What is the magnitude of its acceleration at this point?
A. 9.8 m/s^2
B. −9.8 m/s^2
C. $0 < a < 9.8$ m/s^2
D. zero
E. other (specify)

N11T.8 Two balls have the same size and surface texture, but one is twice as heavy as the other. How many times larger is the terminal speed of the more massive ball falling through air compared to that of the lighter ball?
A. the balls fall with the *same* speed in air
B. $[2]^{1/2}$ times larger
C. 2 times larger
D. 4 times larger
E. other (specify)

N11T.9 Two balls have the same weight and surface texture, but one has twice the diameter of the other. How many times larger is the terminal speed of the smaller ball falling through air compared to that of the bigger ball?
A. the balls fall with the *same* speed in air
B. $[2]^{1/2}$ times larger
C. 2 times larger
D. 4 times larger
E. other (specify)

HOMEWORK PROBLEMS

BASIC SKILLS

N11B.1 A ballistic missile traveling 3200 km between its launch point and final destination freely falls during much of its trajectory, most of which is above the atmosphere. Is this missile a projectile or not? Explain.

N11B.2 A package dropped from an airplane flying at an altitude of 1.0 km will take how long to reach the ground? (Ignore air resistance.)

N11B.3 How long will it take a stone thrown with a vertical velocity of 22 m/s to reach the peak of its trajectory?

N11B.4 If a fireworks rocket has an initial upward speed of 58 m/s when launched, for how long will it coast upward before reaching its peak? (Ignore air resistance and assume that the rocket engines burn out very shortly after launch.)

N11B.5 If an object is launched from the ground with an initial z-velocity of 25 m/s, how much time will pass before it returns to the ground? (Ignore air resistance.)

N11B.6 If an object is launched from a point 10 m above the ground with an initial initial z-velocity of 25 m/s, how much time will pass before it returns to the ground? (Ignore air resistance.)

N11B.7 Estimate the terminal speed for a ping-pong ball whose diameter is $\approx$ 1 inch and whose mass is $\approx$ 5 g. (C for a sphere is roughly 0.5.)

N11B.8 Use the hyperbolic tangent function on your calculator to find an object's downward speed as a fraction of v_T at times $t = \frac{1}{2}v_T/g$, v_T/g, $2v_T/g$, and $3v_T/g$. Are your results consistent with the graph in Figure N11.4?

SYNTHETIC

N11S.1 Estimate the speed at which the drag on a 150-g steel ball becomes equal to about 1% of its weight. (The density of steel is about 7900 kg/m^3, and C for a sphere is about equal to 0.5.)

N11S.2 Your frisbee is lodged in a tree 12 m above your head. How fast would you have to throw a baseball to dislodge the frisbee? (Check your work using a calculation based on conservation of energy. Which is easier to use in this kind of a problem?)*

N11S.3 A police officer is chasing a burglar across a roof top. Both are running at a speed of 6.0 m/s. Before the burglar approaches the edge of the roof, the burglar needs to make a decision about whether to jump the gap to the next building, whose roof is 6.2 m away but 3.5 m lower. Will the burglar make it (assuming that the burglar's initial velocity is 6.0 m/s horizontally when the jump begins)?*

N11S.4 You are standing on a cliff 32 m tall overlooking a flat beach. The edge of the water is 25 m from the base of the cliff. How fast would you have to throw a stone horizontally to reach the water?*

N11S.5 In serving a tennis ball, the server hits the ball horizontally from a height of about 2.2 m. The ball has to travel over the net (0.9 m high) a distance of 12 m away. What is the minimum initial speed that the tennis ball can have if it is to make it over the net?*

N11S.6 You are the pilot of a Coast Guard rescue plane. Your job is to drop a package of emergency supplies to a person floating in the ocean. If your plane is flying directly toward the person at an elevation of 520 m at a speed of 85 m/s (about 190 mi/hr), about how far ahead of the person should you release the package, if it is to land near the person? Assume that the package's motion is not significantly affected by air friction (probably not a very good assumption here).*

N11S.7 A batter hits a fly ball with an initial velocity of 37 m/s at an angle of 32° from the horizontal. How long does the outfielder have to get to the appropriate position to catch the ball?*

N11S.8 During volcanic eruptions, chunks of solid rock can be blasted out of a volcano: these projectiles are called *volcanic blocks*. Imagine that during an eruption of Mt. St. Helens (a volcano in Washington state) a block lands near an observing station that is located 8.4 km east and 1.8 km below the summit. If the block was ejected from the summit at a 35° angle from the horizontal, find **(a)** its initial speed and **(b)** its time of flight.*

*For these starred problems, please make sure that your solution follows the simple-projectile motion problem framework outlined in examples N11.1 and N11.2.

N11S.9 The formula for a parabola in the zx plane is

$$z(x) = Ax^2 + Bx + C \qquad \text{(N11.17)}$$

where A, B, and C are constants. Prove that the trajectory of a simple projectile has this form.

N11S.10 The range R of a projectile is the horizontal distance that it travels between its launch at the time it hits the ground. **(a)** Show that if the projectile is launched from the ground its range will be

$$R = \frac{2v_0^2 \sin\theta\cos\theta}{g} \qquad \text{(N11.18)}$$

(b) Prove that a projectile launched from the ground with a given initial speed v_0 will travel the farthest if it is launched at an angle of 45° with respect to the horizontal. (*Hint*: There exists a trigonometric identity that says that $\sin 2\theta = 2\sin\theta\cos\theta$.)

N11S.11 Integrate the expression for the z-velocity as a function of time for an object falling from rest with drag to find the object's z-position as a function of time (assume that $z(0) = 0$. Use this formula to calculate the approximate distance that a person would fall in 30 s and their speed at the end of that time, and compare it to the results that one would find if there were no air resistance. [*Hint*: Here is your own private, one-entry table of integrals:

$$\int \tanh(bu)\,du = \frac{1}{b}\ln(\cosh bu) \qquad \text{(N11.19)}$$

where b is any constant, u is the variable one is integrating with respect to, ln() is the natural logarithm function, and cosh() is the hyperbolic cosine function. Note that cosh(0) is equal to 1, not zero (check this with your calculator). Also, the terminal velocity for someone falling through air is about 60 m/s.]

RICH-CONTEXT

N11R.1 Abel Knaebble, professional stunt man and amateur physicist, wants to jump the Grand Canyon in his new, super-streamlined motorcycle (which is completely immune to the effects of air friction). He simply plans to ride his motorcycle off one rim, so his initial velocity will be entirely horizontal. Assume that the width of the canyon is 8.0 km at the proposed jump site, that the mass of his fully loaded cycle (including Abel himself) is 380 kg, and that the rim of the canyon from which he will jump is 320 m higher than the far side. How fast will he need to take off to make it to the other side? Discuss some of the impracticalities of this proposed stunt.

N11R.2 An essentially spherical boulder the size of several houses sits precariously on a slope above a village, as shown in Figure N11.5. If this boulder ever gets loose, will it hit the village? (*Hint*: First estimate the speed with which the boulder will roll off the cliff. The total center-of-mass plus rotational kinetic energy of a rolling spherical object is $\frac{7}{10}mv^2$: see chapter C9.)

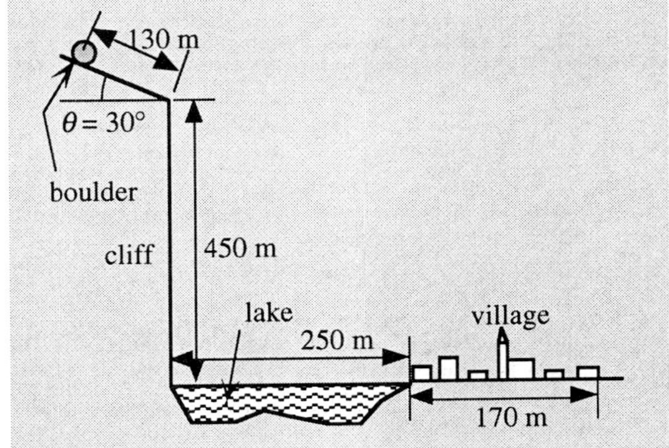

Figure N11.5: A village at risk?

ADVANCED

N11A.1 Prove that the ranges of two projectiles launched from the ground with a given initial speed v_0 at an angle of $45°\pm\phi$ are the same. [*Hint*: $\sin(\pi/2+\phi) = \cos\phi$.]

N11A.2 Imagine that we fire a projectile from the ground into the air with an initial velocity of 30 m/s at an angle of 45° with respect to the horizontal. Assume that the terminal speed of this object is 40 m/s. Use the trajectory construction method described in chapter N5 to predict when and where the object will return to the ground, taking account of the drag force. Compare this to the range the projectile would have in the absence of drag (see problem N11S.10). [*Hints:* I suggest taping two or three sheets of paper lengthwise and using a scale of 1 cm = 2 m and a time interval Δt of 0.4 s. The drag force will always act opposite to the object's velocity. Show that its magnitude is such that $F_D/m = gv^2/v_T^2$. We want to be able to draw the acceleration arrow $\vec{a}\Delta t^2$ for this object: show using Newton's second law that

$$\vec{a}\Delta t^2 = \vec{g}\Delta t^2 + g\Delta t^2 \frac{(v\Delta t)^2}{(v_T\Delta t)^2}(-\hat{v}) \qquad \text{(N11.20)}$$

where $\hat{v}$ is the direction of the velocity. Interpret what this means in terms of how you construct the acceleration arrows at each point, and start constructing the trajectory. Use the previous velocity arrow to in the calculation implied by equation N11.20. Be very careful, making measurements and drawing lines as precisely as you can.]

ANSWERS TO EXERCISES

N11X.1 About 20% and about 0.4% respectively.
N11X.2 About 11°.
N11X.3 It would take an object 1.0 s to fall 4.9 m from rest; answer is independent of the horizontal velocity.
N11X.4 The quadratic formula implies that

$$\text{if } ax^2+bx+c=0 \quad \text{then } x = \frac{-b\pm\sqrt{b^2-4ac}}{2a} \qquad \text{(N11.21)}$$

In the case at hand, $a = -g/2$, $b = v_{0z}$, and $c = z_0$. Plugging these into the above yields equation N11.9.
N11X.5 If z_0 is negative and $2g|z_0| > v_{0z}^2$, then the quantity inside the square root is negative and the equation has no solution. Having z_0 being negative means that the projectile begins below $z = 0$. Conservation of energy means that the projectile in this case must have an initial kinetic energy of $\frac{1}{2}mv_{0z}^2 > mg\,|z_0| \Rightarrow v_{0z}^2 = 2g\,|z_0|$ just to get up to $z = 0$. Therefore, if $v_{0z}^2 < 2g\,|z_0|$, the projectile never makes it up to $z = 0$ and thus there *should be* no solutions to the equation that tells us the times when $z = 0$.
N11X.6 The object reaches terminal velocity when the drag acting on it is equal in magnitude to its weight, so

$$mg = \tfrac{1}{2}C\rho Av_T^2 \quad \Rightarrow \quad v_T^2 = 2mg/C\rho A \qquad \text{(N11.22)}$$

Taking the square root of this yields equation N11.11.
N11X.7 By the equation above, $CA = 2mg/\rho v_T^2$. Estimating $m = 60$ kg, we get $CA = 0.27\ \text{m}^2$. If $C \approx 0.5$, then this means that the cross-sectional area of the person is about 0.54 m^2, which is plausible (a person is about 1.5 m tall and about 0.4 m wide = 0.60 m^2).
N11X.8 If the drag term is zero in equation N11.13 (which is the same thing as saying that the terminal speed goes to infinity), then equation N11.13 becomes

$$\int_{v_z(0)}^{v_z(t)} dv_z = -g\int_0^t dt = -g(t-0) = -gt \qquad \text{(N11.23)}$$

But since $\int du = u$ for any variable u, and v_z is just another variable here, the left-most integral simply becomes v_z evaluated at the limits of the integral:

$$\int_{v_z(0)}^{v_z(t)} dv_z = v_z(t) - v_z(0) \qquad \text{(N11.24)}$$

Plugging this into equation N11.23, we get

$$v_z(t) - v_z(0) = -gt \quad \Rightarrow \quad v_z(t) = -gt + v_z(0) \qquad \text{(N11.25)}$$

N11X.9 Since $g \approx 10\ \text{m/s}^2$, it would take 6 seconds for an object that is freely falling to reach a speed of 60 m/s. If an object falling with drag takes about twice as long to actually reach that terminal speed, then a person would have to fall about 12 s to reach the terminal speed. As a check, note that when $t = 12$ s,

$$\frac{gt}{v_T} = \frac{(10\ \cancel{\text{m}}/\cancel{\text{s}^2})(12\ \cancel{\text{s}})}{60\ \cancel{\text{m}/\text{s}}} = 2 \qquad \text{(N11.26)}$$

which is the place on Figure N11.4 where the object approximately reaches its terminal speed. (Alternatively, we could have just solved $2 = gt/v_T$ for t.)

N12

OSCILLATORY MOTION

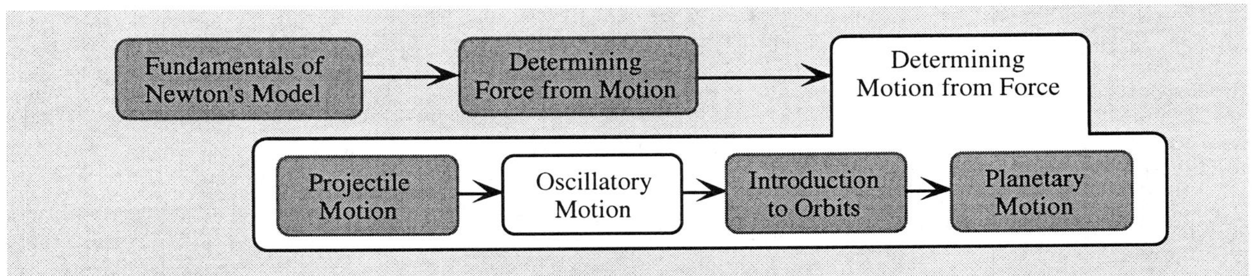

N12.1 OVERVIEW

This is the second chapter in the series of chapters exploring how we can use Newton's second law and knowledge of the forces acting on an object to predict its motion. In this chapter, we explore the very important case of *oscillatory motion*, where an object moves in one dimension in response to an interaction whose potential energy has the form $V(x) = \frac{1}{2}kx^2$.

Oscillatory motion is one of the most common types of motion in physics. An object hanging from the end of a spring or a pendulum swinging back and forth are simple illustrations of this kind of repetitive, back-and-forth motion, but the same mathematics that we can use to describe these motions can also be used to describe oscillatory motion in other contexts, such as the up-and-down motion of an ocean wave, the vibration of a piano string, the time-varying motion of electrons in a TV transmitter, the vibrations of a column of air in a musical instrument, and the oscillations of atoms in a solid. Understanding the oscillation of an object at the end of a spring is the first step toward understanding all of these other motions.

We will begin by exploring the behavior of a model system consisting of an object connected to a fixed point by an ideal spring (a system called the *simple harmonic oscillator*) and then look at a few applications of this model in other contexts. Here is an overview of the sections in this chapter.

N12.2 *A MASS ON A SPRING* describes the basic model and its assumptions, and establishes some conventions.

N12.3 *SOLVING THE EQUATION OF MOTION* shows how we can use Newton's second law to predict the motion of the mass on a spring.

N12.4 *THE OSCILLATOR AS A MODEL* describes why this model is so useful in a wide variety of physical situations.

N12.5 *A MASS HANGING FROM A SPRING* illustrates how we can adapt the model to situations where gravity also acts on the object.

N12.6 *THE SIMPLE PENDULUM* displays another (more complicated) situation where the harmonic oscillator model is clearly just a simplified description the actual motion.

This material is essential background for parts of unit *E*, *Q*, and *T*.

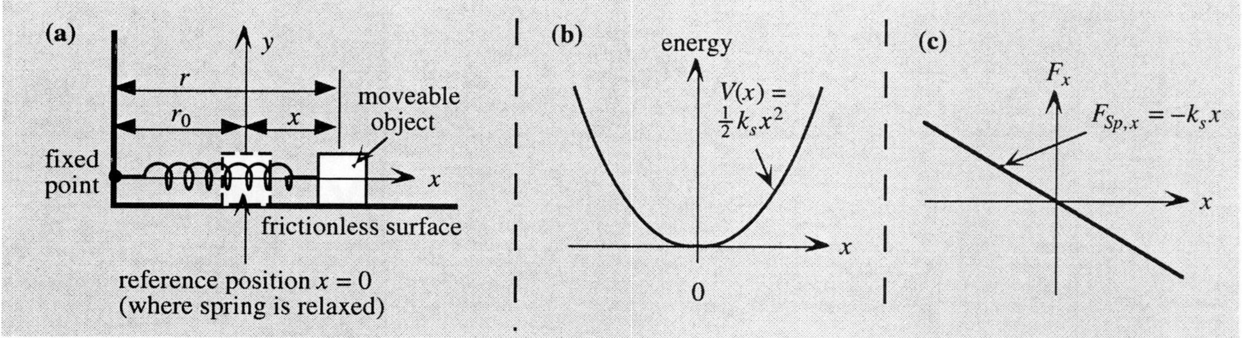

Figure N12.1: (a) The archetype of the simple harmonic oscillator: a moveable object allowed to move in one dimension that is connected to a fixed point by an ideal, massless spring whose relaxed length is r_0. **(b)** A graph of the system's potential energy as a function of *x*. **(c)** A graph of the *x*-force on the moveable object as a function of *x*.

N12.2 A MASS ON A SPRING

Imagine an ideal, massless spring with one end connected to a fixed point (like a wall) and the other connected to a moveable object with mass m, as shown in Figure N12.1a. Assume that the object is free to slide in one dimension on a horizontal frictionless surface. If the surface is truly horizontal, the object's weight will be exactly canceled by the normal force due to its interaction with the surface, and in the absence of friction the net force acting on the object will be the force supplied by the spring.

Conventional definition of reference frame

We conventionally define the x axis to coincide with the line along which the object moves. Let r be the the object's position relative to the fixed point at a given time and let r_0 be the same when the spring is relaxed. If we define our reference frame origin so that $x = r - r_0$, then $x = 0$ corresponds to the object's position when the spring is relaxed, and $|x|$ expresses the distance that the spring is either stretched or compressed.

We saw in chapter C7 that the potential energy function $V(x)$ for an **ideal spring** is

$$V(r) = \tfrac{1}{2}k_s(r - r_0)^2 = \tfrac{1}{2}k_s x^2 \tag{N12.1}$$

where k_s is the **spring constant** that characterizes the spring's stiffness. Knowing the spring's potential energy function, we can use the method described in section N3.5 to find the force that the spring interaction exerts on the object. Equation N3.17 says that the "r component" of this force is given by:

$$\begin{aligned} F_{Sp,r} &= -\frac{dV}{dr} = -\tfrac{1}{2}k_s\frac{d}{dr}(r - r_0)^2 = -\tfrac{1}{2}k_s\frac{d}{dr}(r^2 + 2rr_0 + r_0^2) \\ &= -\tfrac{1}{2}k_s(2r - 2r_0) = -k_s(r - r_0) = -k_s x \end{aligned} \tag{N12.2}$$

The r component of $\vec{F}_{Sp}$ is defined to be the component along the direction that r increases when the object is moved, which in this case is the same as the $+x$ direction. Therefore, we see that the x component of the force exerted on the object by the spring is given by the simple linear formula

$$F_{Sp,x} = -k_s x \tag{N12.3}$$

This equation is called **Hooke's law** after the British scientist (a contemporary of Newton) who first stated it. (Also, if you have ever wondered why the factor of $\frac{1}{2}$ appears in the definition of the potential energy, now you know: putting the factor into equation N12.1 makes equation N12.3 simpler.)

Let's see what Hooke's law tells us about the force exerted by the spring. First and foremost, note that the magnitude $F_{Sp} = |F_{Sp,x}| = |-k_s x| = k_s|x|$ of the spring force is directly proportional to the distance $|x|$ that the spring is stretched. Secondly, note that the force acts in the negative x direction when x is positive and the positive x direction when x is negative: in both cases this tends to push the object back towards $x = 0$. This is completely consistent with what we know qualitatively about the behavior of springs.

Exercise N12X.1: The units of the spring constant k_s were given in chapter C7 as being J/m^2 (these units are appropriate for equation N12.1). What would be the appropriate units for k_s in the context of equation N12.3? Show that your units are equivalent to J/m^2.

Exercise N12X.2: If a spring whose relaxed length is 5.0 cm exerts a force of magnitude 3.0 N when stretched to 8.0 cm, what is its spring constant k_s?

N12.3 SOLVING THE EQUATION OF MOTION

In *any* situation where the net x-force on an object moving in one dimension is given by equation N12.3, Newton's Second Law reads:

$$F_{Sp,x} = ma_x \quad \Rightarrow \quad a_x = \frac{F_{Sp,x}}{m} = \frac{-k_s x}{m} = -\omega^2 x \qquad \text{(N12.4)}$$

where

$$\omega^2 \equiv \frac{k_s}{m} \qquad \text{(N12.5)}$$

(The reason that I defined the *square* of ω to equal k/m will become clear shortly.) Since $a_x = dv_x/dt = d^2x/dt^2$, equation N12.4 becomes:

The harmonic oscillator equation

$$\frac{d^2x}{dt^2} = -\omega^2 x \qquad \text{(N12.6)}$$

This important equation is called the **harmonic oscillator equation**.

How can we find the object's motion in this case? When we studied projectile motion in chapter N5, we simply integrated both sides of Newton's second law twice to find the projectile's position as a function of time (see section N5.4). We cannot do the same thing in this case, because in order to integrate both sides of equation N12.6 with respect to time, we have to know how x depends on time. Unfortunately, this is what we are trying to *find*. As I mentioned in the section on drag in the last chapter, this commonly occurs when we are trying to determine an object's motions from forces, and we ultimately have to find some trick to get around this problem.

The harmonic oscillator equation is an example of a simple **differential equation:** a differential equation sets a function, $x(t)$ in this case, in an equation having terms that also involve derivatives of the function. Differential equations cannot generally be solved in any straightforward manner: often the best method is to make an educated guess as to what the solution is, plug the guess into the differential equation, and check to see whether the guessed solution satisfies the equation. If the guess works, you have solved the equation; if not, you try another guess. (A course in differential equations basically makes you a more intelligent guesser!)

Our trick here: guess the solution and see if it works

So the trick we will use here is that we are just going to guess a possible solution and see if it works. What kind of an intelligent guess can we make in this case? We know that an object on the end of a spring will oscillate, so we expect $x(t)$ to be some function that repeats in time, like $\sin(qt)$ or $\cos(qt)$, where q is some constant. The differential equation tells us that the second time derivative of our $x(t)$ function should be equal to a negative constant times that same function. What function has a second derivative that is equal to a negative number times itself?

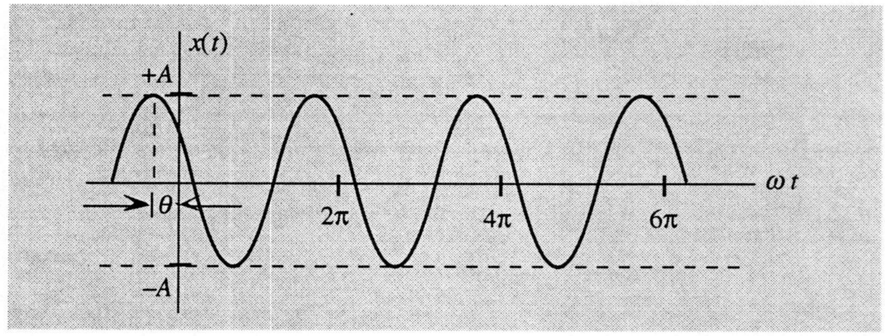

Figure N12.2: The solution to the harmonic oscillator equation is $x(t) = A\cos(\omega t + \theta)$. This function oscillates between the limits $\pm A$, and goes through one complete oscillation every time that ωt increases by 2π. The phase constant θ specifies how far the function is along its cycle at time $t = 0$ (a positive value of θ thus corresponds to shifting the wave crest left from zero by the magnitude of θ). The value of ω is determined by the values of k and m, while the values of A and θ are determined by initial conditions.

Again, $\sin(qt)$ and $\cos(qt)$ both have that characteristic. Note that (see equations N9.26 in the ***Math Skills*** section on the chain rule in chapter N9):

$$\frac{d}{dt}\sin(qt) = q\cos(qt), \quad \frac{d^2}{dt^2}\sin(qt) = \frac{d}{dt}q\cos(qt) = -q^2\sin(qt) \quad \text{(N12.7a)}$$

$$\frac{d}{dt}\cos(qt) = -q\sin(qt), \quad \frac{d^2}{dt^2}\cos(qt) = -\frac{d}{dt}q\sin(qt) = -q^2\cos(qt) \quad \text{(N12.7b)}$$

So if we set $x(t)$ equal to either one of these functions, we could satisfy the harmonic oscillator equation $d^2x/dt^2 = -\omega^2 x$ as long as we identify the constant q as being equal to $\omega = \sqrt{k/m}$.

The equation $x(t) = \sin(\omega t)$ actually can't be right, as $x(t)$ has units of meters while the sine function always produces a unitless number by definition (it is the ratio of two sides of a right triangle). So a viable solution has to be something like $x(t) = A\sin(\omega t)$, where A is a constant with units of meters.

However, as we have already pointed out, something like $x(t) = A\cos(\omega t)$ will *also* be a solution to this equation. The most general solution to the harmonic oscillator equation is in fact

The most general solution to the harmonic oscillator

$$x(t) = A\cos(\omega t + \theta) \quad \text{(N12.8)}$$

where A and θ are constants that turn out to be determined by initial conditions. The object's motion is thus described by a cosine wave that cycles back and forth between the limits $x = +A$ and $x = -A$, as shown in Figure N12.2.

The meaning of the amplitude and phase constants

We call the constant A (which has units of length) the **amplitude** of the oscillation: note that the *total* distance that the oscillating object travels from one extreme to the other is $2A$. We call the constant θ the **phase constant** of the solution: as shown in Figure N12.2, this constant specifies how far along a given cycle the oscillation is at time $t = 0$ (measured from the first peak to the left of the origin). For example, if $\theta = 0$, the object is at $+A$ at $t = 0$; if $\theta = \pi/2$, then the object is at $x = 0$ at $t = 0$, and so on. Different values of θ essentially shift the oscillation back and forth along the ωt axis of Figure N12.2.

Note that if $\theta = \pi/2$, then $\cos(\omega t + \theta)$ is equivalent to $-\sin(\omega t)$ (if you shift the peak to the left of the origin back a full quarter cycle from the origin, you can see that it is like an upside-down sine function). Similarly, if $\theta = \pi$, $\cos(\omega t + \theta) = -\cos\theta$, and if $\theta = 3\pi/2$, then $\cos(\omega t + \theta) = +\sin\theta$. The point is that equation N12.8 embraces *both* the sine and cosine solutions to the harmonic oscillator equation and everything in between. Since we can also always change the sign of the function by choosing the right phase constant θ, we conventionally choose θ so that the amplitude A is positive.

Other quantities associated with the oscillation

The constant $\omega = \sqrt{k_s/m}$ is called the **phase rate** of the oscillation: it specifies the rate at which the quantity $\omega t + \theta$ (which is sometimes called the *phase* of the oscillation) increases with time. Since the object goes through one complete cycle every time that $\omega t + \theta$ increases by 2π, the larger ω is, the more cycles the object will complete in a given time.

The **period** T of the oscillation is defined to be the time that it takes the object to go through one complete oscillation of the cosine function. This will happen when the value of ωt (and thus $\omega t + \theta$) increases by 2π radians, that is:

$$\omega(t+T) = \omega t + 2\pi \qquad \text{(N12.9)}$$

Subtracting ωt from both sides, we see that:

$$\omega T = 2\pi \quad \Rightarrow \quad T = \frac{2\pi}{\omega} = 2\pi\sqrt{\frac{m}{k_s}} \qquad \text{(N12.10)}$$

The **frequency** f of an oscillation is defined to be the number of *cycles* completed per unit time. Since exactly one cycle is completed in time T by definition, the frequency f is given by:

$$f = \frac{1 \text{ cycle}}{T} = \frac{\omega}{2\pi}\text{ cycle} = \frac{\text{cycle}}{2\pi}\sqrt{\frac{k_s}{m}} \qquad \text{(N12.11)}$$

(The Greek letter ν (nu) is also commonly used for frequency, but since this letter is easy to confuse with the letter v used for velocity, I'll avoid it here.) The standard unit for frequency f in the SI system is the **hertz** (abbreviation: Hz), where 1 Hz $\equiv$ 1 cycle/s, whereas the standard SI units for phase rate ω are simply s^{-1} (which can be thought of as being radians/s).

Note that we can determine the oscillation's frequency f (or ω) in terms of the spring constant k_s and the object's mass m ($\omega = \sqrt{k_s / m}$) while we find the oscillation's amplitude A and phase constant θ from initial conditions (as we will see).

Exercise N12X.3: Show by direct substitution that $x(t) = A\cos(\omega t + \theta)$ does indeed solve the harmonic oscillator equation (equation N12.6).

Exercise N12X.4: Verify that $\sqrt{k_s / m}$ has units of s^{-1}.

Exercise N12X.5: An object with a mass of 2.0 kg is attached to a spring with spring constant of 100 N/m. The object oscillates a distance of 10 cm from one extreme to another. What is the amplitude of the oscillation? What is its period? What is its phase rate?

N12.4 THE OSCILLATOR AS A MODEL

Harmonic oscillator model has many applications

The simple harmonic oscillator model is not only useful for describing the behavior of objects connected to springs, it is also a good model for an astounding range of physical systems (from ocean waves to electrical oscillations to atomic vibrations), making it one of the most useful models in all of physics.

We have already discussed part of the reason why this model turns out to be so useful in section C8.5. There, I argued that any time the potential energy function for an interaction has a valley, we can approximate the bottom of that valley by the parabolic harmonic oscillator potential energy function. In this section, I want to present a somewhat different way of saying the same thing.

Characteristics of an oscillating system

The most important characteristics of *any* oscillating object are (1) that it has a position or configuration (called its **equilibrium position**) where the force on it is zero, (2) if it is displaced from that position, it experiences a force (called a **restoring force**) that pushes it *back toward* the equilibrium position, and (3) this force (at least for small displacements) grows in magnitude as the object's displacement increases. Any object satisfying these criteria will oscillate around its equilibrium position if it is displaced and then released.

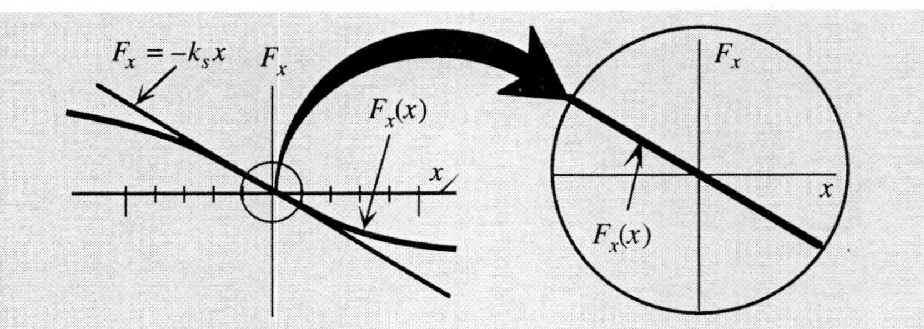

Figure N12.3: Almost *any* function $F_x(x)$ that is negative for $x > 0$ and positive for $x < 0$ can be approximated for small $|x|$ by a straight line $-k_s x$ for an appropriately chosen value of k_s.

Clearly, an object attached to a spring has these characteristics. The special characteristic of a simple harmonic oscillator is that the magnitude of the restoring force is strictly proportional to the distance that the object is displaced. Many oscillating systems do *not* share this characteristic. For example, the force on an atom in a solid is a complicated function of its position as a result of the complicated electrostatic interactions between the atom and its neighbors.

For small displacements, $F_x \approx -k_s x$ in many cases

Even so, calculus tells us that we can approximate the curve of almost any physically reasonable (that is, differentiable) function $F_x(x)$ that is positive for $x < 0$ and negative for $x > 0$ by a straight line $F_x = -k_s x$, where $k_s = -dF_x/dx$ evaluated at $x = 0$. This is a good approximation when x is sufficiently small in magnitude, as shown in Figure N12.3.

This means an object responding to almost *any* restoring force $F_x(x)$ will find that for small displacements $F_x \approx -k_s x$. In this *small oscillation limit*, almost any oscillating object will behave like a simple harmonic oscillator. This is why the harmonic oscillator model is so useful and important.

Exercise N12X.6: Consider an object moving along the x axis under the influence of a force whose x component is $F_x = a[(b-x)^2-b^2]$ where a and b are positive constants. Note that this force component is negative when $b > x > 0$ and positive when $x < 0$, so it qualifies as a restoring force. Argue that for small values of $|x|$, this formula becomes $F_x \approx -k_s x$, and find k_s in terms of a and/or b.

N12.5 A MASS HANGING FROM A SPRING

The simplest way to construct a practical simple harmonic oscillator is to simply hang an object vertically at the end of a spring. This is much easier than constructing the system illustrated in Figure N12.1 (frictionless surfaces are hard to come by). But is a hanging system really the same as that in Figure N12.1?

Net force here includes the force of gravity

The complication here is that the net force on the object is a sum of the vertical component of the spring force and the force of gravity. If we set up our coordinate system so that the x axis is along the direction of oscillation (vertical in this case) and set $x = 0$ to be the object's position when the spring is relaxed, then the net x-force on the object is:

$$F_{\text{net},x} = -k_s x - mg \tag{N12.12}$$

This is *not* the same as the harmonic oscillator force law (equation N12.3).

We can get rid of this term by redefining origin

But it turns out that this force law really does work out to be the harmonic oscillator force law if we choose our reference frame cleverly. Let's choose a reference frame whose x axis is vertical (and let's choose the $+x$ direction to be upward), but let's define $x = 0$ not to be the position of the hanging object when the spring is relaxed, but rather to be its position at *equilibrium*, where the spring tension force exactly balances the object's weight. (The spring will be somewhat stretched at this position!) Let x_R be the position of the object when

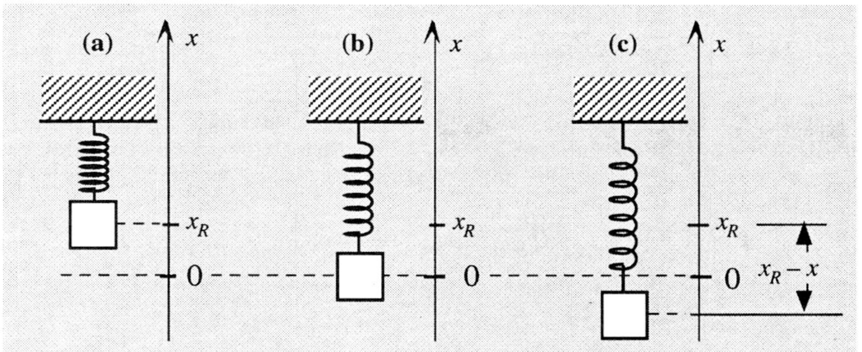

Figure N12.4: A system where an object hangs from a spring. **(a)** The object located the position x_R where the spring is relaxed. **(b)** The position of the object when the spring tension exactly balances the object's weight: this position is defined to be $x = 0$. **(c)** Here, the spring is stretched further: the distance by which the spring is stretched when the object is at x is $x_R - x$.

the spring is relaxed: the distance s by which the spring is stretched when the object is at any given position x is then

$$s = x_R - x \qquad \text{(assuming that } x < x_R\text{)} \tag{N12.13}$$

(see Figure N12.4). Note that since the $+x$ direction is upward, as x becomes more positive, the object is moving *up*, and the distance that the spring is stretched will *decrease*: this is why x is *subtracted* in equation N12.13.

Since the spring will exert an upward force proportional in magnitude to s, the net force acting on the object at position x in this reference frame is:

$$F_{\text{net},x} = +k_s s - mg = +k_s(x_R - x) - mg = k_s x_R - k_s x - mg \tag{N12.14}$$

Note that $k_s s$ is *positive* here because the force is upward (in the $+x$ direction) when the spring is stretched by s. (As a check, note that this formula implies that as the object gets lower, x decreases and the upward force exerted by the spring increases, as we would expect.) At $x = 0$, the net force on the moveable object is supposed to be zero, so equation N12.13 tells us that

$$0 = F_{\text{net},x} = k_s(x_R - 0) - mg \quad \Rightarrow \quad k_s x_R = mg \tag{N12.15}$$

If we subsitute this into equation N12.14, the two constant terms cancel, so

$$F_{\text{net},x} = -k_s x \tag{N12.16}$$

which is the right force law for a simple harmonic oscillator. Choosing the right origin thus enables us to cancel the gravitational force with a constant term in the spring tension force: the resulting net force is directly proportional to x.

So hanging mass behaves like a harmonic oscillator

The point of all of this is that an object hanging from a spring, even though it is not the same thing physically as a mass oscillating *horizontally* back and forth, will *behave* in exactly the same manner. The gravitational force here simply has the effect of displacing the equilibrium point ($x = 0$ in the reference frame we've been using) downward from the position where the spring is relaxed. The hanging oscillator even oscillates at the same frequency that it would if it were horizontal!

EXAMPLE N12.1

Problem: An object of mass 0.52 kg hangs from a spring with spring constant $k_s = 130$ N/m. The object is measured to pass through its equilibrium point with a speed of 1.0 m/s. Define the time when it passes the equilibrium point going upward to be $t = 0$. **(a)** What is the value of θ in equation N12.8? **(b)** What is the amplitude of its oscillation?

Solution **(a)** If we use the usual coordinate system for the vertical oscillator, the object's position is $x = 0$ at $t = 0$ (since it is passing through its equilibrium point at that time). Therefore:

EXAMPLE N12.1 (continued)

$$x(0) = A\cos(\omega \cdot 0 + \theta) = A\cos\theta \tag{N12.17}$$

is equal to *zero* in this case. The cosine function is only zero when $\theta = \pi/2$ or $3\pi/2$, so θ has to be one of these values. We can determine which one by checking the direction of the velocity at $t = 0$. Using the chain rule, we find that:

$$\begin{aligned} v_x &= \frac{dx}{dt} = \frac{d}{dt}A\cos(\omega t+\theta) = -A\sin(\omega t+\theta)\frac{d}{dt}(\omega t+\theta) \\ &= -A\omega\sin(\omega t+\theta) \end{aligned} \tag{N12.18}$$

So at time $t = 0$, we have:

$$v_x(0) = -A\omega\sin(\omega \cdot 0+\theta) = -A\omega\sin\theta \tag{N12.19}$$

Since $v_x(0)$ has to be positive according to the problem statement, $-A\omega\sin\theta$ has to be positive, implying that $\sin\theta$ has to be *negative* (A and ω are positive). Of the choices $\theta = \pi/2$ or $3\pi/2$, only the latter gives $\sin\theta < 0$, so $\theta = 3\pi/2$.

(b) We are given that $v_x(0) = +1.0$ m/s. If we plug $\theta = 3\pi/2$ (which is a phase shift of ¾ of an oscillation) into equation N12.8 and solve for A, we get

$$v_x(0) = -A\omega\sin(\tfrac{3\pi}{2}) = +A\omega \quad \Rightarrow \quad A = \frac{v_x(0)}{\omega} = \frac{v_x(0)}{\sqrt{k_s/m}}$$

$$\Rightarrow \quad A = v_x(0)\sqrt{\frac{m}{k_s}} = (1.0\ \text{m/s})\sqrt{\frac{0.52\ \text{kg}}{130\ \text{N/m}}\left(\frac{1\ \text{N}}{1\ \text{kg}\cdot\text{m/s}^2}\right)} = 0.062\ \text{m} \tag{N12.20}$$

So the amplitude of the oscillation is about 6.2 cm.

EXAMPLE N12.2

Problem: Imagine that an object with mass m is hanging at rest from a spring. You pull the object down a distance D, hold it, and then release it from rest at time $t = 0$. What are A and θ corresponding to these initial conditions?

Solution If we define the origin so that $x = 0$ at the position where the object hangs at rest, then the problem statement implies that

$$x(0) = A\cos(0+\theta) = -D, \qquad v_x(0) = -A\omega\sin(0+\theta) = 0 \tag{N12.21}$$

Now, $-A\omega\sin\theta = 0$ only if $\theta = 0$ or π. Only $\theta = \pi$ yields a negative value for the cosine, which is required to satisfy the first equation. The phase shift is thus half an oscillation here. Plugging $\theta = \pi$ into the first equation, we get

$$-D = A\cos\pi = -A \quad \Rightarrow \quad A = D, \quad \theta = \pi \tag{N12.22}$$

Finding A and θ from initial conditions

These examples illustrate how we can use initial conditions to determine the values of A and θ in the solution $x(t) = A\cos(\omega t+\theta)$. If $x(0) = 0$ or $v_x(0) = 0$, we can quickly determine the phase constant θ, as shown in the examples. Problems where *neither* are zero are more difficult but still solvable when ω (or k_s and m) is known: the trick is to divide equation N12.19 by N12.17 to get

$$-\frac{v_x(0)}{\omega x(0)} = \frac{A\sin\theta}{A\cos\theta} = \tan\theta \quad \Rightarrow \quad \theta = \tan^{-1}\left(\frac{-v_x(0)}{\omega x(0)}\right) \tag{N12.23}$$

Since $\tan\theta = \tan(\theta+\pi)$, there are actually two possible solutions for θ. But if you plug the correct one back into equation N12.17 and solve for A, you should get

$$A = \frac{x(0)}{\cos\theta} > 0 \text{ (since } A \text{ is positive by definition)} \tag{N12.24}$$

EXAMPLE N12.3

Problem: Imagine that an object hanging from a spring oscillates with a period of $T = 2.0$ s. At $t = 0$, it is 5 cm above the equilibrium point and is moving upward at 10 cm/s. What are A and θ for this oscillation?

Solution According to equation N12.23 and $\omega = 2\pi/T$, we have

$$\tan\theta = \frac{-v_x(0)}{\omega x(0)} = \frac{T}{2\pi}\frac{-v_x(0)}{x(0)} = \frac{-(2.0\,\text{s})(10\,\text{cm/s})}{2\pi(5\,\text{cm})} = -\frac{2}{\pi} \qquad \text{(N12.25)}$$

So $\theta = \tan^{-1}(-2/\pi) = -0.57$ or $\pi - 0.57 = 2.57$. The first of these is the correct value because it yields a positive value for A in equation N12.24:

$$A = \frac{x(0)}{\cos\theta} = \frac{5.0\text{ cm}}{\cos(-0.57)} = +5.9\text{ cm} \qquad \text{(N12.26)}$$

Yet another way to find the amplitude from initial conditions is to use conservation of energy. Whenever the cosine in $x(t) = A\cos(\omega t + \theta)$ is equal to ± 1 (which will be at either extreme end of the oscillation), then the object's velocity is equal to zero, since $v_x(t) = -\omega A\sin(\omega t + \theta)$ and $\sin(\omega t + \theta) = 0$ whenever $\cos(\omega t + \theta) = 1$ (we also know that an object will change direction at an extreme point, so $v_x(t)$ must be passing through zero at that time). Therefore, the system's total energy at an extreme point must be

$$E = K + V(x) = 0 + \tfrac{1}{2}k_s x^2 = \tfrac{1}{2}k_s A^2 \qquad \text{(N12.27)}$$

Since energy is conserved, this must be the system's energy at time $t = 0$ too, so

$$\tfrac{1}{2}k_s A^2 = E = \tfrac{1}{2}m[v_x(0)]^2 + \tfrac{1}{2}k_s[x(0)]^2 \qquad \text{(N12.28)}$$

So, if you know $v_x(0)$ and $x(0)$, you can find A this way.

EXAMPLE N12.4

Problem: You see a couple of neighborhood boys bouncing on the hood of your car. When one jumps on the hood, you see the car's front end oscillate with a period of about 1.5 s. After you yell at the boys, you get to wondering about the spring constant k_s of the car's suspension. Estimate k_s from what you saw.

Solution The weight of a typical car ≈ 3000 lbs, corresponding to a mass of roughly 1500 kg. Let's say that the front suspension effectively suspends about half of this mass, or ≈ 800 kg when the ≈ 50-kg mass of the boy is included. According to equation N12.17, the period of oscillation is $T = 2\pi\sqrt{m/k_s}$, so

$$k_s = \frac{4\pi^2 m}{T^2} = \frac{4\pi^2(800\,\text{kg})}{(1.5\,\text{s})^2}\left(\frac{1\text{ N}}{1\,\text{kg}\cdot\text{m/s}^2}\right) = 14{,}000\text{ N/m} \qquad \text{(N12.29)}$$

Here is a way to check this estimate. If you were to sit on the hood (assume that your weight ≈ 700 N ≈ 155 lbs), the hood should sink until the compressed springs exert an upward force equal to your weight, that is, by about $|\Delta x| = \Delta F/k_s = (700\text{ N})/14{,}000\text{ N/m} \approx 0.05\text{ m} = 5\text{ cm}$. This seems about right.

Exercise N12X.7: An 1.0 kg object hangs from a spring whose spring constant is 100 N/m. You take the mass, pull it down 10 cm, and then give it an initial downward speed of 0.50 m/s. What is are the values of A and θ here?

Exercise N12X.8: A 68-kg friend of yours goes bungee jumping, and you watch. You notice that after the jump, the friend oscillates once up and down in about 6.0 s. Estimate the effective spring constant of the bungee cord.

N12.6 THE SIMPLE PENDULUM

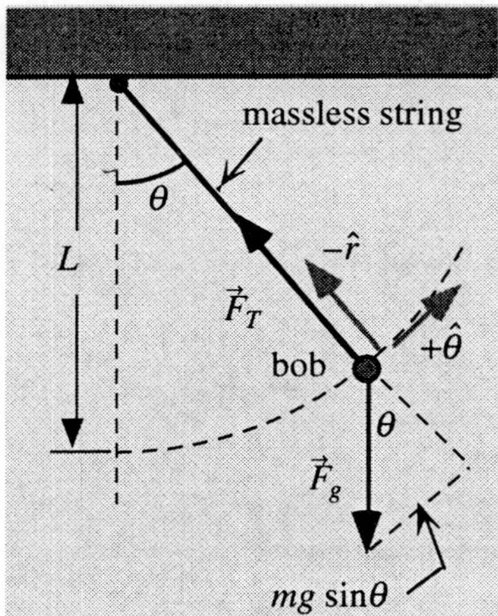

Figure N12.5: The simple pendulum.

A *simple pendulum* is an example of a system that has nothing to do with a mass connected to a spring, and yet (for small oscillations) obeys the same mathematics as a simple harmonic oscillator does. A **simple pendulum** consists of an object (called a **bob**) that swings back and forth at the end of a massless inextensible string of length L tied to a fixed point (see Figure N12.5).

The bob is confined to a circular path by the string, so this is an example of *nonuniform circular motion*. As discussed in chapter N9, if $v_\perp$ is the component of the bob's velocity in the direction of $\hat{\theta}$ (tangential to the circle in the counterclockwise direction), its acceleration at any instant in this situation is

$$\vec{a}(t) = \frac{dv_\perp}{dt}\hat{\theta} - \frac{v^2}{L}\hat{r} \tag{N12.30}$$

According to Newton's second law, this is equal to $\vec{F}_{\text{net}}/m$, where $\vec{F}_{\text{net}}$ is the net force on the bob and m is the bob's mass. The bob is acted on by two forces, a gravitational force due to the earth and the tension force exerted by its interaction with the string. The tension force always acts in the $-\hat{r}$ direction, so only the gravitational force ever has a component in the $\hat{\theta}$ direction. When the string makes an angle θ (defined to be positive counterclockwise) with the vertical, the component of $\vec{F}_g$ in the $\hat{\theta}$ direction is $F_{g,\hat{\theta}} = -mg\sin\theta$, so

$$\frac{dv_\perp}{dt} = \frac{F_{g,\hat{\theta}}}{m} = -g\sin\theta \tag{N12.31}$$

Since $L|d\theta|$ is the distance along the circle the bob travels when its angle changes by $d\theta$, $v_\perp = L d\theta/dt$. (Note that $v_\perp$ is positive when the bob is moving counterclockwise and negative when it is moving clockwise.) Plugging this into equation N12.31, we get

The pendulum equation

$$L\frac{d^2\theta}{dt^2} = -g\sin\theta \quad\Rightarrow\quad \frac{d^2\theta}{dt^2} = -\frac{g}{L}\sin\theta \tag{N12.32}$$

Now for small θ (less than a few tenths of a radian), $\sin\theta \approx \theta$. (Try this on a calculator and see for yourself!). Using this, we get

The low-angle limit is equal to oscillator equation

$$\frac{d^2\theta}{dt^2} = -\frac{g}{L}\theta \;\Rightarrow\; \frac{d^2\theta}{dt^2} = -\frac{g}{L}\theta \quad \text{(for small angles)} \tag{N12.33}$$

Now, compare this equation to equation N12.6: you should see that this equation is mathematically identical to that equation, with θ here playing the role of x there and $\sqrt{g/L}$ here playing the role of ω there. We don't need to solve this differential equation again: all that we need to do is use the answers that we found before, substituting θ for x and $\sqrt{g/L}$ for ω. So the angle of the pendulum bob depends on time as follows (by analogy to equation N12.8):

So the solution is analogous as well

$$\theta(t) \approx A\cos(\omega t + \phi) \quad \text{(for small angles)} \tag{N12.34}$$

where $\omega = \sqrt{g/L}$ here, A expresses the absolute value of the angle at the extreme points and ϕ is the phase constant (note that I am not using θ for this to avoid confusion with the pendulum's angle θ). Its period is

$$T = 2\pi/\omega = 2\pi\sqrt{L/g} \tag{N12.35}$$

Exercise N12X.9: Imagine that a pendulum hangs vertically at rest. We pull the bob aside to an initial angle of $+15°$ and release the bob from rest. It swings with a period of exactly 1.0 s. What are the values of A, ϕ and L here?

SUMMARY

I. A MASS CONNECTED TO A SPRING
   A. Mass m sliding on frictionless surface and connected to an ideal spring
      1. this is the archetype of a *harmonic oscillator*
      2. If x is the direction of motion and $x = 0$ is object's position when the spring is relaxed then we can show from the spring potential energy formula that $F_{Sp,x} = -k_s x$, where k_s is *spring constant*
   B. Any object moving in one dimension whose net x-force is given by $F_x = -k_s x$ is a *simple harmonic oscillator* (SHO)

II. SOLVING THE EQUATIONS OF MOTION
   A. The harmonic oscillator equation is $d^2x/dt^2 = -\omega^2 x$ (N12.6)
      1. $\omega = \sqrt{k_s/m}$ = *phase rate* (units of rad/s = s^{-1}) (N12.7)
      2. This *differential equation* cannot be solved by direct integration
      3. The general method for solving such equations is trial and error!
   B. In this case, the general solution is $x(t) = A\cos(\omega t + \theta)$ (N12.8)
      1. A is a constant with units of length called the *amplitude*
      2. θ (the *phase constant*) specifies where object is in its cycle at $t = 0$
      3. Both A and θ are determined by initial conditions
         [Generally: $\theta = \tan^{-1}(-v_{x0}/\omega x_0)$, $A = x_0/\cos\theta$ (N12.23,24)]
   C. The *period* of the oscillation is $T = 2\pi/\omega$; its *frequency* is $f = \omega/2\pi$.

III. IMPORTANCE OF THE SIMPLE HARMONIC OSCILLATOR
   A. *Any* oscillating object behaves like a SHO *if*
      1. it has a position (*equilibrium*) where net force on object is zero.
      2. it experiences a *restoring force* if it is displaced from equilibrium (it is a "restoring" force if the sign of F_x is opposite to that of x)
      3. $|F_x|$ *increases* as the object's displacement increase (at least for small displacements)
   B. For sufficiently small displacements
      1. almost any "restoring force" becomes approximately $F_x \approx -k_s x$
      2. thus almost any oscillating system will behave like a SHO.

IV. AN OBJECT HANGING FROM A SPRING
   A. Even though net force on object hanging from a spring includes gravity
      1. if we shift our origin so that $x = 0$ is where $F_{net,x} = 0$ (equilibrium)
      2. then the force law in this reference frame does reduce to $F_x = -k_s x$
   B. So an object oscillating vertically on a spring behaves exactly like a simple harmonic oscillator

V. THE SIMPLE PENDULUM
   A. *Simple pendulum*: an object (*bob*) suspended from a fixed point by a inextensible, massless string.
   B. Angle of pendulum $\theta(t)$ obeys $d^2\theta/dt^2 = -[g/L]\sin\theta$ (N12.32)
   C. This is not the harmonic oscillator equation
      1. but if θ is small, $\sin\theta \approx \theta$
      2. the equation of motion becomes $d^2\theta/dt^2 \approx -[g/L]\theta$ (N12.33)
      3. this is equivalent to the harmonic oscillator equation, with θ standing for x, g/L for ω^2
      4. So in low-angle limit, pendulum behaves like harmonic oscillator with $T = 2\pi/\omega = 2\pi\sqrt{L/g}$ (N12.35)
   D. This is an example of system that is *not* a SHO but nonetheless behaves like one for small oscillations

GLOSSARY

force law: an expression that tells how the force exerted on an object by its interaction with another object depends on the separation of the two objects. We can find a force law for any interaction that can be described by a potential energy function that depends on the objects' separation.

ideal spring: a massless spring that exerts a force exactly proportional to the distance that it is extended or compressed from its relaxed length.

harmonic oscillator force law (or **Hooke's law**): $F_x = -k_s x$, where x is the object's position in an appropriate coordinate system.

simple harmonic oscillator: any object moving in one dimension that experiences a net force given by the expression $F_x = -k_s x$, where x is the object's position in an appropriate coordinate system. An object that is connected to a fixed point by an ideal spring and which slides along a horizontal line on a frictionless surface is an example of a simple harmonic oscillator.

spring constant: the constant k in the harmonic oscillator force law. This constant has units of N/m, and expresses the strength of the effective spring in a simple harmonic oscillator.

differential equation: an equation that expresses a relationship between a function and its derivatives.

harmonic oscillator equation: $d^2x/dt^2 = -\omega^2 x$, where $\omega^2 = k_s/m$. This differential equation is the equation of motion for the object in a harmonic oscillator: its solution can be written $x(t) = A\cos(\omega t + \theta)$, where A and θ are constants determined by initial conditions.

amplitude: the constant A in $x(t) = A\cos(\omega t + \theta)$, equal to 1/2 the distance between the limits of an oscillation.

phase constant: the constant θ in the general oscillation solution $x(t) = A\cos(\omega t + \theta)$ that expresses how far along the oscillation is in its cycle at $t = 0$.

period T: the time required to complete one cycle of an oscillation.

frequency f: the number of cycles that an oscillating object completes in a unit time. Measured in **hertz** (where 1 Hz = 1 cycle/s).

phase rate ω: the rate at which the argument (sometimes called the *phase*) of the cosine in $x(t) = A\cos(\omega t + \theta)$ increases with time. One cycle of the oscillation corresponds to a phase change of 2π. The SI units of ω are rad/s = s^{-1}.

equilibrium position: the position of an oscillating object where it experiences zero net force.

restoring force: the force acting on an oscillating object that seeks to return it to its equilibrium position. If the object is moving along the x axis and $x = 0$ at the object's equilibrium position, then a restoring force will have x component $F_x < 0$ if $x > 0$, and $F_x > 0$ when $x < 0$.

simple pendulum: a massive object (called a pendulum **bob**) connected to a fixed point by a massless, inextensible string. A simple pendulum behaves like a harmonic oscillator if the angle of swing is very small.

TWO-MINUTE PROBLEMS

N12T.1 If you double the amplitude of a harmonic oscillator, the oscillator's period
A. decreases by a factor of 2
B. decreases by a factor of $\sqrt{2}$
C. does not change
D. increases by a factor of $\sqrt{2}$
E. increases by a factor of 2
F. other (specify)

N12T.2 If you double the amplitude of a harmonic oscillator, the object's maximum speed does what? (Use the answers for problem N12T.1.)

N12T.3 If you double the spring constant of a harmonic oscillator, the oscillation frequency does what? (Use the answers for problem N12T.1.)

N12T.4 A glider on an air track is connected by a spring to the end of the air track. If it takes 0.30 s for the glider to travel the distance of 12 cm from one turning point to the other, its amplitude is:
A. 12 cm
B. 6 cm
C. 24 cm
D. 36 cm
E. 3.6 cm
F. not enough information!

N12T.5 Consider the glider described in problem N12T.4. Its phase rate is
A. 0.30 s^{-1}
B. 0.15 s^{-1}
C. 0.60 s^{-1}
D. 3.77 s^{-1}
E. 0.096 s^{-1}
F. other (specify)

N12T.6 A glider on an air track is connected by a spring to the end of the air track. If it is pulled 3.5 cm in the $+x$ direction away from its equilibrium point and then released from rest at $t = 0$, what is the phase constant θ?
A. 0
B. $\pi/4$
C. $\pi/2$
D. π
E. $3\pi/2$
F. other (specify)

N12T.7 A glider on an air track is connected by a spring to the end of the air track. If it is pulled 3.5 cm in the $-x$ direction away from its equilibrium point and then released from rest at $t = 0$, what is the phase constant θ?
A. 0
B. $\pi/4$
C. $\pi/2$
D. π
E. $3\pi/2$
F. other (specify)

N12T.8 A mass hanging from the end of a spring has an phase rate of $\omega = 6.3\ s^{-1}$ ($\approx$ 1 cycle/s). Let's define $t = 0$ to be when the mass passes $x = 0$ going up. If its speed as it passes is 1.0 m/s, what is its amplitude A?
A. zero
B. 0.16 m
C. 1.0 m
D. 6.3 m
E. not enough information!
F. other (specify)

N12T.9 To double the period of a pendulum, you need to multiply its length by a factor
A. 1/2
B. 2
C. $\sqrt{1/2}$
D. $\sqrt{2}$
E. 4
F. other (specify)

HOMEWORK PROBLEMS

BASIC SKILLS

N12B.1 An oscillating object repeats its motion every 3.3 s. What is the period of this oscillation? What is its frequency? What is its phase rate?

N12B.2 An object of mass 0.30 kg hanging from a spring is observed to have an oscillation frequency of 2.2 Hz. What is the spring's spring constant k_s?

N12B.3 An object of mass 0.36 kg hanging at the end of a spring oscillates with an amplitude of 4.8 cm and a frequency of 1.2 Hz. What is the spring's spring constant k_s?

N12B.4 A magnesium atom (mass ≈ 24 proton masses) in a crystal is measured to oscillate with a frequency of roughly 10^{13} Hz. What is the effective spring constant of the forces holding that atom in the crystal?

N12B.5 An object of mass 0.30 kg hanging from a spring is pulled 2.5 cm below its equilibrium position and then is released from rest. What is θ for this oscillation in equation N12.8? Explain your reasoning.

N12B.6 A pendulum is observed to swing with a period of 2.0 s. How long is it?

N12S.7 What will be the natural oscillation period of a 30-kg child on a swing whose seat is 3.2 m below the bar where the chains from the seat are attached?

SYNTHETIC

N12S.1 An object of mass 0.25 kg extends a spring by a distance of 5.0 cm when it hangs from the spring at rest. If it is then set in vertical motion, what will be its period of oscillation?

N12S.2 An object of mass 0.30 kg hanging from a spring is lifted 2.5 cm above its equilibrium position and is dropped from rest. What is the amplitude of the subsequent oscillation? Explain your reasoning.

N12S.3 Any real spring has mass. Do you think that this mass would make the actual period of a real harmonic oscillator longer or shorter than the period predicted by equation N12.10? Explain your reasoning.

N12S.4 An object of mass 1.0 kg hanging from the end of a spring (whose spring constant is 120 N/m) is observed at $t = 0$ to pass downward through position $x = 2.5$ cm traveling at a speed of 0.60 m/s. **(a)** What is the amplitude of oscillation? **(b)** What is the phase constant θ for this oscillation? Please explain your reasoning.

N12S.5 An object of mass 0.60 kg hangs from the end of a spring. Imagine that the object is lifted upward and held at rest at the position where the spring is not stretched. The object is then released. It is observed that the lowest point in the object's subsequent oscillation is 12 cm below the point where it was released. **(a)** What is the amplitude of the oscillation? **(b)** What is the spring constant of the spring? **(c)** What is the object's maximum speed? Please explain your reasoning and show your work for each step.

N12S.6 When a 65-kg friend of yours sits on a trampoline, your friend sinks about 45 cm below the trampoline's normal level surface. If your friend were to bounce gently on the trampoline (never leaving its surface), what would be your friend's period of oscillation? (*Hint*: Model the trampoline as if it were a harmonic oscillator. Do you think that this will be a good model?)

N12S.7 If you stand on a pogo stick, you note that its spring-loaded foot is pushed in about 18 cm. If you bounce *gently* up and down on it (so that the foot never leaves the ground), what is your approximate oscillation frequency?

N12S.8 Do you think that a pendulum swinging through a large angle will have a longer or shorter period than the period predicted by equation N12.35? Explain your reasoning carefully. [Possibly helpful hints: consider extreme cases, or compare $\sin\theta$ to θ.]

N12S.9 The net x-force on an oscillating object moving in the x-direction is given by $F_x = -b(x - a^2x^3)$, where $b = 120$ and $a = 5$ in appropriate SI units. **(a)** What are the units of a and b? **(b)** Show that for small oscillations, this force law reduces to the harmonic oscillator force law $F_x \approx -k_s x$, where k_s depends on the value of a and/or b. **(c)** What is the frequency of such small oscillations? **(d)** How small is "small"? For what range of x will $F_x = -k_s x$ to within 1%?

N12S.10 Imagine that the x-force on an oscillating object in a certain situation is given by $F_x = -a\sin(bx)$, where $a = 100$ and $b = 10$ in appropriate SI units. **(a)** What are the SI units of a and b? **(b)** Show that for small x, this force law reduces to $F_x \approx -k_s x$, where k_s can be calculated from a and b. **(c)** Find the frequency of small oscillations.

N12S.11 We can actually determine the motion of a harmonic oscillator from *conservation of energy* instead of solving the harmonic oscillator equation. Here's how.
(a) Show that a bit of manipulation of the conservation of energy formula for the harmonic oscillator yields

$$A^2 = x^2(t) + [v_x(t)/\omega]^2 \qquad \text{(N12.36)}$$

where $A = 2E/k_s$ and $\omega = \sqrt{k_s/m}$ as usual.
(b) Argue that this means that at any instant of time, we can find an angle ψ such that

$$x(t) = A\cos\psi \quad \text{and} \quad v_x(t) = -\omega A\sin\psi \qquad \text{(N12.37)}$$

Thus these expressions give x and v_x for *all* times, with ψ depending in some unknown way on time. (*Hint*: $|x|$ will always be less than A, so we can *define* $x(t) = A\cos\psi$. Then solve equation N12.36 for the other term.)
(c) Use the definition $v_x = dx/dt$ and the chain rule to show that the most general possible expression for ψ is

$$\psi = \omega t + \theta \qquad \text{(N12.38)}$$

where θ is some constant. Therefore

$$x(t) = A\cos\psi = A\cos(\omega t + \theta) \;\; !! \qquad \text{(N12.39)}$$

RICH-CONTEXT

N12R.1 How do you measure the mass of an astronaut in orbit (you can't just use a scale!)? For the Skylab program, NASA engineers designed a Body Mass Measuring Device (BMMD). This is essentially a chair of mass m mounted on a spring with a carefully measured spring constant $k = 605.6$ N/m. (The other end of the spring is connected to the Skylab itself, which has a mass much larger than the astronaut or the chair, and so remains essentially fixed as the astronaut oscillates.) The period of oscillation of the empty chair is measured to be 0.90149 s. When an astronaut is sitting in the chair, the period is 2.12151 s. What is the mass of the astronaut? Please describe your reasoning! (Adapted from Halliday and Resnick, *Fundamentals of Physics*, 3/e, New York: Wiley, 1988, p. 324.)

N12R.2 Known mobster and gambler Larry the Loser is found dead hanging at about the 6-story level from a bungee cord tied to the top of the 14-story (42-m) Prudential Building in downtown Chicago. Larry has a mass of 90 kg, and when the bungee cord is cut down, it is found to have a relaxed length of 5 stories. Detective Lestrade from the Chicago Police Department thinks that Larry must have been bungee-jumping from the top of the building and just unluckily hit the ground, but *you* know that Larry was *murdered*. What is your evidence? [*Note*: a bungee cord will not exert any force on the jumper until the jumper has fallen a distance equal to the bungee cord's relaxed length. Also, the force a bungee cord exerts actually exceeds $k_s|x|$ in magnitude when it is stretched a lot.]

ADVANCED

N12A.1 Imagine that the force law expressing the interaction between a hydrogen atom and a chlorine atom in an HCl molecule is approximately

$$F_x = -a[(b/r)^2 - (c/r)^3] \qquad \text{(N12.40)}$$

where F_x is the x-force on the hydrogen atom, a is a constant having units of force and b and c are constants having units of distance and r is the distance measured from the chlorine atom (which is so massive compared to the hydrogen atom that we will consider it fixed). **(a)** What is the equilibrium position r_0 for the hydrogen atom (express your answer in terms of b and c)? **(b)** Define $x = r - r_0$, and show that for small x, this force law becomes approximately $F_x = -k_s x$. (One approach: find the first few terms of a Taylor-series expansion of the exact force law. Another approach: use the approximation $[1+q]^n \approx 1+nq$ if $q << 1$.) **(c)** What is the frequency of small oscillations of the hydrogen atom in terms of its mass m, and a, b, and c?

ANSWERS TO EXERCISES

N12X.1 The appropriate units for k_s in equation N12.3 are N/m. Since 1 J = 1 N·m, this is the same as J/m².
N12X.2 100 N/m.
N12X.3 If we define $u = \omega t + \theta$, then $x(t) = A\cos u$, and

$$\frac{d}{dt}A\cos u = -A\sin u\frac{du}{dt} = -\omega A\sin u, \qquad \text{(N12.41)}$$

$$\text{since } \frac{du}{dt} = \frac{d}{dt}(\omega t + \theta) = \omega + 0 = \omega \qquad \text{(N12.42)}$$

Taking the time derivative again we get

$$\frac{d^2x}{dt^2} = -\omega A\frac{d}{dt}\sin u = -\omega^2 A\cos u = -\omega^2 x \qquad \text{(N12.43)}$$

So this solution does indeed solve the SHO equation.
N12X.4 The quantity k_s/m has the units

$$\text{units}\left(\frac{k_s}{m}\right) = \frac{\cancel{\text{N}}/\cancel{\text{m}}}{\cancel{\text{kg}}}\left(\frac{1\,\cancel{\text{kg}}\cdot\cancel{\text{m}}/\text{s}^2}{1\,\cancel{\text{N}}}\right) = \frac{1}{\text{s}^2} \qquad \text{(N12.44)}$$

So $\sqrt{k_s/m}$ indeed does have units of s⁻¹.
N12X.5 The amplitude is half the total distance between extremes or 5 cm. The period is $T = 2\pi\sqrt{m/k_s} = 0.90$ s. Its phase rate is $\sqrt{k_s/m} = 7.1$ s⁻¹.
N12X.6 Multiplying out the square, we find that

$$F_x = a(x^2 - 2bx + b^2 - b^2) = ax^2 - 2abx \qquad \text{(N12.45)}$$

For $x << b$, the first term will be very small compared to the second, and we can neglect it, leaving

$$F_x \approx -2abx = -k_s x, \quad \text{where } k_s = 2ab \qquad \text{(N12.46)}$$

N12X.7 Using equation N12.23 and N12.24, we get

$$\theta = \tan^{-1}\left(\frac{-v_x(0)}{\omega x(0)}\right) = \tan^{-1}\left(\frac{-(-0.5\,\cancel{\text{m}}/\cancel{\text{s}})}{(10\,\cancel{\text{s}}^{-1})(-0.1\,\cancel{\text{m}})}\right)$$
$$= \tan^{-1}(-0.5) = -0.46 \qquad \text{(N12.47)}$$

As a fraction of an oscillation, this result corresponds to shifting the cosine wave by $0.46/2\pi = 0.073 = 7.3\%$ of a complete oscillation. But the result for the phase shift θ could *also* be $\pi - 0.46 = 2.68$ (43% of an oscillation), since $\tan\theta = \tan(\theta+\pi)$. It is fact the latter value that gives the correct positive value for the amplitude:

$$A = \frac{x(0)}{\cos\theta} = \frac{-0.1\text{ m}}{\cos(2.68)} = +0.11\text{ m} \qquad \text{(N12.48)}$$

N12X.8 Since $T = 2\pi\sqrt{m/k_s}$, we have

$$k_s = \frac{4\pi^2 m}{T^2} = \frac{4\pi^2(68\,\cancel{\text{kg}})}{(6.0\,\cancel{\text{s}})^2}\left(\frac{1\text{ N}}{1\,\cancel{\text{kg}}\cdot\text{m}/\cancel{\text{s}^2}}\right) = 75\,\frac{\text{N}}{\text{m}} \qquad \text{(N12.49)}$$

N12X.9 Since $0 = v_x(0) = Ld\theta/dt = -LA\sin(0+\phi) = -LA\sin\phi$, ϕ must be 0 or π. Only the first give $\cos\phi > 0$, which we need since $\theta(0) = A\cos\phi = +15°$, so we must have $\phi = 0$ and $A = +15°/\cos(0) = +15° = \pi/6$ radians. Since for the pendulum $\omega = \sqrt{g/L}$, we have

$$T = \frac{2\pi}{\omega} = 2\pi\sqrt{\frac{L}{g}} \Rightarrow L = \frac{T^2 g}{4\pi^2} \qquad \text{(N12.50)}$$

Plugging in the numbers we get $L = 0.25$ m.

N13

INTRODUCTION TO ORBITS

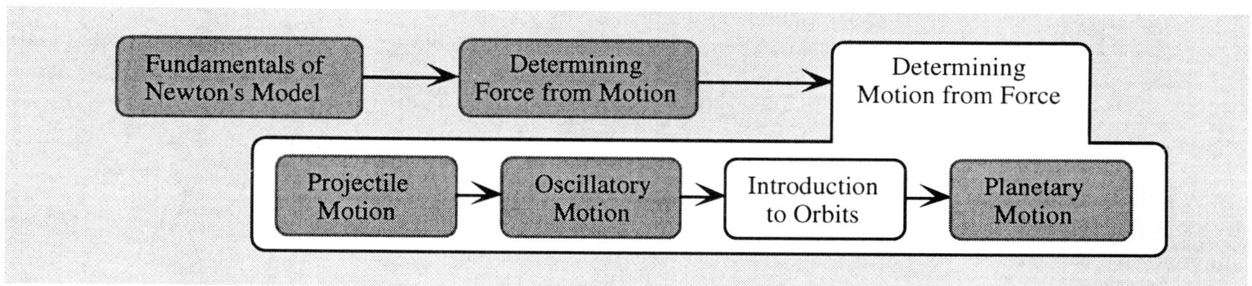

N13.1 OVERVIEW

In the previous chapters of this unit, we have been exploring the application of Newton's laws of motion to a variety of essentially terrestrial situations. Here at the end of the unit, we will finally turn to the problem of celestial physics. In this chapter and the next, we will explore how to use what we learned about the gravitational force in chapter N13 to predict the motion of the planets and other celestial objects. We will see that just as Newton's laws accurately describe the motion of terrestrial objects, they also accurately predict the motion of the planets. Newton's laws thus provide a *universal* model for (macroscopic) mechanics.

In this chapter, we will discuss some of the history of our understanding of planetary motion and lay some foundations for the last chapter, where we will explore Newton's predictions regarding planetary motion in depth. Here is an overview of the sections in this chapter.

N13.2 *KEPLER'S LAWS.* Fifty years before Newton first studied the problem of planetary motion, Johannes Kepler stated three empirical laws that seemed to describe the motion of the planets (assuming the heliocentric model of the solar system) according to the best data available. Any theory of planetary motion, therefore, has to explain these laws. This section discusses Kepler's Laws in their historical context.

N13.3 *ORBITS AROUND A MASSIVE PRIMARY* discusses the simplifications that result if one object in an interacting pair is much more massive than the other.

N13.4 *KEPLER'S SECOND LAW* shows that Kepler's second law follows directly from the law of conservation of angular momentum.

N13.5 *CIRCULAR ORBITS AND KEPLER'S THIRD LAW.* The orbits of many celestial objects are nearly circular. This section explores what Newton's laws say about orbits that we know are nearly circular, and proves Kepler's third law for circular orbits.

N13.6 *CIRCULAR ORBIT PROBLEMS* shows how to solve a circular orbit problems using a variety of examples.

N13.7 *BLACK HOLES AND DARK MATTER* shows how physicists have recently used Kepler's third law to amass evidence for the existence of giant black holes and unseen dark matter.

N13.2 KEPLER'S LAWS

In the year 1600, Johannes Kepler came to Prague to join the research staff at an observatory operated by Tycho Brahe, an astronomer who was both the official "imperial mathematician" of the Holy Roman Empire and a friend of Galileo. The following year, Brahe died, and Kepler succeeded him as "imperial mathematician" and as director of the observatory. His new position gave him complete access to Tycho Brahe's extraordinary collection of careful astronomical observations of the planets, the result of a lifetime of work by perhaps the greatest naked-eye astronomer who ever lived. (The telescope was not invented until about 1610.)

Kepler's laws

In 1609, Kepler published a work entitled *Astronomia Nova* (New Astronomy) in which he stated two empirical laws that seemed to be consistent with Brahe's planetary observations. In modern language, these laws state that

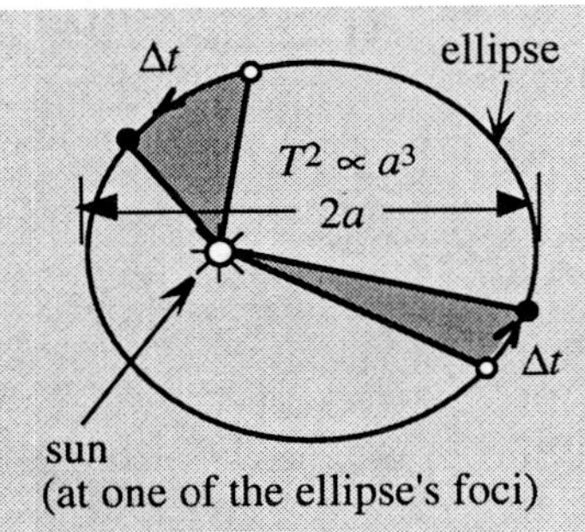

Figure N13.1: Kepler's laws illustrated. The gray regions show the area swept out by a line connecting the planet and the sun during equal time intervals Δt during different parts of the orbit: their areas are equal by Kepler's second law.

1. The orbits of the planets are *ellipses*, with the sun at one focus.
2. The line from the sun to a planet sweeps out *equal areas in equal times*.

(I'll define the *focus* of an ellipse in the next chapter.) Kepler offered no theoretical explanation for these laws: he simply presented them as being *descriptive* of planetary orbits, according to Brahe's observational data. These laws represented a rather radical departure from the accepted wisdom of the time: up to then, it was assumed by most astronomers that the motions of the planets could be described in terms of combinations of uniform circular motions (though it was becoming clear that such schemes had to be extraordinarily complex to fit the best observational data available).

Ten years later, Kepler published in his *Harmonice Mundi* (Harmonics of the World) a third empirical law:

3. The square of a planet's period T (the time that it takes to complete one orbit) is proportional to the cube of the semimajor axis a of its orbit.

(The **semimajor axis** of an ellipse is defined to be half of the width measured across the widest part of the ellipse.) The three laws stated above are known as **Kepler's Three Laws of Planetary Motion.**

Kepler's laws are empirical

Again let me emphasize that these laws are entirely empirical: they do not so much explain as *describe* the motion of the planets. However, because Kepler supported them so carefully with observational data of extraordinary quality, these laws became widely known and accepted in spite of their radical character. Newton's triumph was to *explain* these laws.

Newton offered an explanation of these laws

For more than six decades after the publication of the last of these laws, the scientific community was unable to say anything about why these laws should be true. Isaac Newton's incredible triumph was to show that each of these laws follow directly from his second law and the assumption that the force of gravity between two objects depends on the inverse square of the distance separating them. In other words, Newton offered an *explanation* of Kepler's laws in terms of physical principles that applied equally well to terrestrial motion: one simply had to accept that the planets were endlessly falling around the sun.

Let me emphasize how radical *this* suggestion was at the time! Before Newton, scholars had believed that the laws of physics pertaining to the motion of heavenly bodies (where unceasing motion in approximate circles seemed to be the rule) were completely distinct from the laws pertaining to terrestrial motion (where objects generally come to rest rather quickly). It took Newton's genius not only to see that this very credible division between celestial and terrestrial physics was in fact not necessary at all (which, granted, was *beginning* to occur to others as well), but also to provide a complete theoretical perspective that unified terrestrial and celestial physics in a manner that demonstrably *worked*. The simplicity, beauty, and extraordinary predictive power of Newton's ideas were so compelling that it brought the physics community to its first real consensus on

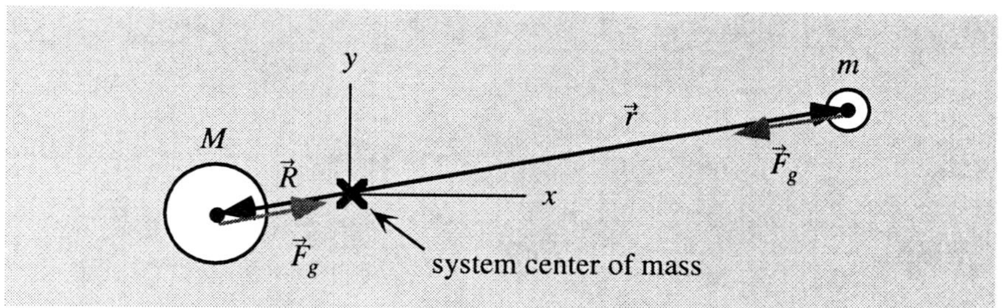

Figure N13.2: An isolated system of two objects interacting gravitationally. Note that if we define the origin of our reference frame to be attached to the system's center of mass, the positions of the objects are opposite. The gravitational force exerted on each object by the interaction points directly toward the other.

a grand theoretical structure for physics. This first consensus (as discussed in chapter C1) in some sense marks the birth of physics as a scientific discipline.

Our goal: prove Kepler's laws follow from principles we have studied

Our goal in this chapter and the next is to prove that Kepler's laws are a consequence of the laws of mechanics that we have been studying. We will not quite follow the same path that Newton did in proving this: the laws of conservation of energy and angular momentum give us more powerful tools than even Newton had at his disposal. (Using these laws will enable us to do in a few pages what it took Newton scores of pages to show in the *Principia*.)

N13.3 ORBITS AROUND A MASSIVE PRIMARY

The first step toward understanding what Newton's laws say about planetary motion is to take advantage of the simplifications that result when (1) our system of interest consists of a very massive object interacting with a much lighter object, and (2) we choose the origin of our reference frame to be the system's center of mass.

General situation: isolated pair of interacting objects

Consider an isolated system consisting of an object of mass M interacting with a smaller object of mass m (see Figure N13.2). If the system is *really* isolated, then its center of mass will move at a constant velocity and thus can be used as the origin of an inertial reference frame. In practical situations, it actually is more likely that the system is *freely falling* in some external gravitational field (for example, the earth and moon falling around the sun), but we can *still* treat the system's center of mass as the origin of an inertial reference frame if we ignore the external gravitational field (as we saw in chapter N10).

Results in a frame based on the system's CM

If we define the center of mass of the system to be the origin, then the definition of the center of mass means that

$$0 = \frac{(m\vec{r}+M\vec{R})}{M+m} \quad\Rightarrow\quad M\vec{R} = -m\vec{r} \quad\Rightarrow\quad \vec{R} = -\frac{m}{M}\vec{r} \qquad \text{(N13.1)}$$

This implies that in *all* circumstances (no matter how the two objects move and/or interact with each other) the positions $\vec{R}$ and $\vec{r}$ of the objects relative to the center of mass are opposite and have magnitudes that are strictly proportional to each other: if r gets bigger or smaller than so does R (proportionally).

Taking the time derivative of both sides of equation N13.1, we get

$$\vec{V} = -\frac{m}{M}\vec{v} \quad\Rightarrow\quad V = \frac{m}{M}v \qquad \text{(N13.2)}$$

This means that

$$\tfrac{1}{2}MV^2 = \tfrac{1}{2}M\left(\frac{m}{M}v\right)^2 = \left(\frac{m}{M}\right)\tfrac{1}{2}mv^2 \quad\Rightarrow\quad K_M = \frac{m}{M}K_m \qquad \text{(N13.3)}$$

where K_M and K_m are the kinetic energies of the massive and light objects respectively. Thus (as we've seen before), the kinetic energy of the more massive object in this frame is smaller than that of the lighter object by the factor m/M.

We can also express the angular momentum $\vec{L}_M$ of the larger object around the system's center of mass in terms of $\vec{L}_m$ as follows

$$\vec{L}_M \equiv \vec{R} \times M\vec{V} = \left(-\frac{m}{M}\right)\vec{r} \times M\left(-\frac{m}{M}\right)\vec{v} = +\frac{m}{M}(\vec{r} \times m\vec{v}) = \frac{m}{M}\vec{L}_m \quad \text{(N13.4)}$$

The system's total angular momentum around the center of mass is thus

$$\vec{L}_{\text{tot}} = \vec{L}_M + \vec{L}_m = \left(\frac{m}{M}+1\right)\vec{L}_m \quad \text{(N13.5)}$$

The system's total angular momentum around its center of mass will be conserved *if and only if* the light object's angular momentum is conserved around the center of mass.

Implications when $M >> m$

Equations N13.1 through N13.5 apply to *any* isolated (or freely falling) system of two interacting objects described in a reference frame whose origin is the system's center of mass. If in addition, we have $M >> m$, then

1. The position of the massive object is $\vec{R} \approx 0$ (by N13.1)
2. This object is essentially at *rest* at the origin: $\vec{V} \approx 0$ (by N13.2)
3. Its kinetic energy is negligible: $K_M \approx 0$ (by N13.3)
4. Its angular momentum is negligible: $\vec{L}_M \approx 0$ (by N13.4)
5. The objects' separation ≈ distance of lighter object from the origin

In such a case, the massive object (which we call the **primary** of this system under these circumstances) is essentially at rest at the origin, and the lighter object (which we call a **satellite**) orbits it. The primary then provides an essentially fixed origin for the gravitational force exerted on the satellite.

This approximation is very useful in the solar system

This approximation holds very well in the solar system. Even Jupiter's mass is more than 1000 times smaller than the sun's mass, and the earth's mass is more like 330,000 times smaller. Similarly, the moons that orbit the major planets typically have masses much smaller than their primary: even our own moon, which is the second largest moon in the solar system compared to its primary (after the Pluto/Charon system), has 81 times less mass than the earth.

The fact that planetary masses are so small compared to the sun has another important implication. When computing the orbit of one planet, we can ignore the gravitational effects of the others (to an excellent degree of approximation): the sun is so much more massive than anything else that the gravitational force that it exerts is by far the greatest influence on each planet's motion. Therefore we can pretend in our analysis that each planet orbits the sun as if it were alone.

Exercise N13X.1 The radius of the earth's orbit is about 1.5×10^{11} m on the average. What is the distance between the sun and the center of mass of the earth/sun system? How does this compare to the sun's radius (700,000 km)?

N13.4 KEPLER'S SECOND LAW

Equations N13.4 and N13.5 imply that the angular momentum of *either* object in an isolated interacting pair is *separately* conserved. This has two important consequences: (1) the object's orbit lies in a plane, and (2) its position vector sweeps out equal areas in equal times.

Orbit of either object must lie in fixed plane in space

The first of these consequences follows from the definition of the angular momentum, which says that $\vec{L}$ for either object is defined to be $\vec{L} \equiv \vec{r} \times m\vec{v}$. By definition of the cross product, this means that $\vec{L}$ is always perpendicular to $\vec{r}$. But if $\vec{L}$ has a fixed orientation in space (because it is conserved) then the object's position vector $\vec{r}$ must always lie in the fixed plane perpendicular to $\vec{L}$. Therefore the object's orbit lies in a certain fixed plane.

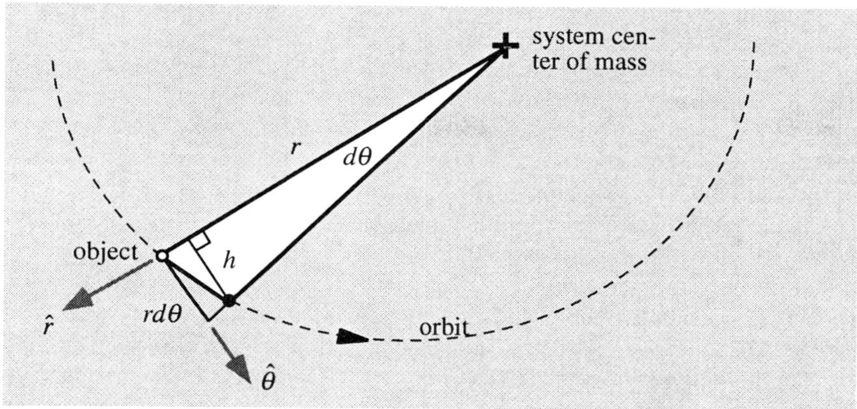

Figure N13.3: The area swept out by the line between the object and the system's center of mass as the object moves a tiny angle $d\theta$ around the center of mass is shown in white (the object's initial position is the white dot and its final position is the black dot). This area is almost equal to $\frac{1}{2}rh$, and h in turn is approximately equal to the arclength $r\,d\theta$.

Position vector of either object sweeps out equal areas in equal times

Kepler's second law follows from the conservation of angular momentum as follows. Imagine that in an infinitesimal time dt, the object moves a certain infinitesimal angle $d\theta$ (shown greatly exaggerated in Figure N13.3) as it moves in its orbit around the system's center of mass. The area swept out by the line between the object and the system's center of mass is the white pie-slice in Figure N13.3. If $d\theta$ is very small, the shape of the slice is very nearly triangular, so its area is $dA \approx \frac{1}{2}$(base)·(height). Now, as shown on the drawing, the height h of the triangle is very nearly equal to the arclength $r\,d\theta$, and this approximation gets better and better as $d\theta \to 0$. So if $d\theta$ is small,

$$dA \approx \tfrac{1}{2}r(r\,d\theta) = \tfrac{1}{2}r^2\,d\theta \qquad \text{(N13.6)}$$

Dividing both sides by dt and taking the limit as dt and $d\theta$ go to zero, we get

$$\frac{dA}{dt} = \tfrac{1}{2}r^2\frac{d\theta}{dt} \qquad \text{(N13.7)}$$

But according to equation C13.8, $L = \text{mag}(\vec{L})$ for the object is

$$L = mr^2\omega = mr^2\frac{d\theta}{dt} \qquad \text{(N13.8)}$$

This means that

$$\frac{dA}{dt} = \frac{1}{2}r^2\frac{d\theta}{dt} = \frac{1}{2m}\left(mr^2\frac{d\theta}{dt}\right) = \frac{L}{2m} \qquad \text{(N13.9)}$$

Therefore, since L is conserved, dA/dt is constant: *the object's radius vector sweeps out equal areas in equal times*. Note that this applies to *either* object, independent of the objects' relative masses or the nature of their interaction.

Application to Kepler's second law

Now, Kepler's second law actually says that the line between the planet and the *sun* sweeps out equal areas in equal times, not the line between the planet and the system's *center of mass*. But if the sun is much more massive than the planet, it essentially *is* located at the system's center of mass, and Kepler's second law is essentially correct.

We can express $d\theta/dt$ as a function of L, m, and r

For future reference, note that because the planet's angular momentum is conserved, its phase speed at any time can be expressed as a function of its radius r at that time: solving equation N13.8 for $d\theta/dt$, we get

$$\frac{d\theta}{dt} = \frac{L}{mr^2} \qquad \text{(N13.10)}$$

We will use this result in the next chapter.

N13.5 CIRCULAR ORBITS AND KEPLER'S THIRD LAW

We will assume $M >> m$ from now to the end of unit

From now on through the rest of the unit, we will assume the primary-satellite approximation, where the mass of the primary is much greater than the mass of the satellite. To simplify notation, when we refer to the satellite's kinetic energy and/or angular momentum from now on, we will drop the subscripts on K_m and $\vec{L}_m$ and simply use K and $\vec{L}$.

Circular orbits at constant speed are possible

If the primary is much more massive than its satellite, then it is possible for the satellite to follow an essentially circular orbit at constant speed *around the primary*. Let us see whether such an orbit is consistent with what we know about uniform circular motion and the gravitational interaction.

We know that an object moving in a circular trajectory at a constant speed is accelerating toward the center of its circular path, and that the magnitude of its acceleration is $a = v^2/R$ = a *constant*, where R is the radius of the object's orbit and v is its constant orbital speed. Newton's second law tells us that this acceleration must be caused by some force that is directed toward the center of the satellite orbit and which has a *constant* magnitude.

The gravitational force $\vec{F}_g$ exerted on the satellite by its gravitational interaction with its primary satisfies these criteria. It is directed toward the center of the satellite's trajectory (to the extent that we can consider the massive object to be at rest). According to equation N3.18, its magnitude is given by Newton's **law of universal gravitation** $F_g = GMm/r^2$ (where r is the distance between the centers of the satellite and primary and m and M are their masses): we derived this in section N3 from the gravitational potential energy formula. Therefore, in the case of a truly circular orbit (where $r = R$ = constant) the magnitude of the force will be GMm/R^2 = a constant, as needed.

This means that it is at least *plausible* that the gravitational force exerted on the satellite due to its interaction with the primary can hold the satellite in a circular orbit. This does not mean that orbits *always* have to be circular: indeed, orbits generally are *not* circular (as we will see in the next chapter). But it does mean that a circular orbit is a possibility. In fact the orbital radii of most major objects in the solar system are constant to within a few percent. Most artificial satellites orbit the earth with approximately circular orbits as well. Therefore, this "special case" of all the kinds of orbits possible is in fact approximately applicable to a wide variety of realistic situations.

Implications of Newton's second law for circular orbits

Let's see what Newton's second law can tell us quantitatively about such orbits. Assuming that the gravitational force is the *only* force acting on the satellite, then $F_{net} = F_g = GMm/R^2$. Therefore, the magnitude of Newton's second law tells us that

$$ma = F_{net} = \frac{GMm}{R^2} \tag{N13.11}$$

Dividing both sides by m and plugging $a = v^2/R$ into this equation, we find that:

$$\frac{v^2}{R} = \frac{GM}{R^2} \quad \Rightarrow \quad v = \sqrt{\frac{GM}{R}} \tag{N13.12}$$

This equation expresses the orbital speed of the satellite in a circular orbit in terms of the mass of the primary and the radius of the orbit. (Note that this is suitably constant.)

We can use this information to determine how long it will take the satellite to go once around its circular orbit. Let us refer to the time required for the satellite to complete one orbit as being the **period** of the orbit T. Since T is the time it takes the satellite to travel a distance $2\pi R$ at a constant speed of v,

$$T = \frac{2\pi R}{v} \tag{N13.13}$$

If you square equation N13.13 and plug in equation N13.12, you can show that

$$T^2 = \frac{4\pi^2}{GM} R^3 \qquad \text{(N13.14)}$$

Exercise N13X.2: Verify equation N13.14.

Kepler's third law for circular orbits

Kepler's third law states that *the square of a planet's period is proportional to the cube of its semimajor axis.* When an orbit is circular, its semimajor axis (half of the distance measured across the widest part of the orbit) *is* its radius R. So we see here that in a few lines we have *derived* Kepler's third law (for the special case of circular orbits, anyway). Moreover, this derivation even gives us the constant of proportionality between T^2 and R^3!

Exercise N13X.3: The earth orbits the sun in an approximately circular orbit once a year. If it were four times as far from the sun, how long would it take to orbit once?

N13.6 CIRCULAR ORBIT PROBLEMS

We can use equations N13.12 and N13.14 to answer many questions about orbiting objects and their primaries. In this section, we'll explore some examples of how the equations derived in the previous section can be applied to problems involving nearly circular orbital motion in the solar system. Since these examples involve circularly "constrained" motion (though only because the orbits *happen* to be roughly circular, not because they are required to be), we can easily adapt the framework we used for constrained motion problems for use with these these problems. (To save space, though, the examples omit the framework outline in the margin.)

Differences between circular orbit and general constrained motion problems

Doing circular orbit problems is much like doing the constrained motion problems we did in chapter N9, with some important modifications. For example, we do not need to use Newton's laws in component form (equations N13.12 or N13.14 will probably be more helpful), and this means that we don't really need to define reference frame axes. (We will assume that the origin of our reference frame is at the center of the massive primary.) A force diagram of an orbiting object is also optional (since it would almost always show only one force vector). The important issues to be addressed in a *conceptual representation* section are the validity of the circular orbit approximation and the assumption that the primary is indeed much more massive than the satellite.

A plan for solving circular orbit problems

So a terse outline of our adaptation of the constrained motion framework for orbital motion problems might look as follows

A. *pictorial representation*
   1. draw a sketch of the situation
   2. label it with appropriate symbols
   3. list knowns and unknowns

B. *conceptual representation*
   1. describe the two interacting objects
   2. make sure that one is very massive
   3. and that the other's orbit is circular (or is assumed to be circular)

C. *math representation:* apply equations N13.12 or N13.14

D. *evaluation:* Check result for correct sign, correct units and a plausible magnitude (as usual)

The examples on the following pages illustrate solutions following this format as well as some of the many applications of equations N13.12 through N13.14.

EXAMPLE N13.1
Time for a Shuttle Orbit

Problem: Imagine that the space shuttle *Atlantis* is in a circular orbit at an altitude of 250 km above the Earth's surface. What is its orbital speed? What is its orbital period?

Solution Figure N13.4 shows a sketch and a list of knowns and unknowns.

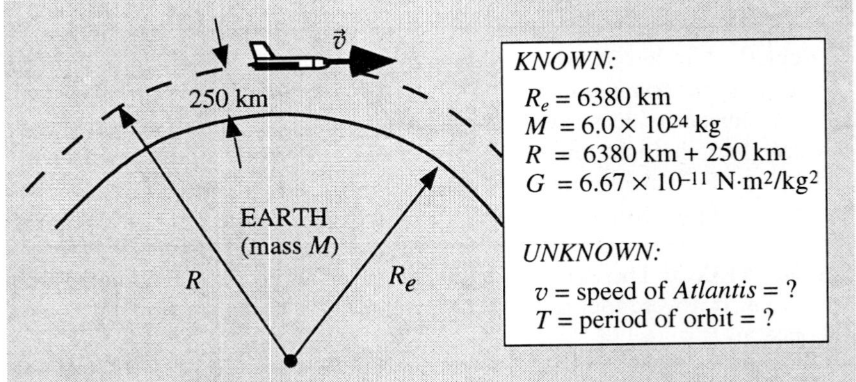

Figure N13.4: The *pictorial representation* section of the solution to the space shuttle problem.

According to the problem statement, the *Atlantis* is following a circular orbit around the earth, and the earth is much more massive than the *Atlantis*, so equations N13.12 and N13.14 should apply. Equation N13.12 tells us that the speed of a circularly orbiting object is:

$$v = \sqrt{\frac{GM}{R}} = \sqrt{\frac{(6.67\times10^{-11}\ \text{N}\cdot\text{m}^2/\text{kg}^2)(6.0\times10^{24}\ \text{kg})}{(6{,}380{,}000\ \text{m} + 250{,}000\ \text{m})}\left(\frac{1\ \text{kg}\cdot\text{m/s}^2}{1\ \text{N}}\right)}$$

$$= 7{,}770\ \text{m/s} = 7.8\ \text{km/s} \tag{N13.15}$$

Once we have the speed, it is easier to use equation N13.13 rather than equation N13.14. The orbit's period is thus

$$T = \frac{2\pi R}{v} = \frac{2\pi(6380\ \text{km} + 250\ \text{km})}{7.77\ \text{km/s}}\left(\frac{1\ \text{min}}{60\ \text{s}}\right) = 89.4\ \text{min} \tag{N13.16}$$

Note that this is a little bit longer than the result we found in Example N9.5.3. This is plausible: since the shuttle orbits some distance above the earth's surface, the force of gravity on the orbiting shuttle will be a bit smaller than it would be on the earth's surface. This means that the shuttle must move a bit more slowly if its acceleration is to match the reduced force.

EXAMPLE N13.2
How to Weigh Jupiter

Problem: Ganymede, the largest moon of Jupiter, has a nearly circular orbit whose radius is 1.07 Gm. Measurements show that the moon goes around Jupiter once every 7 days, 3 hours, and 43 minutes. What is Jupiter's mass?

Solution Figure N13.5 shows a sketch and lists knowns and unknowns.

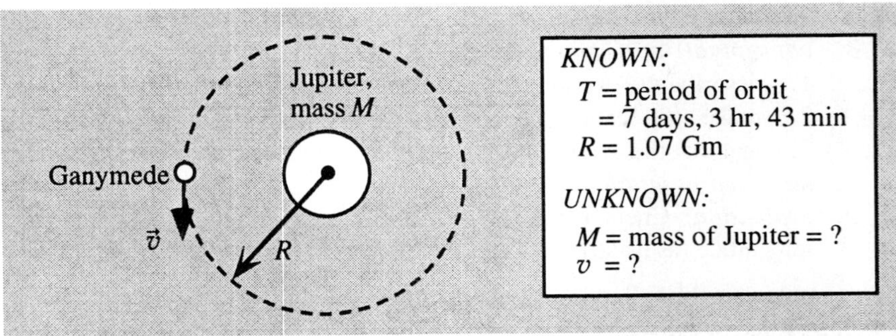

Figure N13.5: The *pictorial representation* section of the solution for the Ganymede problem.

EXAMPLE N13.2 (continued)

The interacting objects here are Ganymede and Jupiter. We are told that Ganymede's orbit is circular and given its radius R and period T, and we can safely assume that the Ganymede is much less massive than Jupiter. Since we know R and T, we can solve equation N13.14 for the mass M of Jupiter, as follows

$$T^2 = \frac{4\pi^2}{GM}R^3 \quad \Rightarrow \quad M = \frac{4\pi^2 R^3}{GT^2} \tag{N13.17}$$

The period T expressed in seconds is

$$\begin{aligned} T &= 7\,\text{d}\left(\frac{24\,\text{h}}{1\,\text{d}}\right)\left(\frac{3600\,\text{s}}{1\,\text{h}}\right) + 3\,\text{h}\left(\frac{3600\,\text{s}}{1\,\text{h}}\right) + 43\,\text{min}\left(\frac{60\,\text{s}}{1\,\text{min}}\right) \\ &= 6.18 \times 10^5\,\text{s} \end{aligned} \tag{N13.18}$$

Plugging this into equation N13.17, we get

$$\begin{aligned} M &= \frac{4\pi^2(1.07\times 10^9\,\text{m})^3}{(6.67\times 10^{-11}\,\text{N}\cdot\text{m}^2/\text{kg}^2)(6.18\times 10^5\,\text{s})^2}\left(\frac{1\,\text{N}}{1\,\text{kg}\cdot\text{m/s}^2}\right) \\ &= 1.9 \times 10^{27}\,\text{kg} \end{aligned} \tag{N13.19}$$

Note that we can determine the mass of an object by observing the motion of a smaller object orbiting it (this is in fact the standard method of determining the mass of astronomical objects).

EXAMPLE N13.3

Problem: Astronomical measurements show that Jupiter orbits the sun in a nearly circular orbit once every 11.86 years. How does the radius of Jupiter's orbit compare with that of the earth?

Solution Figure N13.6 shows a sketch and list of knowns and unknowns.

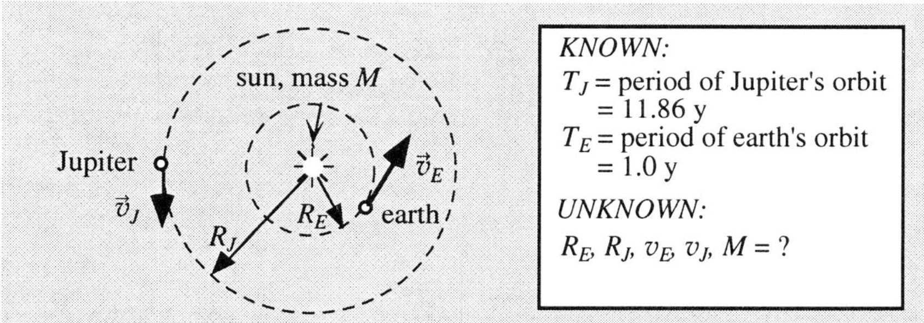

Figure N13.6: The *pictorial representation* section of the solution for the Jupiter problem.

In this problem, Jupiter and earth are both orbiting the sun, which is much more massive than either planet. We are ignoring the comparatively tiny interaction between earth and Jupiter. To do this problem, we have to *assume* that both the earth's orbit and Jupiter's orbit are circular (which is pretty closely true). If this is so, then equation N13.14 applies to both orbits, so

$$T_J^2 = \frac{4\pi^2}{GM}R_J^3 \quad \text{and} \quad T_E^2 = \frac{4\pi^2}{GM}R_E^3 \tag{N13.20}$$

Since we are really looking for the ratio R_J/R_E, the fastest way to find that ratio is to divide the first expression by the second:

$$\frac{T_J^2}{T_E^2} = \frac{(4\pi^2/GM)R_J^3}{(4\pi^2/GM)R_E^3} = \frac{R_J^3}{R_E^3} \quad \Rightarrow \quad \frac{R_J}{R_E} = \left(\frac{T_J}{T_E}\right)^{2/3} = \left(\frac{11.86\,\text{y}}{1.0\,\text{y}}\right)^{2/3} \tag{N13.21}$$

So $R_J = (11.86)^{2/3}R_E = 5.20R_E$, a result consistent with other measurements.

EXAMPLE N13.4
The Rutherford Atom

Problem: In 1911, Ernest Rutherford proposed (on the basis of certain scattering experiments performed by himself and his graduate students) that all atoms consist of a tiny, massive, and positively charged nucleus surrounded by lightweight orbiting electrons. Each electron was assumed to be held in its orbit by the force of electrostatic attraction (which has a magnitude of $F_e = k\,|q_1 q_2|\,/\,r^2$ according to problem N11S.1) between the electron's negative charge $-e$ (where $e = 1.6 \times 10^{-19}$ C) and the nucleus' positive charge. We now know that this model is a tremendous oversimplification, but it was an important step toward understanding the structure of atoms.

Let's consider hydrogen, the simplest possible atom. In this model a hydrogen atom would consist of a single electron orbiting a single proton. The proton has a mass about 1836 times greater than that of the electron and has a charge of $+e$. For the sake of argument, let's assume that the electron's orbit is approximately circular and has a radius equal to the atom's measured radius of about 0.54 Å, or 5.4×10^{-11} m. What is the approximate orbital speed of such an electron according to this model?

Solution Figure N13.7 shows a sketch of the situation and a list of knowns (using the values of k and m on the inside front cover) and unknowns.

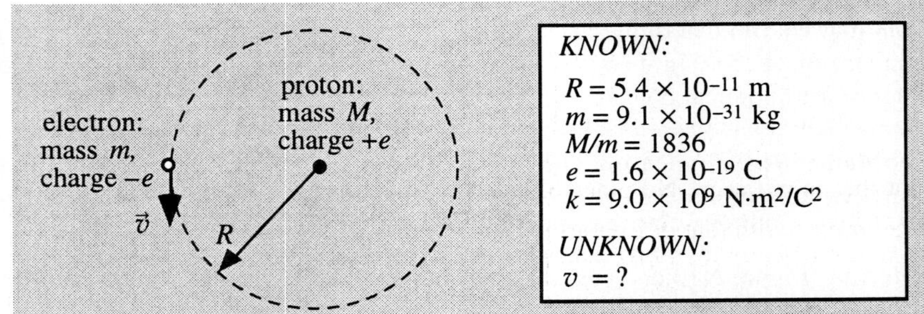

Figure N13.7: The *pictorial representation* section of the solution for the Rutherford atom.

Since the proton is so much more massive than the electron, we can consider the proton to be at rest at the system's center of mass. In this case, the force that causes the electron's acceleration in its circular orbit is not a gravitational force but an electrostatic attraction force (whose magnitude is $F_e = ke^2/R^2$ in this case). Newton's second law for this electron thus implies that (since the electron's acceleration in its circular orbit is $a = v^2/R$)

$$m\frac{v^2}{R} = F_{\text{net}} = F_e = \frac{ke^2}{R^2} \qquad \text{(N13.22)}$$

Solving this for v, we get

$$v = \sqrt{\frac{ke^2}{mR}} = \sqrt{\frac{(9.0\times10^9\ \cancel{\text{N}}\cdot\text{m}^2/\cancel{\text{C}^2})(1.6\times10^{-19}\ \cancel{\text{C}})^2}{(9.1\times10^{-31}\ \cancel{\text{kg}})(5.4\times10^{-11}\ \cancel{\text{m}})}\left(\frac{1\ \cancel{\text{kg}}\cdot\cancel{\text{m}}/\text{s}^2}{1\ \cancel{\text{N}}}\right)}$$

$$= 2.2\times10^6\ \cancel{\text{m/s}}\left(\frac{c}{3.0\times10^8\ \cancel{\text{m/s}}}\right) = 0.0073c = \frac{c}{140} \qquad \text{(N13.23)}$$

Interestingly, independent measures of the electron's orbital speed (having to do with the magnetic field produced by the circulating charge and relativistic effects) yield approximately the same result, so even though this model may be oversimplified, it must be on to something!

The point of this problem is that circular orbit problems don't have to be limited to astronomical situations!

N13.7 BLACK HOLES AND DARK MATTER

In late May of 1994, H. C. Ford and R. J. Harms announced that they had discovered a giant black hole at the center of the galaxy known as M87. This was the first time that anyone had presented an apparently iron-clad argument for the existence of a giant black hole at the center of any galaxy.

A black hole is an object so dense that light cannot escape it. So how can we hope to "discover" an object that by definition doesn't emit any light? If you can find something orbiting the black hole and you can measure the orbit's period and radius, you can compute the black hole's mass using Kepler's third law!

Ford and Harms used the repaired Hubble Telescope to take spectra from very small regions of a gas cloud near the center of M87. By measuring the doppler shifts of the spectra obtained, they were able to determine that gas that was a radius $R = 60$ lyr ($\approx 5.7 \times 10^{17}$ m) from the center of the cloud was orbiting the center with a speed of 450 km/s. If we assume that the orbit is roughly circular, $T = 2\pi R/v$. Plugging the latter equation into Newton's version of Kepler's third law and solving for M, we get

$$\frac{4\pi^2 R^3}{GM} = T^2 = \left(\frac{2\pi R}{v}\right)^2 \Rightarrow M = \frac{4\pi^2 R^3 v^2}{G(2\pi R)^2} = \frac{Rv^2}{G} \qquad \text{(N13.24)}$$

Plugging $R = 5.7 \times 10^{17}$ m and $v = 450$ km/s we get 1.7×10^{39} kg, which is about 10^9 solar masses (1 solar mass = 2.0×10^{30} kg). This is just one of many bits of evidence cited by Ford and Harms.

How do we know that this is a black hole? No other explanation works! A star cluster containing a billion stars within a radius of 60 lyr would emit lots of light that is not seen. No plausible model *other* than the black hole model explains how a billion solar masses could fit within a sphere 60 lyr in radius and yet be consistent with all the other available data taken by Ford and Harms.

For at least a decade, astronomers also have been collecting evidence that 90% or more of the mass of the universe is *dark matter*, that is, matter that is *not* in stars that emit light or dust clouds that emit radio signals, reflect or obstruct light. Consistent evidence for this dark matter comes from a variety of studies of the orbital motion of stars within galaxies, satellite galaxies around large galaxies, and galaxies in galactic clusters. Many of these studies use Newton's version of Kepler's third law to determine the mass of the unseen matter by its gravitational effects.

In one recent study, D. N. C. Lin, B. F. Jones and A. R. Klemona carefully measured the transverse movement of the *Large Magellanic Cloud* (LMC) relative to background objects. The LMC is a small galaxy that is a companion to and presumably orbits our own much larger galaxy. Lin and his collaborators measured the transverse velocity of the LMC to be about 200 km/s. Since the LMC is about 170,000 lyr (1.6×10^{21} m) from our galaxy, if it were in a circular orbit, it would imply that the mass of our galaxy is

$$M = \frac{Rv^2}{G} = \frac{(1.6\times10^{21}\ \text{m})(200{,}000\ \text{m/s})^2}{6.67\times10^{-11}\ \text{N}\cdot\text{m}^2/\text{kg}^2}\left(\frac{1\ \text{N}}{1\ \text{kg}\cdot\text{m/s}^2}\right) = 1.0\times10^{42}\ \text{kg} \qquad \text{(N13.25)}$$

which is about 500 billion solar masses (a more accurate calculation based on the LMC's actual trajectory puts the estimate at more like 600 billion solar masses). Visible matter in our galaxy amounts to about 100 billion solar masses.

What is this dark matter? No one knows! Very recent experiments have strongly suggested that it *cannot* be dwarf stars, large planets, or black holes, mounting evidence suggest that this matter is *not* even ordinary matter made of protons, neutrons, and electrons. Part of the point here is that we would not even know that this major fraction of the universe exists if it weren't for Newton's form of Kepler's third law!

SUMMARY

I. KEPLER'S LAW
  A. Origin of Kepler's laws
    1. Johannes Kepler became director of Prague observatory in 1601
    2. This gave him access to Tycho Brahe's accurate observations
    3. Kepler invented his laws to describe the implications of this data
    4. These laws are completely empirical
  B. A statement of Kepler's laws
    1. planets orbit in ellipses, with sun at one focus
    2. a line from the sun to planet sweeps out equal areas in equal times
    3. $T^2 \propto a^3$, where T = period of orbit, a = semimajor axis (a = half of the ellipse's greatest width)
  C. Newton's triumph was to offer an *explanation* of these empirical laws using principles that were applicable to terrestrial physics as well

II. ORBITS AROUND A MASSIVE PRIMARY
  A. Consider an isolated (or freely falling) system of two interacting objects
    1. let the objects have masses M and m
    2. a reference frame whose origin is the system's CM will be inertial
      a) if the frame is isolated, CM moves at a constant velocity
      b) if the frame is freely falling, we ignore external gravitational interactions and treat the frame as *if* it were inertial
  B. If we have $M >> m$
    1. we call the massive object the *primary*, the other its *satellite*
    2. we can make the following approximations in the CM frame
      a) the primary is essentially at *rest* at the origin: $\vec{R} \approx 0$, $\vec{V} \approx 0$
      b) its kinetic energy is negligible: $K_M \approx 0$
      c) its angular momentum is negligible: $\vec{L}_M \approx 0$
      d) the objects' separation ≈ the satellite's distance from the origin
  C. The approximation that $M >> m$ works well in the solar system

III. KEPLER'S SECOND LAW
  A. Conservation of angular momentum implies that for *either* object
    1. the object's orbit around the CM lies in a plane, because
      a) $\vec{L} = \vec{r} \times m\vec{v}$ implies that $\vec{r} \perp \vec{L}$
      b) $\vec{L}$ is fixed in space, so $\vec{r}$ must lie in fixed plane $\perp \vec{L}$
    2. the object's position vector sweeps out equal areas in equal times
      a) if object moves infinitesimal angle $d\theta$ in time dt
      b) area swept out is $dA \approx \frac{1}{2} r^2 d\theta$
      c) magnitude of object's angular momentum is $L = mr^2\, d\theta/dt$
      d) so $dA/dt = L/2m$ = constant
  B. Kepler's second law: a line connecting the *sun* (not CM) and the planet sweeps out area at constant rate, but if $M >> m$, sun's position ≈ CM

IV. CIRCULAR ORBIT PROBLEMS
  A. From now on, we will assume $M >> m$
    1. so the primary basically provides fixed source of gravitational field
    2. we will use $\vec{L}$ and K to refer to the *satellite's* AM and KE.
  B. Circular orbits are possible (if gravity is the only force on the object) (the constant inward force provides the constant inward $\vec{a}$ needed)
  C. Applying Newton's second law to circular orbit of radius R, we get
    1. a constant orbital speed = $v = [GM/R]^{1/2}$ (N13.12)
    2. $T^2 = (4\pi^2/GM)R^3$, where T is orbital period (N13.14)
       (This is Kepler's third law for the case of circular orbits)
  D. Circular orbit problems are like constrained motion problems, except
    1. we use eqn. N13.12 or14 or the *magnitude* of Newton's 2nd law
    2. this means that setting up coordinate axes is not crucial
    3. the net force is the object's weight so a force diagram is not needed

GLOSSARY

ellipse: a specific mathematical curve that looks like a flattened circle (we will define this term more carefully in the next chapter).

semimajor axis a (of an ellipse): half of the distance measured across the ellipse at its widest point.

period T (of an orbit): the time that it takes an object to complete exactly one orbit around the system's center of mass.

Kepler's three laws of planetary motion: empirical laws devised by Johannes Kepler in the early decades of the 1600's to describe the observed motion of the planets. These laws state that:

1. the planets move in ellipses, with the sun at one focus (the focus of an ellipse will be defined in chapter N14)
2. the line from the sun to the planet sweeps out equal areas in equal times
3. the period of the orbit is proportional to the 3/2 power of its semimajor axis.

primary: the more massive object in an isolated, two-object interacting system where the mass of one object is much larger than the other.

satellite: the lighter object in an isolated, two-object interacting system where the mass of one object is much larger than the other.

TWO-MINUTE PROBLEMS

N13T.1 Kepler's second law implies that as a planet's distance from the sun *increases* in an elliptical orbit, its orbital speed

A. increases
B. decreases
C. remains the same
D. it depends

N13T.2 The sun's mass is about 1000 times that of Jupiter, and the radius of Jupiter's orbit is about 1100 times the sun's radius. The center of mass of the sun-Jupiter system is inside the sun. (T or F)

N13T.3 Two stars, one with radius r and the other with radius $3r$ orbit each other so that their centers of mass are $5r$ apart. Assume that the stars have the same uniform density. The center of mass of this system is inside the larger star (T or F).

N13T.4 The speed of a satellite in a circular orbit of radius R around the earth is 3.0 km/s. The speed of another satellite in a different circular orbit around the earth is half this value. What is the radius of that satellite's orbit?

A. $4R$
B. $2R$
C. $\sqrt{2}\,R$
D. $R/\sqrt{2}$
E. $R/2$
F. $R/4$
T. other (specify)

N13T.5 A satellite orbits the earth once every 2.0 h. What is the orbital period of another satellite whose orbital radius is 4.0 times larger?

A. 4.0 hours
B. 8.0 hours
C. 16 hours
D. 64 hours
E. other (specify)

N13T.6 The radius of Saturn's orbit is 9.53 times that of the Earth. What is the period of Saturn's orbit (assuming that it is nearly circular)?

A. 9.53 y
B. 29 y
C. 91 y
D. 866 y
E. other (specify)

N13T.7 The radius of the earth's (almost circular) orbit around the sun is 150,000,000 km, and it takes a year for the earth to go around the sun. Imagine that a certain satellite goes in an almost circular orbit of radius 15,000 km around the earth (this radius is 10,000 times smaller than the earth's orbital radius around the sun). What is the period of this orbit?

A. 10^{-6} year
B. 10^{-4} year
C. 10^{-3} year
D. 10^{6} year
E. these periods are not related in any simple way

HOMEWORK PROBLEMS

BASIC SKILLS

N13B.1 A satellite in a circular orbit around the earth has a speed of 3.00 km/s. What is the radius of this orbit?

N13B.2 The circular orbit of a satellite going around the earth has a radius of 10,000 km. What is the satellite's orbital speed?

N13B.3 *Geostationary* satellites are placed in a circular orbit around the Earth at such a distance that their orbital period is exactly equal to 24 h (this means that the satellite seems to hover over a certain point on the Earth's surface). What is the radius of such a geocentric orbit?

N13B.4 What is the orbital speed of the moon? The radius of the moon's orbit is roughly 382 Mm.

N13B.5 What is the earth's speed as it orbits the sun?

N13B.6 Triton is the largest moon of Neptune (roughly the same size as the Earth's moon). It orbits Neptune once every 5.877 days at a distance of roughly 354 Mm from Neptune's center. What is the mass of Neptune?

N13B.7 The radius of Neptune's nearly circular orbit is about 30 times larger than that of the earth's orbit. What is the period of Neptune's orbit?

N13B.8 The radius of Venus' nearly circular orbit is about 30 times larger than that of the earth's orbit. What is the period of Venus' orbit?

N13B.9 A neutron star is an astrophysical object having a mass of roughly 3.0×10^{30} kg (about 1.4 times the mass of the sun) but a radius of only about 12 km. If you are in a circular orbit of radius 320 km (about 200 mi), how long would it take you to go once around the star?

SYNTHETIC

N13S.1 Is the center of mass of the earth/moon system inside the earth?

N13S.2 Imagine that the sun, earth, and Jupiter are aligned so that all three are in a line. What is the magnitude of the gravitational force exerted by Jupiter on the earth compared to that exerted by the sun on the earth?

N13S.3 What is the magnitude of the gravitational force exerted on the earth by the moon compared to that exerted by on the earth by the sun?

N13S.4 During a certain 5-day time period, the line connecting a comet with the sun changes angle by about 3.2°. Assume that the comet's distance from the sun is roughly 130 million km during this time. During a 5-day time period some time later, the angle of this line changes by 0.80°. How far is the comet from the sun now? Explain.

N13S.5 Consider a light object and a massive object connected by a spring. Both objects are floating in deep space. Will the path of the light object around the massive object lie in a plane? Will it obey Kepler's second law?

N13S.6 What would be the orbital speed of the electron in a hydrogen atom if the interaction between the proton and electron were gravitational instead of electrostatic? (Assume that the electron's orbit is still circular and its radius is still 5.4×10^{-11} m.)

N13S.7 Imagine that two objects with masses M and m are connected by a spring with zero relaxed length, so that the attractive force that each exerts on the other is $F = k_s r$. Assume that $M >> m$, and that the satellite orbits in a circular orbit of radius R around its primary. Find an expression (analogous to Kepler's third law) that gives the period of the orbit T as a function of the orbital radius R, the spring constant k^s, and whatever else you need.

N13S.8 What if the magnitude of the force of gravity exerted by one object on another were given by

$$F_g = \frac{GMm}{r^3} \qquad \text{(N13.26)}$$

where r is the separation between the two objects? How would we have to rewrite Kepler's third law?

N13S.9 In Newton's time, the distance between the Earth and the moon was known to be about 60 times the radius of the earth. According to the law of universal gravitation, about how many times smaller is the earth's gravitational field strength at the radius of the moon's orbit than it is on the earth's surface? Use this information (rather than the mass of the earth) to estimate the period of the moon's orbit in days. How does this compare with the moon's actual orbital period? Please show your work. (This was one of the ways that Newton supported his inverse-square law of gravitation.)

N13S.10 In a certain binary star system, a small red star with a mass ≈ 0.22 solar masses orbits a bright white-hot star with a mass ≈ 4.2 solar masses. (These masses are estimated from the stars' color and luminosity.) The red star is observed to eclipse the other every 482 days. What is the approximate distance between these stars? What assumptions (if any) do you have to make?

RICH-CONTEXT

N13R.1 Consider a spherical asteroid made mostly of iron (whose density is 7.9 g/cm^3) whose radius is 22 km. Could you run fast enough to put yourself in orbit around this asteroid?

N13R.2 Imagine that you wanted to put a satellite in such a circular orbit that it appeared on the western horizon every Monday morning at 6 am and was never at any other time on the western horizon. What radius should the orbit have? (Don't forget to account for the earth's rotation!)

ADVANCED

N13A.1 Consider an isolated two-object system interacting gravitationally such that M is not necessarily much greater than m. Assume that the smaller object's orbit around the system's center of mass is circular and has radius r. Argue that both objects have the same orbital period T around the system's center of mass and that

$$T^2 = \frac{4\pi^2 D^3}{G(M+m)} \qquad \text{(N13.27)}$$

where D is the separation of the two objects.

ANSWERS TO EXERCISES

N13X.1 According to equation N13.1, if the distance between the earth and system's center of mass is r (which is then also the radius of the earth's orbit) and the distance between the sun and the system's center of mass is R, then

$$R = \frac{m}{M}r = \frac{6.0\times 10^{24}\ \text{kg}}{2.0\times 10^{30}\ \text{kg}}(1.5\times 10^{11}\ \text{m}) = 450{,}000\ \text{m} = 450\ \text{km} \qquad \text{(N13.28)}$$

(Note that you can look up the earth and sun's masses in the inside front cover.) This is nearly 2,000 times smaller than the sun's radius.

N13X.2 Plugging equation N13.13 into N13.12, we get

$$T = \frac{2\pi R}{\sqrt{GM/R}} = 2\pi R\sqrt{\frac{R}{GM}} = 2\pi\sqrt{\frac{R^3}{GM}} \qquad \text{(N13.29)}$$

Squaring both sides of this gives us equation N13.14.

N13X.3 According to equation N13.14, $T \propto R^{3/2}$. If we increase R by a factor of 4, then $R^{3/2}$ increases by a factor of $(\sqrt{4})^3 = 2^3 = 8$. So the orbit would last 8 years.

N14

PLANETARY MOTION

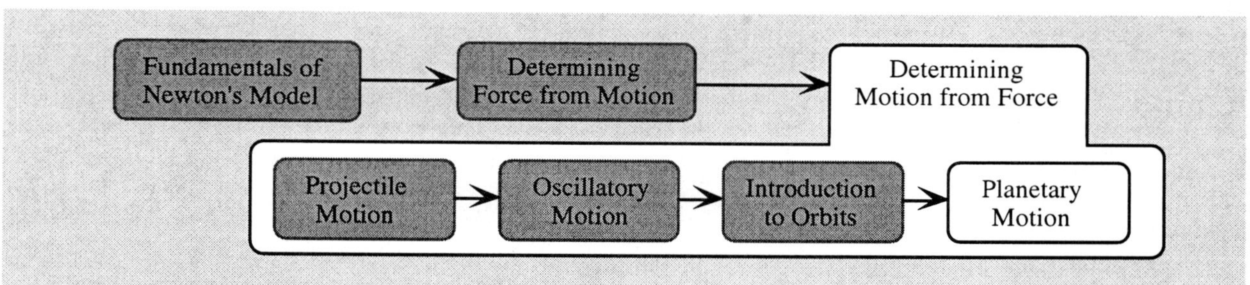

N14.1 OVERVIEW

In this final chapter of the unit, we will bring many of the tools that we have developed in this unit (and the previous unit) to bear on the problem that represents the crowning triumph of the newtonian synthesis. We will prove here that Kepler's first and third laws of planetary motion are in fact mathematical *consequences* of Newton's mechanics and the law of universal gravitation, showing how Newton's theory successfully embraces celestial as well as terrestrial physics.

N14.2 *CONIC SECTIONS* reviews the mathematical definition and properties of conic sections in general and ellipses in particular, providing the background we need for the rest of the chapter.

N14.3 *RADIAL ACCELERATION* uses the methods of chapter N9 to find the radial component of an object's acceleration at any point in an orbit: it is this component that is linked to the gravitational force by Newton's second law.

N14.4 *KEPLER'S FIRST LAW* shows how we can prove Kepler's first law (that planets orbit in ellipses) using Newton's second law.

N14.5 *AN ORBIT'S ECCENTRICITY* looks at how the eccentricity of an orbit is connected to the system's total energy E.

N14.6 *IMPLICATIONS* explores the link that our proof implies about the relationship between the system's energy and the shape of the orbit.

N14.7 *KEPLER'S THIRD LAW* proves that Kepler's third law holds for elliptical orbits as well as circular orbits.

N14.8 *SOLVING ORBIT PROBLEMS* illustrates how we can use conservation principles to determine many things about realistic orbits.

This chapter concludes this unit on newtonian mechanics. In this unit, you have seen some of the tremendous analytical power and range of Newton's great vision, and perhaps have come to understand why his theory dominated physics for more than two centuries.

Newtonian physics, however, is *not* the whole story. In the early part of the 1900s, newtonian ideas began to unravel. In the next unit, we will study the theory of special relativity, the first great post-newtonian theory. Even so, as we study that theory, we will see the great debt that *all* physical theories (even modern ones) owe to Newton, who first showed how it could be done.

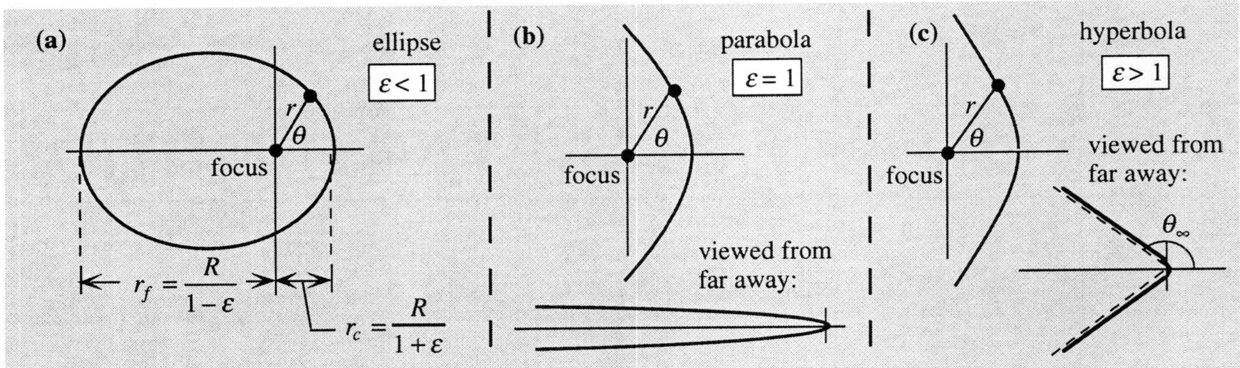

Figure N14.1: The three main types of curves described by equation N14.1.

N14.2 CONIC SECTIONS

The core of the Newtonian synthesis was Newton's proof that his three laws of mechanics and the law of universal gravitation imply that the orbits of planets necessarily are *ellipses,* consistent with Kepler's surprising empirical first law. Newton's proof was in fact even more general than this: he was able to show that the path of *any* object moving around a massive primary under the influence of gravity alone (even if the object, unlike a planet, were to approach the primary from infinity and then return to infinity) must be a *conic section,* a general class of curves for which ellipses are a subset. Our goal in this chapter is to see how this proof works.

What is a conic section?

Mathematically, a **conic section** is defined to be a curve whose points obey the following equation:

$$\frac{1}{r} = \frac{1}{R}(1+\varepsilon\cos\theta) \quad \text{or} \quad r = \frac{R}{1+\varepsilon\cos\theta} \tag{N14.1}$$

where $\vec{r}$ is the position vector of a given point on the curve relative to a fixed point called the curve's **focus**, $r = \text{mag}(\vec{r})$, θ is the angle that $\vec{r}$ makes with a line that connects the focus to the nearest point on the curve, R is an arbitrary positive constant (with units of distance) that specifies the size of the curve, and ε is an arbitrary positive (and unitless) constant called the curve's **eccentricity** that specifies something about the shape of the curve. Such a curve is called a *conic section* because it happens to specify the curve that results when you take the intersection of an arbitrarily oriented infinite plane and the surface of an infinite cone.

Types of conic sections

If $\varepsilon = 0$, the equation of the curve given by equation N14.1 simply becomes $r = R$, and so the curve is a **circle**. If $0 < \varepsilon < 1$, then the curve is an **ellipse**: the curve's distance r from the focus oscillates between a minimum value r_c at angle $\theta = 0$ to a maximum value r_f at $\theta = \pi$, where

$$r_c = \frac{R}{1+\varepsilon} \quad \text{and} \quad r_f = \frac{R}{1-\varepsilon} \quad \text{(for an ellipse)} \tag{N14.2}$$

(r_c stands for "r of the closest point to the focus" and r_f for "r of the farthest point from the focus"). Figure N14.1a shows what an ellipse looks like.

If $\varepsilon = 1$, then the curve is closest to the focus at $\theta = 0$ ($r_c = \frac{1}{2}R$), but the distance to the curve increases to infinity as θ approaches π: such a curve is a **parabola** (which is like an infinitely stretched ellipse: see Figure N14.1b).

If $\varepsilon > 1$, then the closest point on the curve is still at $\theta = 0$, but there now exists an angle $\theta_\infty < \pi$ such that $\varepsilon\cos\theta_\infty = -1$: the curve will go to infinity at that angle. For angles between θ_∞ and π, r in equation N14.1 would be *negative*, which is absurd for a distance, so the curve is actually undefined for angles in this range. Similarly, the curve goes to infinity at the angle $-\theta_\infty$ and is undefined for angles between $-\theta_\infty$ and $-\pi$. Therefore in this case the curve is only defined for angles θ such that $|\theta| < \theta_\infty$. This kind of curve is a **hyperbola** (see Figure N14.1c).

Exercise N14X.1: What is θ_∞ when $\varepsilon = 3/2$?

Note that only the circle and ellipse close on themselves: the parabola and hyperbola come in from infinity and go back out to infinity. Therefore, only circles or ellipses are appropriate orbits for planets, which by definition are bodies that go repetitively around the sun.

N14.3 RADIAL ACCELERATION

One of the most important characteristics of the gravitational force acting on a planet is that it is always directed exactly toward the sun. Newton's second law tells us that the planet's acceleration must also point in this direction. It will therefore help us in determining the planet's motion if we can determine what the component of an orbiting object's acceleration in this direction is.

Steps in determining the acceleration's radial component

We can determine this fairly easily if we draw on the techniques that we developed to analyze circular motion in chapter N9. We saw in the last chapter that a planet's orbit must lie in a plane. Assume that we have defined our reference frame so that the orbit of the object in question lies in the xy plane. We can then represent the object's position at any instant of time as follows:

$$\vec{r}(t) = \begin{bmatrix} r\cos\theta \\ r\sin\theta \end{bmatrix} = r\begin{bmatrix} \cos\theta \\ \sin\theta \end{bmatrix} = r\hat{r}, \text{ where } \hat{r} \equiv \begin{bmatrix} \cos\theta \\ \sin\theta \end{bmatrix} \tag{N14.3}$$

where θ is the angle that the object's position vector $\vec{r}$ makes with the x axis at that instant (see Figure N14.2). As we saw in chapter N9, $\hat{r}$ is a unit vector that points directly away from the origin. As the object moves, θ changes, and thus the direction of $\hat{r}$ changes: in fact

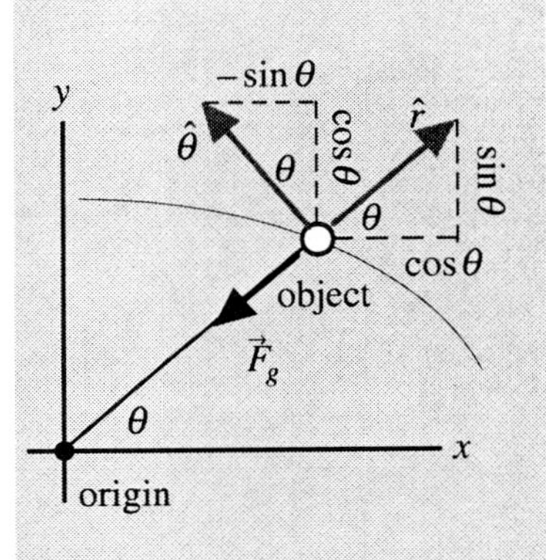

Figure N14.2: A sketch showing the directions of the unit vectors $\hat{r}$ and $\hat{\theta}$ at a given instant of time. If the primary is at the origin, the force of gravity on the object points in the $-\hat{r}$ direction.

$$\frac{d\hat{r}}{dt} = \frac{d}{dt}\begin{bmatrix} \cos\theta \\ \sin\theta \end{bmatrix} = \begin{bmatrix} -\sin\theta \\ \cos\theta \end{bmatrix}\frac{d\theta}{dt} = \hat{\theta}\frac{d\theta}{dt}, \text{ where } \hat{\theta} \equiv \begin{bmatrix} -\sin\theta \\ \cos\theta \end{bmatrix} \tag{N14.4a}$$

where $\hat{\theta}$ is a unit vector pointing perpendicular to $\hat{r}$ on the counterclockwise side (see Figure N14.2). Similarly, we saw in chapter N9 that

$$\frac{d\hat{\theta}}{dt} = -\hat{r}\frac{d\theta}{dt} \tag{N14.4b}$$

Now we are ready to calculate an object's acceleration in terms of these radial and perpendicular unit vectors. Equation N14.3 tells us that the object's position vector at any time is given by $\vec{r}(t) = r\hat{r}$, where both the length r and the direction $\hat{r}$ may change with time as the object moves. Therefore, according to the product rule of derivatives and equations N14.4, we have

$$\vec{v}(t) = \frac{d}{dt}(r\hat{r}) = \frac{dr}{dt}\hat{r} + r\frac{d\hat{r}}{dt} = \frac{dr}{dt}\hat{r} + r\hat{\theta}\frac{d\theta}{dt} \tag{N14.5}$$

You can take the time-derivative of this once again and use both the product and sum rules to show that

The acceleration of an object in terms of $\hat{r}$ and $\hat{\theta}$.

$$\vec{a}(t) = \left[\frac{d^2r}{dt^2} - r\left(\frac{d\theta}{dt}\right)^2\right]\hat{r} + \left[2\frac{dr}{dt}\frac{d\theta}{dt}\hat{\theta} + r\frac{d^2\theta}{dt^2}\right]\hat{\theta} \tag{N14.6}$$

Exercise N14X.2: Use equations N14.5 and N14.4 to fill in the missing steps leading to equation N14.6.

Exercise N14X.3: Check that for the case of uniform circular motion (where r = constant and $d\theta/dt$ = constant), equation N14.6 says that the object's acceleration points toward the origin and has a magnitude of $a = v^2/r$. [*Hint*: In the case of circular motion, $v = |rd\theta|/dt$.]

Definition of the radial and perpendicular components of the acceleration

At any given instant of time, the $\hat{\theta}$ and $\hat{r}$ directions define two perpendicular coordinate axes in space, just like the y and x axes do. The part of the acceleration vector that multiplies the $\hat{r}$ unit vector thus defines the **radial component** a_r of the acceleration; the part that multiplies $\hat{\theta}$ is the **perpendicular component** $a_\perp$ (analogous to x and y components respectively).

Now, the law of universal gravitation implies that the force exerted on a satellite of mass m orbiting a primary of mass M points directly toward the primary and has a magnitude of $F_g = GMm/r^2$. If the primary is at the origin of our reference frame, then this means that this force acts entirely in the $-\hat{r}$ direction. So the radial component of Newton's second law in this case implies that

The radial component of Newton's second law

$$-\frac{GMm}{r^2} = F_{g,r} = F_{\text{net},r} = ma_r = m\left[\frac{d^2r}{dt^2} - r\left(\frac{d\theta}{dt}\right)^2\right] \tag{N14.7}$$

The gravitational force has *no* component in the perpendicular direction, so the component of Newton's second law in that direction tells us that

The perpendicular component of Newton's second law

$$0 = F_{\text{net},\perp} = ma_\perp = m\left[2\frac{dr}{dt}\frac{d\theta}{dt} + r\frac{d^2\theta}{dt^2}\right] \tag{N14.8}$$

This component of Newton's second law actually implies that angular momentum $L = mr^2(d\theta/dt)$ is conserved (see problem N14S.11), so we will simply use conservation of angular momentum in what follows. It is the radial component that determines the shape of the orbit.

N14.4 KEPLER'S FIRST LAW

In this section, we will prove Kepler's first law, the most subtle but important of Kepler's three laws. The proof itself is important for several reasons, not just because it shows how Newton's second law and the law of universal gravitation are sufficient to explain Kepler's first law (which was the triumph that launched physics as a science and is the foundation of the "great idea" of this unit). This proof is also the crowning illustration in this unit of how we can use knowledge about forces to predict motion, and is a good example as well of how the harmonic oscillator model can pop up in the most unusual contexts (something that seems to happen again and again in physics).

As is generally the case when we are trying to determine an object's motion from its forces, we have to use some kind of trick to find the answer. In this particular case, we will use three tricks: (1) we will use conservation of angular momentum to eliminate $d\theta/dt$ from equation N14.7, (2) we will reexpress this equation in terms of the variable $u = 1/r$ to make it simpler, and (3) we will draw on what we learned about the solution to the harmonic oscillator equation in chapter N12 to find the solution here.

The proof here is longer than most that appear in this book. The appropriate way to study such a proof is to first go through the derivation step by step (working the exercises), making sure that you understand how each step follows from the last. Then reread the derivation and try to construct in your mind the big picture of how the derivation flows (the marginal comments may help).

First trick: eliminating $d\theta/dt$

In chapter N13, we saw that conservation of angular momentum means that

$$\frac{d\theta}{dt} = \frac{L}{mr^2} \tag{N14.9}$$

where m is the planet's mass (see equation N13.10). If we plug this into equation N14.7, we find that $r(d\theta/dt)^2 = r(L/mr^2)^2 = L^2/m^2r^3$. The radial component of Newton's second law in this case thus states that

$$\frac{d^2r}{dt^2} - \frac{L^2}{m^2r^3} = a_r = \frac{F_{g,r}}{m} = -\frac{GM}{r^2} \qquad \text{(N14.10)}$$

The second trick: Change variables from r and t to $u = 1/r$ and θ

In principle, could use this equation to find out how r varies with time. However, our goal here is not to find the function $r(t)$ but rather the function $r(\theta)$ that describes the shape of the planet's orbit. It turns out that we can more easily solve the problem by finding not $r(\theta)$ but $u(\theta)$ where

$$u \equiv \frac{1}{r} \quad \text{which means that} \quad r \equiv \frac{1}{u} \qquad \text{(N14.11)}$$

It is not *obvious* that making this substitution will help, but the fact that inverse factors of r appear often both in equation N14.10 *and* in equation N14.1 makes it worth a try. Let's see if we can convert the d^2r/dt^2 term in equation N14.10 into something involving u and θ instead of r and t. By the chain rule,

$$\frac{dr}{dt} = \frac{dr}{du}\frac{du}{d\theta}\frac{d\theta}{dt} = \left(\frac{d}{du}\frac{1}{u}\right)\frac{du}{d\theta}\frac{d\theta}{dt} = -\frac{1}{u^2}\frac{du}{d\theta}\frac{Lu^2}{m} = -\frac{L}{m}\frac{du}{d\theta} \qquad \text{(N14.12a)}$$

where I have used both equations N14.9 and N14.11 in simplifying this result. In a similar manner, *you* can use this equation show that

$$\frac{d^2r}{dt^2} \equiv \frac{d}{dt}\left(\frac{dr}{dt}\right) = \frac{d\theta}{dt}\frac{d}{d\theta}\left(\frac{dr}{dt}\right) = -\frac{L^2}{m^2r^2}\frac{d^2u}{d\theta^2} \qquad \text{(N14.12b)}$$

Exercise N14X.4: Verify equation N14.12b.

If we now plug this into Newton's second law (equation N14.10) and express all factors of r in terms of factors of u, we get:

$$-\frac{L^2u^2}{m^2}\frac{d^2u}{d\theta^2} - \frac{L^2u^3}{m^2} = -GMu^2 \Rightarrow \frac{d^2u}{d\theta^2} + u = \frac{GMm^2}{L^2} \qquad \text{(N14.13a)}$$

Now, if we define a *new* variable

$$w(\theta) \equiv u(\theta) - \frac{GMm^2}{L^2} = \frac{1}{r} - \frac{GMm^2}{L^2} \qquad \text{(N14.14)}$$

you should be able to show that equation N14.13a can be rewritten as follows:

$$\frac{d^2w}{d\theta^2} = -w \qquad \text{(N14.13b)}$$

Exercise N14X.5: Verify equation N14.13b.

The third trick: compare with the harmonic oscillator equation

Now, compare this equation to equation N12.6 (page 165). You can see that equation N14.13b has the same *form* as the harmonic oscillator equation, with w and θ here playing the roles of x and t there, and with ω^2 there being 1 here. This means that the *solution* to equation N14.13b should be the same as the solution to equation N12.6 with the same substitutions made:

$$w(\theta) = A\cos(\theta + \delta) \qquad \text{(N14.15a)}$$

where A and δ are as yet undetermined constants. Using equation N14.14 and defining $\varepsilon \equiv AL^2/GMm^2$, you can show that we can rewrite this as:

The general equation for a satellite's orbit around a massive primary

$$\frac{1}{r} = \frac{1}{R}[1+\varepsilon\cos(\theta+\delta)] \quad \text{where} \quad R \equiv \frac{1}{GM}\left(\frac{L}{m}\right)^2 \tag{N14.15b}$$

Exercise N14X.6: Verify equation N14.15b.

Exercise N14X.7: Show by direct substitution that the solution in equation N14.15a satisfies equation N14.13b.

In equation N14.15b, the phase constant δ simply determines the angle at which the planet has its smallest radius (that is, the angle at which the cosine is maximum). If we agree that we will *always* orient our reference frame so that the point of closest approach corresponds to angle $\theta = 0$, then $\delta = 0$. If you now compare equation N14.15b with equation N14.1, you can see that we have just *proved* that the trajectory of an orbiting object must indeed be a conic section! Since the ellipse is only conic section appropriate for a repeating *planetary* orbit (as discussed in the first section), we see that planets move in ellipses. Q.E.D.

N14.5 AN ORBIT'S ECCENTRICITY

Having done this, it is instructive to see how the constant ε in equation N14.15b is linked to total *energy* of the orbiting body. Consider the point of closest approach r_c, which (if we take $\delta = 0$) is at $\theta = 0$. According to equation N14.15b, we have:

$$\frac{1}{r_c} = \frac{GMm^2}{L^2}(1+\varepsilon) \tag{N14.16}$$

Just before the point of closest approach, the orbiting body has been moving closer to the origin; just after the point of closest approach, it is moving away from the origin. Therefore, the body's radial velocity $v_r \equiv dr/dt$ is going through zero at that point and its speed is equal to the absolute value of the perpendicular component $v_\perp$ of its velocity (that is, the component of $\vec{v}$ in the $\hat{\theta}$ direction). Conservation of energy at this point thus tells us that

$$E = \tfrac{1}{2}mv_\perp^2 - \frac{GMm}{r_c} \tag{N14.17a}$$

We saw in chapter N9 that $v_\perp = r(d\theta/dt)$, so if we substitute $v_\perp = r(d\theta/dt) = r(L/mr^2) = L/mr$ into this equation and also substitute in the right side equation N14.16 for r_c, we get

$$\begin{aligned}
E &= \frac{L^2}{2mr_c^2} - \frac{GMm^2}{r_c} = \frac{L^2}{2m}\left(\frac{GMm^2(1+\varepsilon)}{L^2}\right)^2 - GMm^2\left(\frac{GMm^2(1+\varepsilon)}{L^2}\right) \\
&= \frac{G^2M^2m^3}{2L^2}(1+\varepsilon)^2 - \frac{G^2M^2m^3}{L^2}(1+\varepsilon) \\
&= \frac{G^2M^2m^3}{2L^2}\left[1+2\varepsilon+\varepsilon^2-2-2\varepsilon\right] = -\frac{G^2M^2m^3}{2L^2}(1-\varepsilon^2)
\end{aligned} \tag{N14.17b}$$

If you now solve this equation for ε, you can show that

How eccentricity is related to E/m and L/m

$$\varepsilon = \left[1+\frac{1}{G^2M^2}\left(\frac{2E}{m}\right)\left(\frac{L}{m}\right)^2\right]^{1/2} \tag{N14.18}$$

Exercise N14X.8: Verify equation N14.18

Equation N14.18, in conjunction with equation N14.15b (with $\delta = 0$)

$$\frac{1}{r} = \frac{1}{R}[1+\varepsilon\cos\theta] \quad \text{where} \quad R \equiv \frac{1}{GM}\left(\frac{L}{m}\right)^2 \tag{N14.19}$$

completely determines the orbiting body's trajectory in terms of its total energy per mass E/m and its angular momentum per mass L/m. Note that we never really need to know the object's mass; we can determine the values of E/m and L/m if we know the value of GM, the body's speed v, the perpendicular component of its velocity $v_\perp$, and its distance from the origin r at any point along its orbit, since $2E/m = v^2 - 2GM/r$, and $L/m = rv_\perp$.

We can completely determine an orbit's shape if we know $GM, v, v_\perp$, and r at any instant of time

N14.6 IMPLICATIONS

Our derivation of the orbit equation N14.19 does not specify that the orbiting object has to be a planet: it applies equally well to *any* satellite orbiting a comparatively massive primary. Therefore, this equation applies to moons orbiting planets, artificial satellites orbiting the earth, unpowered spaceships passing by planets and so on.

Equation N14.19 applies to *any* object orbiting a massive primary

Our proof of Kepler's first law yields another unsuspected byproduct: equation N14.18 specifies a simple but nonobvious link between the orbit's eccentricity and the system's total energy E:

The link between energy and an orbit's shape

If $E < 0$,	then $\varepsilon < 1$,	orbit is *elliptical*	
If $E = 0$,	then $\varepsilon = 1$,	orbit is *parabolic*	(N14.20)
If $E > 0$,	then $\varepsilon > 1$,	orbit is *hyperbolic*	

This does makes sense if you think about it, though. Since an object's gravitational potential energy $V(r) = -GMm/r$ is always negative (with the usual definition of the reference separation), having a total energy $E \geq 0$ means that E is larger than V at all values of r, implying that our object will have kinetic energy $K > 0$ even at very large r. The most important distinction between elliptical orbits and parabolic and hyperbolic orbits is that in the latter cases, the object comes in from $r = \infty$ and returns. This would not be possible unless $E \geq 0$.

There is also an interesting link between an *elliptical* orbit's energy and the quantity known as the ellipse's *semimajor axis*. An ellipse's **semimajor axis** a is defined to be half its widest diameter. Figure N14.3 shows that this quantity is equal to half the sum of r_c and r_f:

The link between energy and an elliptical orbit's size

$$a \equiv \tfrac{1}{2}(r_c + r_f) \tag{N14.21}$$

Equation N14.19 implies that $r = (L/m)^2/GM(1+\varepsilon\cos\theta)$ the smallest possible r is at angle $\theta = 0$ and the largest at angle $\theta = \pi$, so

$$a = \tfrac{1}{2}\left[\frac{(L/m)^2}{GM(1+\varepsilon)} + \frac{(L/m)^2}{GM(1-\varepsilon)}\right] = \frac{(L/m)^2}{2GM}\left[\frac{1}{1+\varepsilon} + \frac{1}{1-\varepsilon}\right] \tag{N14.22}$$

With the help of equation N14.17b and a bit of algebra, you can show that

$$E = -\frac{GMm}{2a} \tag{N14.23}$$

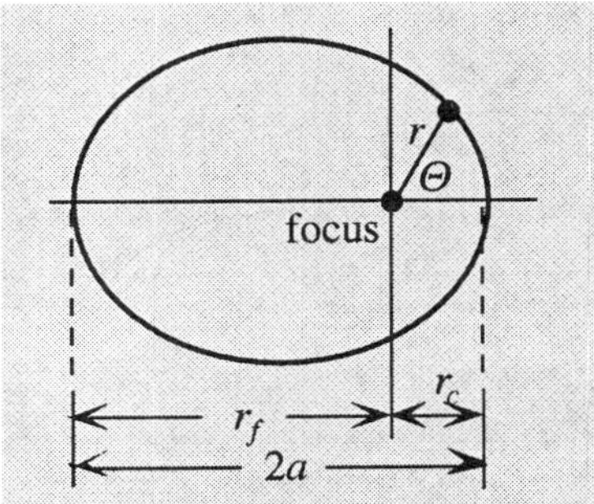

Figure N14.3: The connection between an ellipse's semimajor axis a and the distances to the orbit's closest and farthest points r_c and r_f, respectively.

Exercise N14X.9: Verify equation N14.23.

This can be a very useful equation in a number of contexts. For example, one can use this equation to calculate the change in an object's total energy that is required to change an object's orbit from one elliptical shape to another.

The following example illustrates how we might use the information provided by equation N14.20 in a practical situation.

EXAMPLE N14.1

Problem: Imagine that a strange object is discovered 22 AU (3.3 Tm) from the sun. (*Note*: Astronomers often express distances in the solar system in **astronomical units**, where 1 AU is defined to be the average radius of earth's orbit = 1.5×10^{11} m.) Measurements show that the component of the object's velocity toward the sun is 11 km/s and the magnitude of its perpendicular velocity component is 2.0 km/s. Is this a previously undiscovered member of the solar system, or is it an interloper from outside the solar system? How close will it get to the sun?

Solution The value of GM for the sun is

$$GM = \left(6.67\times10^{-11}\frac{\text{N}\cdot\text{m}^2}{\text{kg}^2}\right)(2.0\times10^{30}\text{ kg})\left(\frac{1\text{ kg}\cdot\text{m/s}^2}{1\text{ N}}\right)$$

$$= 1.33\times10^{20}\text{ m}^3/\text{s}^2 \tag{N14.24}$$

The object's speed is $v = [(11\text{ km/s})^2 + (2.0\text{ km/s})^2]^{1/2}$ according to the pythagorean theorem. The value of $(2E/m) = v^2 - GM/r$ for this object is thus

$$\left(\frac{2E}{m}\right) = (11{,}000\text{ m/s})^2 + (2{,}000\text{ m/s})^2 - \frac{2(1.33\times10^{20}\text{ m}^{3\to2}/\text{s}^2)}{3.3\times10^{12}\text{ m}}$$

$$= +4.4\times10^7\text{ m}^2/\text{s}^2 \tag{N14.25}$$

Since this is positive, the orbit is hyperbolic, so the object must be from *outside* the solar system (assuming that it is unpowered). Ominous! We also have

$$(L/m) = rv_\perp = (3.3\times10^{12}\text{ m})(2000\text{ m/s}) = 6.6\times10^{15}\text{m}^2/\text{s} \tag{N14.26}$$

Therefore

$$R = \frac{1}{GM}\left(\frac{L}{m}\right)^2 = \frac{(6.6\times10^{15})^2\text{ m}^4/\text{s}^2}{1.33\times10^{20}\text{ m}^3/\text{s}^2} = 3.3\times10^{11}\text{ m} \tag{N14.27a}$$

$$\varepsilon = \left[1+\frac{(4.4\times10^8\text{ m}^2/\text{s}^2)(6.6\times10^{15}\text{ m}^2/\text{s})^2}{(1.33\times10^{20}\text{ m}^3/\text{s}^2)^2}\right]^{1/2} = 1.44 \tag{N14.27b}$$

$$r_c = \frac{R}{1+\varepsilon} = \frac{3.3\times10^{11}\text{ m}}{2.44} = 1.35\times10^{11}\text{ m} \tag{N14.27c}$$

Note that r_c is just inside the earth's orbital radius of 1.5×10^{11} m.

Values of GM for various objects

Exercise N14X.10: Show that GM for the earth is 4.0×10^{14} m^3/s^2. (Similarly GM for the moon is about 4.9×10^{12} m^3/s^2, and GM for Jupiter is about 1.3×10^{17} m^3/s^2.)

N14.7 KEPLER'S THIRD LAW

We are now in a position to derive Kepler's third law for elliptical orbits. We found in the last chapter that the planet's radius vector sweeps out area at a constant rate of $dA/dt = L/2m$ (equation N13.9). If we integrate both sides of this expression from $t = 0$ to $t = T$ (where T is the time required for the planet to complete exactly one orbit), we get (since $L/2m$ is a constant)

$$\int_0^T \frac{dA}{dt}dt = \int_0^T \frac{L}{2m}dt \quad\Rightarrow\quad A = \frac{L}{2m}T \tag{N14.28}$$

where A is the total area swept out by the radius vector in time T, which is simply the area enclosed by the planet's elliptical orbit. It turns out that the area of an ellipse whose equation is $r = R/(1+\varepsilon\cos\theta)$ is $A = \pi R^{1/2}a^{3/2}$, where a is the semimajor axis of the ellipse (see problem N14S.13). Using this, equation N14.19, and N14.28, you can show that

$$T = \frac{2\pi}{\sqrt{GM}}a^{3/2} \quad \text{or} \quad T^2 = \frac{4\pi^2}{GM}a^3 \qquad \text{(N14.29)}$$

Kepler's third law for elliptical orbits

This equation says that *the square of a planet's period is proportional to the cube of its semimajor axis:* this is Kepler's third law. Again, this law actually applies to *any* object in an elliptical orbit around a massive primary.

Exercise N14X.11: Verify equation N14.29.

Derivation determines the constant of proportionality

Equation N14.29 provides an important piece of information that the verbal statement of Kepler's third law does not: *it specifies the constant of proportionality in terms of the GM.* This actually makes it possible to compute the numerical value for the period of an object's orbit if you know *GM* for its primary. Alternatively, if we know the period and semimajor axis of an object's orbit, we can use equation N14.29 to find the mass of its primary.

EXAMPLE N14.2

Problem: A certain asteroid in an elliptical orbit is a distance r_c = 3.5 AU from the sun at the closest point in its orbit and a distance r_f = 4.5 AU at the farthest. What is the period of its orbit in years?

Solution The semimajor axis of the orbit is $a = \frac{1}{2}(r_c + r_f)$ = 4.0 AU. The semimajor axis of the earth's orbit $a_e \approx$ 1 AU (by definition). Since $T \propto a^{3/2}$,

$$\frac{T}{T_e} = \left(\frac{a}{a_e}\right)^{3/2} = \left(\frac{4.0\ \cancel{\text{AU}}}{1.0\ \cancel{\text{AU}}}\right)^{3/2} = 8.0 \qquad \text{(N14.30)}$$

Since the period of the earth's orbit T_e = 1 y by definition, the period of the asteroid's orbit must be 8.0 y.

N14.8 SOLVING ORBIT PROBLEMS

In many practical situations, we can determine the characteristics of elliptical orbits directly from conservation of energy and angular momentum without having to deal with the orbit equation N14.19. The purpose of this section is to discuss how this is done.

Consider an object in an elliptical orbit around a massive object with mass *M*. At the closest and farthest points of the orbit, the orbiting object's velocity is entirely perpendicular to the radial direction (see Figure N14.4), since at these extreme points, the radial component of the object's velocity is switching from being outward to inward (or vice versa) so its value passes through zero at the extreme point. This means that $v_\perp = v$ at these extreme points, implying that the magnitude of orbiting object's angular momentum is simply $L = mrv$ at these points. Therefore, conservation of angular momentum and conservation of energy for these points implies that

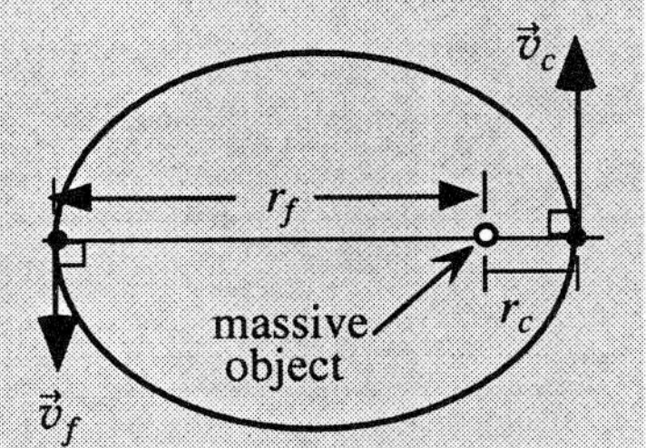

Figure N14.4: An orbiting object's velocity vectors are perpendicular to the radial direction at the orbit's extreme points.

$$\left(\frac{L}{m}\right) = r_c v_c = r_f v_f \qquad \text{(N14.31a)}$$

$$\left(\frac{2E}{m}\right) = v_c^2 - \frac{2GM}{r_c} = v_f^2 - \frac{2GM}{r_f} \qquad \text{(N14.31b)}$$

Conservation equations linking quantities at extreme points

Assuming that we know *GM*, these equations represent two equations in the four unknowns r_c, r_f, v_c, and v_f. Given any *two* of these four unknowns, we can use these equations to determine the other two.

Extreme-point data characterize elliptical orbit

The four extreme-point values r_c, r_f, v_c, and v_f in turn provide essentially everything we need to know about an elliptical orbit. Knowing these values, we can use equations N14.31 and equations N14.18 and N14.19 to determine its eccentricity, the value of R, and the value of r at any angle. We can also calculate the orbit's semimajor axis $a = \frac{1}{2}(r_c + r_f)$, and use a and equation N14.29 to find the orbit's period.

How to solve an orbit problem involving such data

So to solve a typical problem where you are given information about one or both of the extreme points of an orbit, all that you need to do is go through the following steps:

1. sketch the orbit and label the four quantities r_c, r_f, v_c, and v_f,
2. identify the two that are known and the two that are unknown,
3. write down the two conservation laws (equations N14.31),
4. express the conservation laws in unitless form,
5. solve the equations for the unknown quantities, and
6. check that the sign, units, and magnitudes of the answers make sense.

Why work with equations in unitless form?

Step 4 needs somewhat more explanation. Long experience shows that in orbit problems (and in many other situations where complicated computations are necessary) it helps to express equations in unitless form, usually by dividing through by whatever combination of known quantities yields a unitless result. To be specific, let's define the unitless ratios $q \equiv r_f / r_c$ (this will always be greater than 1) and $u \equiv v_f / v_c$ (this will always be less than 1). In this context, it is useful to note that $2GM/r$ and v^2 have the same units, so $2GM / rv^2$ is also a unitless number. Using unitless variables is useful because (1) it saves writing when we are dealing with complicated numerical quantities with convoluted units, (2) it usually ends up making numerical quantities have reasonable magnitudes, and (3) it almost inevitably makes the algebra easier to do and more transparent to read. All this generally makes a problem *much* easier (trust me!).

This process is illustrated in the following examples.

EXAMPLE N14.3

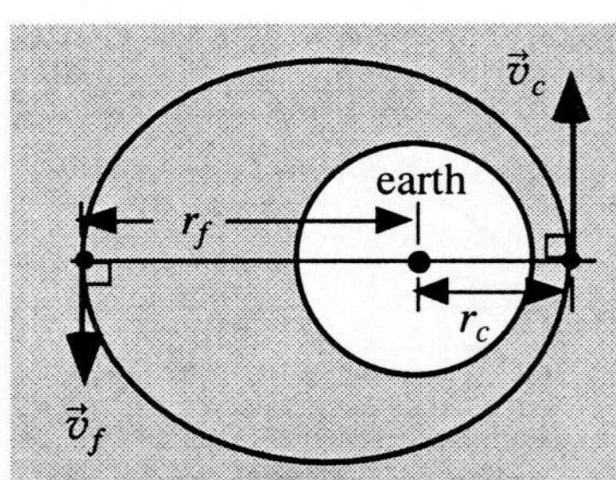

Figure N14.5: The situation for example N14.3.

Problem: At the point of its orbit closest to the earth, a satellite is 200 km above the earth's surface and is traveling at a speed of 10.0 km/s. Find the farthest distance that this satellite gets from the earth, and determine its period.

Solution See Figure N14.5 for a sketch of the orbit. We are given that at the closest point in its orbit, the satellite is 200 km from the earth's surface, and thus 200 km + 6380 km = 6580 km = r_c from the earth's center, and that its speed v_c = 10.0 km/s at this point. We want to find r_f.

Conservation of the satellite's angular momentum implies that

$$r_c v_c = r_f v_f \tag{N14.32}$$

We can make this equation unitless if we divide both sides by the known quantities $r_c v_c$: the equation then becomes

$$1 = qu \quad \text{where} \quad q \equiv r_f / r_c, \quad u \equiv v_f / v_c \tag{N14.33}$$

The law of conservation of energy tells us that

$$v_c^2 - \frac{2GM}{r_c} = v_f^2 - \frac{2GM}{r_f} \tag{N14.34}$$

We can make *this* equation unitless by dividing both sides by v_c^2:

$$1 - \frac{2GM}{r_c v_c^2} = u^2 - \frac{2GM}{r_f v_c^2} \tag{N14.35}$$

EXAMPLE N14.3 (continued)

The term $2GM/r_c v_c^2$ is an entirely known quantity we can call b:

$$b \equiv \frac{2GM}{r_c v_c^2} = \frac{2(3.99 \times 10^{14}\ \text{m}^3/\text{s}^2)}{(6,580,000\ \text{m})(10,000\ \text{m/s})^2} = 1.21 \text{ (unitless!)} \quad \text{(N14.36)}$$

(See exercise N14X.9 for the value of GM for the earth.) Note that b has a nice, reasonable magnitude instead of being something awful like 3.99×10^{14} m^3/s^2.) The term $2GM/r_f v_c^2$ can be expressed in terms of b and q if we multiply top and bottom by r_c:

$$\frac{2GM}{r_f v_c^2} = \frac{2GM}{r_f v_c^2}\left(\frac{r_c}{r_f}\right) = \left(\frac{2GM}{r_c v_c^2}\right)\left(\frac{r_c}{r_f}\right) = \frac{b}{q} \quad \text{(N14.37)}$$

Plugging these into equation N14.35, we get the unitless equation

$$1 - b = u^2 - \frac{b}{q} \quad \text{(N14.38)}$$

Now we can solve the unitless equations N14.33 and N14.38 for our unknowns u and q. What we *really* want to find is q, because that gives us r_f in terms of r_c. Equation N14.33 implies that $u = 1/q$, so if we plug this into equation N14.38, we get an equation entirely in terms of q:

$$1 - b = \frac{1}{q^2} - \frac{b}{q} \quad \text{(N14.39)}$$

Multiplying both sides by q^2 and rearranging (remembering that $b > 1$), we get the quadratic equation

$$(b - 1)q^2 - bq + 1 = 0 \quad \text{(N14.40)}$$

We can solve this using the quadratic formula:

$$q = \frac{b \pm \sqrt{b^2 - 4(b-1)}}{2(b-1)} = \frac{b \pm (2-b)}{2(b-1)} = \frac{1}{b-1} \text{ and } 1 \quad \text{(N14.41)}$$

The two solutions for q in a problem like this yield *both* extreme points of the orbit (since $r_f = r_c$ trivially satisfies equations N14.32 and N14.34), so *one* of the answers for q will always be 1 (this is a useful check on one's algebra!). The other solution is the interesting one:

$$q = \frac{1}{b-1} = \frac{1}{0.21} = 4.76 = \frac{r_f}{r_c} \quad \text{(N14.42a)}$$

$$\text{so } r_f = 4.76 r_c = 4.8(6580 \text{ km}) = 31{,}300 \text{ km} \quad \text{(N14.42b)}$$

(Note that equation N14.33 says that $v_f / v_c = u = 1/q = 0.21$, so $v_f = 0.21 v_c$ = 2.1 km/s, in case you wanted to know.)

The orbit's semimajor axis $a = \frac{1}{2}(r_c + r_f)$ = 18,900 km. Plugging this into Kepler's third law (equation N14.29), we get:

$$T = \sqrt{\frac{4\pi^2(1.89 \times 10^7\ \text{m})^3}{(3.99 \times 10^{14}\ \text{m}^3/\text{s}^2)}} = 2.6 \times 10^4\ \text{s}\left(\frac{1\ \text{h}}{3600\ \text{s}}\right) = 7.2\ \text{h} \quad \text{(N14.43)}$$

EXAMPLE N14.4

Problem: A satellite orbits the earth in an elliptical orbit such that its closest point is a distance r_c = 10,000 km from the earth's center, and its farthest point is r_f = 40,000 km from the earth's center. What is its speed at the closest point?

Solution Here we are given r_c and r_f and want to find v_c. Conservation of the satellite's angular momentum implies that

$$r_c v_c = r_f v_f \quad \Rightarrow \quad u \equiv \frac{v_f}{v_c} = \frac{r_c}{r_f} = \frac{1}{q} = \frac{10{,}000 \text{ km}}{40{,}000 \text{ km}} = \frac{1}{4} \qquad \text{(N14.44)}$$

Conservation of energy implies that:

$$v_c^2 - \frac{2GM}{r_c} = v_f^2 - \frac{2GM}{r_f} \qquad \text{(N14.45a)}$$

Dividing both sides by v_c^2, defining $b = 2GM / r_c v_c^2$ again, and again multiplying top and bottom of the second term by r_c / r_c, we get the unitless equation:

$$1 - \frac{2GM}{v_c^2 r_c} = \left(\frac{v_f}{v_c}\right)^2 - \frac{2GM}{v_c^2 r_c}\frac{r_c}{r_f} \quad \Rightarrow \quad 1 - b = u^2 - \frac{b}{q} \qquad \text{(N14.45b)}$$

Here we don't have to use the quadratic equation: we just plug in the known values of u and q from equation N14.44:

$$1 - b = \frac{1}{16} - \frac{b}{4} \quad \Rightarrow \quad \frac{15}{16} = \frac{3b}{4} \quad \Rightarrow \quad b = \frac{5}{4} = \frac{2GM}{r_c v_c^2} \qquad \text{(N14.46)}$$

Solving for the unknown v_c, we get

$$v_c = \sqrt{\frac{8GM}{5r_c}} = \sqrt{\frac{8(3.99 \times 10^{14} \text{ m}^{3\,2} / \text{s}^2)}{5(10{,}000{,}000 \text{ m})}} = 7{,}990 \text{ m/s} \qquad \text{(N14.47)}$$

This is plausible: orbital speeds near the earth are generally roughly 10 km/s.

An alternative approach to this problem would have been to use the fact that $E = -GMm / 2a$ (equation N14.23). Since the system's energy is conserved

$$v_c^2 - \frac{2GM}{r_c} = \frac{2E}{m} = -\frac{GM}{a} \quad \Rightarrow \quad v_c = \sqrt{GM\left(\frac{1}{r_c} - \frac{1}{a}\right)} \qquad \text{(N14.48)}$$

Since everything on the right side of the last equality is known, we can just plug in numbers and crank out the answer (you should get the same result as before).

SUMMARY

I. CONIC SECTIONS
   A. The general formula for a *conic section*: $r = R/(1+\varepsilon\cos\theta)$ (N14.1)
      1. r is the distance between a given curve point and the curve's *focus*
      2. θ is the angle that the line connecting the point to the focus makes with the line that connects the focus to the closest point on the curve
      3. R is a constant that specifies the scale of the curve
      4. ε is a constant called the curve's *eccentricity*
   B. The shape of the curve is determined by the value of ε
      1. if $\varepsilon = 0$, we have a *circle*
      2. if $0 < \varepsilon < 1$, we have an *ellipse*
      3. if $\varepsilon = 1$, we have a *parabola*
      4. if $\varepsilon > 1$, we have a *hyperbola*
   C. Quantities that describe ellipses
      1. r_c = distance from focus to closest point = $R/(1+\varepsilon)$ (N14.2)
      2. r_f = distance from focus to farthest point = $R/(1-\varepsilon)$
      3. semimajor axis $a \equiv \frac{1}{2}(r_f + r_c) = R/(1-\varepsilon^2)$ (N14.21)
      4. area $A = \pi R^{1/2}a^{3/2}$

II. PROVING KEPLER'S FIRST LAW
   A. Finding the acceleration in terms of $\hat{r}$ and $\hat{\theta}$
      1. Using techniques from chapter N9 and the product rule, we get

$$\vec{a}(t) = \left[\frac{d^2r}{dt^2} - r\left(\frac{d\theta}{dt}\right)^2\right]\hat{r} + \left[2\frac{dr}{dt}\frac{d\theta}{dt}\hat{\theta} + r\frac{d^2\theta}{dt^2}\right]\hat{\theta} \qquad \text{(N14.6)}$$

      2. Newton's second law implies that the first quantity in brackets is equal to $-GMm/r^2$ and the second is equal to zero
         a) the latter relation leads to conservation of angular momentum
         b) the former determines the particular shape of the curve
   B. The remainder of the proof takes advantage of three tricks
      1. the definition of angular momentum $\Rightarrow d\theta/dt = L/mr^2$ (N14.9)
      2. rewriting everything in terms of $u = 1/r$ turns out to help
      3. comparing the resulting differential equation describing the orbit with the harmonic oscillator equation gives us the solution:
         $r = R/(1+\varepsilon\cos\theta)$ with $R \equiv (L/m)^2/GM$ (N14.15)

III. CONNECTIONS WITH ENERGY
   A. A straightforward calculation using the definition of energy leads to

$$\varepsilon = \left[1 + (2E/m)(L/m)^2/GM\right]^{1/2} \qquad \text{(N14.18)}$$

   B. This implies a connection between total energy and an orbit's shape
      1. if $E < 0$ then $\varepsilon < 1$ and the orbit is elliptical
      2. if $E = 0$ then $\varepsilon = 1$ and the orbit is parabolic
      3. if $E > 0$ then $\varepsilon > 1$ and the orbit is hyperbolic
   C. Another interesting link is $E = -GMm/2a$ (N14.23)
   D. One can determine an orbit completely knowing GM, L/m, and $2E/m$

IV. KEPLER'S THIRD LAW
   A. Argument
      1. $dA/dt = L/2m$ (from N13) implies that ellipse's area $A = (L/2m)T$
      2. comparing this with formula for A gives $T^2 = 4\pi^2a^3/GM$
   B. (As we saw in the last chapter) we can use this to find a primary's mass

V. SOLVING ORBIT PROBLEMS
   A. If we know any two of r_c, r_f, v_c, or v_f for an elliptical orbit
      1. we can use conservation of L and E to find the other two
      2. knowing all four plentifully describes an elliptical orbit
   B. A framework for solving an orbit problem:
      1. Draw a picture, label r_c, r_f, v_c, and v_f, and identify the unknowns
      2. Write down the conservation-law equations (equations N14.41)
      3. Express in unitless form using $u \equiv v_f/v_c$ and $q \equiv r_f/r_c$
      4. solve for unknowns and check result

GLOSSARY

conic section: the name describing a family of curves such that the distance r from a point on the curve to a fixed point (called the **focus**) is is given by $r = R/(1+\varepsilon\cos\theta)$, where θ is the anglebetween the line connecting the focus and the point in question and the line between the focus and the closest point on the curve. R and ε are arbitrary positive constants; ε called the curve's **eccentricity**. (These curves happen to be the intersection between an arbitrarily oriented infinite plane and the surface of an infinite cone.)

circle: a conic section with $\varepsilon = 0$. Such a curve has constant radius $r = R$.

ellipse: a conic section with $0 < \varepsilon < 1$. An ellipse is a closed curve, having a well defined radius for all angles θ around the focus.

parabola: a conic section with $\varepsilon = 1$. A parabola is an open curve that goes to infinity at $\theta = \pi$.

hyperbola: a conic section with $\varepsilon > 1$. A hyperbola is an open curve that goes to infinity at angles $\pm\theta_\infty$, where $\cos\theta_\infty = -1/\varepsilon$..

semimajor axis a (of an ellipse): half of the widest possible diameter = $\frac{1}{2}(r_c + r_f)$, where r_c is the radius of the point on the ellipse that is closest to the focus, and r_f is the radius of point that is furthest.

astronomical unit (AU): A unit of distance that astronomers often use to describe distances in the solar system. 1 AU ≡ the average distance between the earth and the sun = 1.5×10^{11} m = 150 million kilometers.

TWO-MINUTE PROBLEMS

N14T.1 According to the text, an orbiting object having energy $E = 0$ follows a parabolic path. When we specify the numerical value for an object's energy, we are implicitly comparing it to the energy the object has in some reference situation. So, when we say that $E = 0$ in this case, we are comparing the energy of the orbiting object to the energy of an object with the same mass that is
A. at rest at the primary's center
B. at rest on the primary's surface
C. in a circular orbit just above the primary's surface
D. at rest at $r = \infty$
E. other (specify).

N14T.2 An object is discovered near the earth with values of $2E/m = 2.12 \times 10^7 \text{ m}^2/\text{s}^2$ and $L/m = 7.8 \times 10^{10} \text{ m}^2/\text{s}$. This object is in what kind of orbit around the earth?
A. elliptical B. parabolic C. hyperbolic

N14T.3 The eccentricity of this object's orbit is
A. 1.08 D. 0.90
B. 1.81 E. 1.35
C. 0.81 F. other (specify)

N14T.4 A comet is discovered in an elliptical orbit around the sun. Its closest distance from the sun is 1.0 AU and measurements of its speed at this distance imply that its greatest distance from the sun is 15 AU. About how many years will pass between the comet's closest approaches to the sun?
A. 8 years D. 226 years
B. 16 years E. 512 years
C. 64 years F. other (specify)

N14T.5 The relationship $r_c v_c = r_f v_f$ for an elliptical orbit follows from the fact that the satellite's velocity is perpendicular to the radius at the two extreme points and
A. conservation of angular momentum
B. conservation of energy
C. conservation of momentum
D. Newton's second law
E. other properties of ellipses
F. the relationship is false

N14T.6 At general points in a general orbit the distance r the object is from the primary's center is inversely proportional to its speed v. This statement is
A. would be true if v were changed to $|v_\perp|$
B. true for elliptical orbits but not for orbits where $\varepsilon \geq 1$
C. would be true if "inversely" were changed to "directly"
F. is completely false: there is no such relationship
T. is completely true as stated.

HOMEWORK PROBLEMS

BASIC SKILLS

N14B.1 An object in an elliptical orbit around the earth has an angular speed around the earth of 5.0 radians/h when it is a certain distance from the earth's center. When it is 2.5 times as far from the earth, what is its angular speed?

N14B.2 An object is discovered near the earth with values of E, L, and m such that $2E/m = -8.2 \times 10^7 \text{ m}^2/\text{s}^2$ and $L/m = 7.8 \times 10^{10} \text{ m}^2/\text{s}$. Find the eccentricity of this orbit, the radius of the closest point of the orbit to the earth, and classify the orbit as elliptical, parabolic or hyperbolic.

N14B.3 A space probe near the earth has values of E, L, and m such that $L/m = 7.8 \times 10^{10} \text{ m}^2/\text{s}$ and $2E/m = 0$. Find the eccentricity of this orbit, the radius of the closest point of the orbit to the earth, and classify the orbit as elliptical, parabolic or hyperbolic.

N14B.4 An asteroid orbiting the sun has a semimajor axis of 3.0 AU. What is the period of its orbit in years?

N14B.5 A space probe is put in an orbit around the sun that is 0.5 AU from the sun at the closet and 2.0 AU at the farthest. What is its period in years?

N14B.6 Satellite A travels around the earth in a circular orbit of radius R. Another satellite orbits in an elliptical orbit such that is R from the earth at its closest point and $3R$ from the earth at its farthest. How does the period of satellite B compare to that of satellite A? Show your work.

N14B.7 Imagine that you are in an orbit around the Earth whose most distant point from the earth is five times farther from the earth's center than the closest point. If your speed is 10.0 km/s at the closest point, what is your speed at the farthest?

N14B.8 A satellite orbits the earth in such a way that its speed at its point of closest approach is roughly three times its speed at the most distant point. How many times more distant from the earth's center is the far point than the close point?

SYNTHETIC

N14S.1 Consider a proton (which has a positive electrical charge) interacting electrostatically with a massive nucleus (which also has positive charge). Ignore the effects of any electrons in the vicinity. Will the path of the proton as it moves around the nucleus lie in a fixed plane? Will it obey Kepler's second law? Can the path possibly be an elliptical orbit? Carefully explain your answers. (Assume that the proton obeys the laws of newtonian mechanics.)

N14S.2 A small asteroid is discovered 14,000 km from the earth's center moving at a speed of 9.2 km/s. Can you tell from the information provided whether this asteroid is in an elliptical or hyperbolic orbit around the earth? Is the direction of its velocity vector important in determining this? Please explain.

N14S.3 A mysterious object is sighted by astronomers moving at a speed of 21 km/s at a distance of 3.8 AU from the sun (1 AU = Earth's orbital radius = 1.5×10^{11} m). Can you tell from the information provided whether this object is in an elliptical or hyperbolic orbit around the sun? Is the direction of its velocity vector important in determining this? Please explain.

N14S.4 Imagine that you are in a circular orbit of radius R = 7500 km around the earth. You'd like to get to a geostationary space station whose circular orbit has a radius of $3R$, so you'd like to put yourself in an elliptical orbit whose closest point to the earth has radius R and whose most distant point has radius $3R$ (such an orbit is called a *Hohmann transfer* orbit, and represents the lowest-energy way to get from one circular orbit to another). By what percent would you have to increase your speed at radius R to put yourself in this transfer orbit?

N14S.5 Imagine that you are an astronaut in a circular orbit of R = 6500 km around the earth. **(a)** What is your orbital speed? **(b)** Say that you fire a rocket pack so that in a very short time, you increase your speed in the direction of your motion by 20%. What are the characteristics of your new orbit?

N14S.6 A certain satellite is in an orbit around the earth whose nearest and farthest points from the earth's center are 7,000 km and 42,000 km respectively. Find the satellite's orbital speed at its point of closest approach.

N14S.7 An asteroid is in an orbit around the sun whose closest point to the sun has a radius R and whose most distant point has a radius $9R$, where R is equal to the radius of the earth's orbit = 1 AU = 1.5×10^{11} m. **(a)** What is the asteroid's speed when it is closest to the sun? **(b)** How does this compare with the earth's orbital speed? **(c)** What is the period of this orbit, in years?

N14S.8 A new comet is discovered 6.6 AU from the sun (1 AU = earth's orbital radius = 1.5×10^{11} m) moving with a speed of 17 km/s. At that time, its velocity vector makes an angle of 174.3° with respect to its position vector. **(a)** Is this comet in a hyperbolic orbit? **(b)** What will be its distance from the sun at the point of closest approach?

N14S.9 Some recent space probes have made several hyperbolic passes past the Earth to help give them the right direction and speed to go to their final destination. Imagine that one such probe has a speed of 12 km/s at a distance of 650,000 km from the Earth. If the angle that its velocity vector makes with its position at that time is 177°, how near will it pass by the Earth at its point of closest approach?

N14S.10 Argue that for a given value of angular momentum L, the smallest (most negative) energy that an orbiting object can have is

$$E = -\frac{G^2M^2m^3}{2L^2} \qquad \text{(N14.49)}$$

and that the orbit in this case will be circular.

N14S.11 Show that equation N14.8 (the perpendicular component of Newton's second law) is equivalent to

$$0 = \frac{1}{r}\frac{d}{dt}\left[r^2\frac{d\theta}{dt}\right] \qquad \text{(N14.50)}$$

(show that doing the derivative in the equation above yields what we have in equation N14.8). Argue from this that the orbiting object's angular momentum L is a constant independent of time.

N14S.12 Draw a trajectory diagram (see section N5.7) for a satellite orbiting the earth whose point of closest approach is 7500 km from the center of the earth and whose entirely tangential speed at that point is 8.0 km/s. Draw the diagram to a scale of 1 cm = 1000 km and use a time step Δt = 5 min = 300 s. Plot half the orbit, and also plot selected points from the ellipse predicted by equations N14.18 and N14.19 for the sake of comparison. [*Hint*: you should find that the length of the acceleration arrow for a given r is $GM\Delta t^2/r^2 = (35.9\text{ cm}^3)/r^2$ in this case.]

N14S.13 You can find the area of an ellipse by dividing it up into a large number of pie-slices of infinitesimal angular width $d\theta$. According to section N13.4 (see equation N13.6) the area of each slice in the limit that $d\theta \to 0$ is

$$dA = \tfrac{1}{2}r(r\,d\theta) = \tfrac{1}{2}r^2 d\theta \qquad \text{(N14.51)}$$

To find the total area of the ellipse, we need to add up all of the slices around the ellipse. In the limit that $d\theta \to 0$, the sum becomes an integral

$$A = \int_0^{2\pi} \tfrac{1}{2}r^2 d\theta = \tfrac{1}{2}R^2 \int_0^{2\pi} \frac{d\theta}{(1+\varepsilon\cos\theta)^2} \qquad \text{(N14.52)}$$

Look this up in any table of definite integrals (that is, integrals with specified upper and lower limits). Please give the reference. Then show use equation N14.22 to show that $a = R/(1-\varepsilon^2)$, and use this to arrive at $A = \pi R^{1/2}a^{3/2}$, as stated below equation N14.28.

RICH-CONTEXT

N14R.1 You are the commander of the starship *Execrable*. You are currently in a standard orbit around a class-M planet whose mass is 4.4×10^{24} kg and whose radius is 6100 km. Your current circular orbit around the planet has a radius of R = 50,000 km. Your exobiologist wants to get in closer (say to an orbital radius of 10,000 km, or $R/5$) to look for signs of life. Your planetary geologist wants to stay at the current radius so that the entire face of the planet can be scanned with the sensors at once. Because you are tired of the bickering, you decide to put the *Execrable* into an elliptical orbit whose minimum distance from the planetary center is $R/5$ and whose farthest point is R from the same. Your navigational computers are down again (of course), so you have to compute by hand how to insert yourself in this new orbit. Your impulse engines are capable of causing the ship to accelerate at a rate of 1 g = 9.8 m/s^2. In what direction should you fire your engines, relative to your current direction of motion (forward or backward)? For how many seconds should you fire them?

N14R.2 You'd like to put your spaceship into a Hohmann transfer orbit between earth and Mars (that is, and elliptical orbit whose closest point to the sun is the same as the earth's orbital radius and whose farthest point is the same as Mars' orbital radius: see problem N14S.4). **(a)** If your spaceship is initially traveling around the sun at the same distance as the earth is and with the same orbital speed, by what factor will you need to increase your speed to put you into the transfer orbit? **(b)** How long will it take you to get to Mars along this orbit?

ADVANCED

N14A.1 Here is another way to prove Kepler's first law. First, use the chain rule to show that

$$\frac{dr}{dt} = \frac{dr}{d\theta}\frac{L}{mr^2} \tag{N14.53}$$

Then show that the definition of energy, equation N14.5 and equation N14.9 imply that

$$\left(\frac{dr}{d\theta}\right)^2\left(\frac{L}{mr^2}\right)^2 + \left(\frac{L}{mr}\right)^2 - \frac{2GM}{r} = \frac{2E}{m} \tag{N14.54}$$

Multiply both sides by r^2, take the square root of both sides, and rearrange things so that you get an equation of the form $dr/d\theta = f(r)$, where f is some function of r. You can rearrange this to read:

$$\frac{dr}{f(r)} = d\theta \tag{N14.55}$$

You can now take the indefinite integral of both sides of this equation to get θ as a function of r, which you can invert to find r as a function of θ. Look up the integral in an integral table (please give a reference to your source) and do the inversion. Check that your final function $r(\theta)$ is equivalent to equation N14.1 for a conic section.

ANSWERS TO EXERCISES

N14X.1 $\theta_\infty = \cos^{-1}(-2/3) = 2.30 = 132°$.

N14X.2 According to the definition of acceleration

$$\vec{a}(t) = \frac{d\vec{v}}{dt} = \frac{d}{dt}\left(\frac{dr}{dt}\hat{r} + r\hat{\theta}\frac{d\theta}{dt}\right) \tag{N14.56}$$

According to the product rule, this becomes (N14.57)

$$\vec{a}(t) = \frac{d^2r}{dt^2}\hat{r} + \frac{dr}{dt}\frac{d\hat{r}}{dt} + \frac{dr}{dt}\hat{\theta}\frac{d\theta}{dt} + r\frac{d\hat{\theta}}{dt}\frac{d\theta}{dt} + r\hat{\theta}\frac{d^2\theta}{dt^2}$$

Using equations N14.4 then yields (N14.58)

$$\vec{a}(t) = \frac{d^2r}{dt^2}\hat{r} + \frac{dr}{dt}\hat{\theta}\frac{d\theta}{dt} + \frac{dr}{dt}\hat{\theta}\frac{d\theta}{dt} + r(-\hat{r})\left(\frac{d\theta}{dt}\right)^2 + r\hat{\theta}\frac{d^2\theta}{dt^2}$$

Regrouping terms then leads to equation N14.6.

N14X.3 If r = constant, then $dr/dt = 0$ and $d^2r/dt^2 = 0$. If $d\theta/dt$ = constant, then $d^2\theta/dt^2 = 0$. If we plug these things into equation N14.6, only one term survives:

$$\vec{a}(t) = -r\left(\frac{d\theta}{dt}\right)^2\hat{r} = -\frac{1}{r}\left(r\frac{d\theta}{dt}\right)^2\hat{r} = -\frac{v^2}{r}\hat{r} \tag{N14.59}$$

according to the hint in the problem. This equation states that the acceleration points in the $-\hat{r}$ direction and has a magnitude of $a = v^2/r$.

N14X.4 According to equation N14.12b, (N14.60)

$$\frac{d^2r}{dt^2} = \frac{d\theta}{dt}\frac{d}{d\theta}\left(\frac{dr}{dt}\right) = \frac{d\theta}{dt}\frac{d}{d\theta}\left(-\frac{L}{m}\frac{du}{d\theta}\right) = -\frac{d\theta}{dt}\frac{L}{m}\frac{d^2u}{d\theta^2}$$

Since $d\theta/dt = L/mr^2 = Lu^2/m$, this becomes

$$\frac{d^2r}{dt^2} = -\frac{Lu^2}{m}\frac{L}{m}\frac{d^2u}{d\theta^2} = -\frac{L^2u^2}{m^2}\frac{d^2u}{d\theta^2} \tag{N14.61}$$

N14X.5 Since GMm/L^2 is a constant (N14.62)

$$\frac{dw}{d\theta} = \frac{d}{d\theta}\left(u - \frac{GMm^2}{L^2}\right) = \frac{du}{d\theta} \Rightarrow \frac{d^2w}{d\theta^2} = \frac{d^2u}{d\theta^2}$$

Substituting this and $u = w + GMm^2/L^2$ into equation N14.13a, we get

$$\frac{d^2w}{d\theta^2} + w + \frac{GMm^2}{L^2} = \frac{GMm^2}{L^2} \tag{N14.63}$$

Canceling the constant terms and adding $-w$ to both sides yields equation N14.13b.

N14X.6 When we substitute equation N14.14 into equation N14.15a, we get

$$\frac{1}{r} - \frac{GMm^2}{L^2} = A\cos(\theta+\delta)$$

$$\Rightarrow \frac{1}{r} = \frac{GMm^2}{L^2} + A\cos(\theta+\delta)$$

$$= \frac{GMm^2}{L^2}\left(1 + \frac{AL^2}{GMm^2}\cos\theta\right) = \frac{GMm^2}{L^2}(1+\varepsilon\cos\theta)$$

with the definition of ε suggested. (N14.64)

N14X.7 If w is as stated in equation N14.15a, then

$$\frac{dw}{d\theta} = \frac{d}{d\theta}[A\cos(\theta+\delta)] = -A\sin(\theta+\delta)$$

$$\frac{d^2w}{d\theta^2} = \frac{d}{d\theta}\left(\frac{dw}{d\theta}\right) = \frac{d}{d\theta}[-A\sin(\theta+\delta)]$$

$$= -A\cos(\theta+\delta) = -w, \text{ as claimed.} \tag{N14.65}$$

N14X.8 Multiplying both sides of equation N14.17b by $2L^2/G^2M^2m^3$, we get

$$\varepsilon^2 - 1 = \frac{2EL^2}{G^2M^2m^3} = \frac{1}{G^2M^2}\left(\frac{2E}{m}\right)\left(\frac{L}{m}\right)^2 \tag{N14.66}$$

Adding 1 to both sides and taking the square root yields equation N14.18.

N14X.9 According to equation N14.22, we have:

$$a = \frac{(L/m)^2}{2GM}\left[\frac{1}{1+\varepsilon} + \frac{1}{1-\varepsilon}\right] \tag{N14.67}$$

$$\text{But} \quad \frac{1}{1+\varepsilon} + \frac{1}{1-\varepsilon} = \frac{1-\varepsilon+1+\varepsilon}{(1+\varepsilon)(1-\varepsilon)} = \frac{2}{1-\varepsilon^2} \tag{N14.68}$$

$$\text{So} \quad \frac{1}{a} = \frac{GM}{(L/m)^2}(1-\varepsilon^2) \Rightarrow 1-\varepsilon^2 = \frac{(L/m)^2}{GMa} \tag{N14.69}$$

Plugging this into equation N14.17b, we get

$$E = -\frac{G^2M^2m^3}{2L^2}\frac{(L/m)^2}{GMa} = -\frac{GMm}{2a} \tag{N14.70}$$

N14X.10 This is simply a matter of plugging in numbers (use the masses given on the inside front cover).

N14X.11 According to equation N14.28, the area of the ellipse is $A = (L/2m)T$. We are given that $A = \pi R^{1/2}a^{3/2}$; equation N14.19 tells us that $R = (L/m)^2/GM$. So

$$\frac{L}{2m}T = A = \pi R^{1/2}a^{3/2} = \pi\sqrt{\frac{L^2a^3}{m^2GM}}$$

$$\Rightarrow T = \pi\frac{2m}{L}\sqrt{\frac{L^2a^3}{m^2GM}} = 2\pi\sqrt{\frac{a^3}{GM}} \tag{N14.71}$$

This is the first equation in N14.29; squaring this equation yields the second.

INDEX TO UNIT *N*